T0136865

Signals and Communication Technology

More information about this series at http://www.springer.com/series/4748

George Pavlidis

Mixed Raster Content

Segmentation, Compression, Transmission

 Springer

George Pavlidis
Athena Research Center
University Campus at Kimmeria
Xanthi
Greece

ISSN 1860-4862   ISSN 1860-4870   (electronic)
Signals and Communication Technology
ISBN 978-981-10-9716-4   ISBN 978-981-10-2830-4   (eBook)
DOI 10.1007/978-981-10-2830-4

Printed on acid-free paper

This Springer imprint is published by Springer Nature
The registered company is Springer Nature Singapore Pte Ltd.
The registered company address is: 152 Beach Road, #22-06/08 Gateway East, Singapore 189721, Singapore

Preface

—...δεῖν γὰρ ἐπ' ἐκείνων καὶ αὐτὸν παρακελεύεσ-
θαι τὸν Πυθαγόραν ζητεῖν ἐξ ἐλαχίστων καὶ
ἁπλουστάτων ὑποθέσεων δεικνύναι τὰ ζητούμενα...

Pythagorians

Since the day that Johann Gutenberg invented the modern typography in Germany in 1450 we are inundated with reams of paper documents. Although the manufacturing of paper was known in China since the second century and high-quality paper was already available in the fifth century, the availability of printed material began to increase significantly only after the invention of Gutenberg, who introduced the metallic elements in typography.

In less than 50 years after the printing of the first book, more than 1,700 printing machines were used in Europe, through which 40,000 different print jobs had already been completed (Johnson 1973). By 1975, about 400 years later, it was estimated that approximately 50,000,000 printing works had been completed, while according to the 1998 estimate there are over 10^{12} documents in the world with a provision for doubling every 3 years (Open Archive Systems 1998). During the beginning of the twenty-first century the trend was still mainly oriented towards the production and reproduction of printed documents at a rate faster than the rate of digitization.

Nowadays, the main issues regarding the printed documents are not related with the processes of paper or document production, but rather with the huge physical storage requirements. On the other hand, the storage requirements for digital images are significantly lower. Indicatively, in a 4.7 GB DVD one may store about 4,320 uncompressed A4 pages digitized at a resolution of 300 dpi in binary form (1 Bits Per Pixel (bpp)—black and white), or about 540 gray-level images (8 bpp), or about 180 full color (24 bpp). This DVD including the protective sleeve requires about 160 cubic centimeters or 1.6×10^{-4} m^3 of physical storage space. A simple comparison to the respective paper volume for the same amount of material (using a typical 80 gr. paper), which gives an estimate of about 432×10^{-4} m^3 for the case of the 4,320 pages, shows that the digitized material requires 270 times less storage volume.

In order to cope with the storage issue that arises from the huge amount of produced digital documents by improving the data compression ratio, numerous compression methods for digitized documents have been studied and developed over the recent years, either case-specific or generic. It is noticeable that by applying an efficient method of digital data compression, a volume compression for 2,000,000 books of 200 pages to less than 0.2 m^3 can be achieved.

Even though these numbers are impressive, this is not only a storage issue. With the advent of the new communication technologies through the pervasive presence of the Internet and the mobile networks, this issue is directly reflected on the management and transmission of the digitized information that a color document represents. Considering the data archival, document digitization contributes significantly to more efficient data management, since search and retrieval can be faster and more intuitive (using semantics). The ability to efficiently transmit digitized documents through wired or wireless networks could boost a fast building trend towards tele-working, tele-cooperation and true cybernetics (digital governance) and civil services. These highlight one of the main issues in information transmission, transmission speed. The provided table shows indicative transmission times for an ISO A4 page digitized at 300 dpi (which is about 25 MB uncompressed) through LAN, DSL, ISDN and PSTN technologies, reported in bpsec and Bps.

Channel bandwidth (approximate true speed)	Approximate transmission time
PSTN 28.8 Kbps (0.0035 MBps)	2 h
PSTN 33.6 Kbps (0.0041 MBps)	1.5 h
PSTN 56 Kbps (0.0068 MBps)	1 h
ISDN 128 Kbps (0.0156 MBps)	25 min
DSL 384 Kbps (0.0469 MBps)	10 min
Dedicated Line 2 Mbps (0.25 MBps)	1.5 min
LAN 10 Mbps (1.25 MBps)	20 s
LAN 100 Mbps (12.5 MBps)	2 s

Apparently, in all cases except the fast LAN networks (or the modern high bandwidth WiFi and mobile networks), transmission times can be rather disappointing and, in some cases, impractical. A straightforward way to tackle with this issue is to reduce the amount of data being transferred using a data compression mechanism. The application of compression is not only an economic solution—since it is an algorithmic response to the problem and, in most cases, does not demand any infrastructural changes—it is also the only meaningful choice towards the quantification of the information within a transmitted message and a distinction between the useful part of the information and the inherent redundancy (since, in most cases, data compression exploits statistical characteristics of the data to reduce the redundancy). Otherwise, the only alternative would be to change all network infrastructures to meet the constantly growing bandwidth demands, which would eventually bring higher investments and, consequently, higher costs for the consumers. Considering also the fact that digital image compression has already a long history of success and can be considered a mature technological approach to solve problems of the modern society, its choice as a solution to this problem seems most appealing. Indicatively, the application of JPEG image compression (which is not the most appropriate method of MRC compression, though) on a typical ISO A4 page digitized at 150 dpi and 24 bpp (true color) can lead to a reduction from 6 MB to 200 KB attaining a 1:30 compression ratio with unnoticeable quality degradation (30.7 dB PSNR) and a data transmission time reduction through a 56 Kbps channel from 15 min to 30 s.

A digitized document is a digital image that represents an 'exact' replica of an original document (Lynn 1990). Digital images can be considered 'superior' to printed documents for a number of reasons, including that they can be efficiently stored, efficiently and nonlinearly searched and retrieved, reproduced and transmitted with accuracy or controlled distortion.

The treatment of digitized documents as generic digital images is not efficient for data compression. This is due to the fact that the digitized documents include mixed visual content with largely differentiating characteristics, including parts with text, parts with graphics and parts with natural images in shades of gray or various colors. This is why they are usually referenced as mixed raster content or MRC (ITUT 1999). Each of the individual parts of such mixed images exhibits different statistical and semantic characteristics and need to be treated differently to achieve efficient data compression. For decades these image characteristics have been the center of attention in the field of image compression, and specific methods have been discovered (and continue being developed) to treat each of the cases. In recent years there has also been an attempt to identify these different parts in mixed images and apply the appropriate data compression method to each identified part. MRC compression specialists have to deal with notions such as segmentation and layered representation, sparse matrix compression, error resilience during transmission through noisy and congested networks, reconstruction of lost data, etc. All these notions represent the central considerations in this treatise, so methods to treat them are presented in detail, based on extended research over a number of years (Pavlidis and Chamzas 2005; Pavlidis et al. 2001, 2002, 2003, 2004, 2005; Politou et al. 2004), which departed significantly from other approaches basically regarding the overall scope.

Most of the works in the field of MRC processing and compression start by adopting some significant assumptions for the type of the characteristics in the images that are to be processed. In most cases, the compression of MRC refers to binary images (1 bpp) as they usually represent black text characters on white background. In addition, in many cases segmentation of MRC is studied without considering compression or even the efficiency in transmission or error correction in noisy environments. These considerations are detailed by a study of segmentation of color MRC images in order to decorrelate their basic structural elements into at least three layers for their independent compression, while keeping a 'loose' compatibility with the ITU Recommendations T.44 (ITUT 1999) for MRC.

This is complemented by a study on compression methods, which by exploiting segmentation operate in an optimal way in the different layers of information targeting an improvement of the total compression ratio and perceived quality. Another approach that completes the picture is presented as a study and simulation of efficient transmission of compressed MRC images through noisy and congested channels, in order to analyze and propose reconstruction mechanisms in an integrated management framework.

The presentation of the main concepts in this treatise is divided into two main parts, the first of which includes introductory chapters to cover the scientific and technical background aspects, whereas in the second part there is a set of research and development approaches in MRC proposed by the author to tackle the issues in MRC segmentation, compression and transmission from a novel point of view. The first chapter reviews *the color theory* and *the mechanism of color vision* in humans; the scope is to form a stable ground for a color representation theory and to identify the main characteristics and limitations of visual perception. The *information theory and data coding* is introduced in the second chapter, and a brief presentation of the most widely used compression methods is included to set the background for data compression and to highlight the complexity involved in dealing with MRC. In this chapter, a contribution of the author is included that involves the study of the new image compression standard, JPEG2000, which led to the development of a novel system for progressive transmission of color images in cultural multimedia databases. The third chapter introduces the *segmentation of images* through an extensive literature review, which highlights the differences regarding the approaches to tackle MRC segmentation. Different approaches are being presented in order to emphasize that segmentation of images is a case-specific (or context-specific) mechanism and there is not one-for-all solution available. In the second part, and specifically in the fourth chapter, the *segmentation of color images for optimized compression* is introduced and analyzed, including a multi-layered decomposition and representation of MRC and the processes that may be adopted to optimize the coding rates of those different layers. A nonlinear projections method is presented, in which the background layer in the multi-layered representation is optimized, to tackle one of the major issues in MRC image compression, that is, the data filling. In the final chapter the *segmentation of color images for optimized transmission* is introduced and analyzed, including two distinctive approaches that are based on segmenting the image data into significant and complementary; the first, a global approach based on the application of multi-layered coding and progressive transmission using typical coding standards and labeling of data as significant and additional, in addition to applying a segmentation method based on the clustering of data from a differential image; the second and latter, the exploitation of the JPEG2000 standard for color image transmission using a selective data segmentation into significant and complementary parts, which provides the functionality to apply cost policies.

Xanthi, Greece   George Pavlidis

Contents

Acronyms

AC	Arithmetic Coding
API	Application Programming Interface
ASCII	American Standard Code for Information Interchange
bpp	Bits Per Pixel
Bps	Bytes Per Second
bps	Bits Per Sample (P. vi, 55, 123)
bps	Bits Per Second (P. 55, 61, 64, 114, 132, 152)
CALIC	Context-based Adaptive Lossless Image Coder
CDF	Cumulative Distribution Function
CIE	Commission Internationale d'Eclairage
CIELAB	CIELab color system
CMF	Color Matching Functions
CPRL	Component-Position-Resolution-Layer
CT	Classification Trees
CWT	Continuous-time Wavelet Transform
dB	Decibel
DCT	Discrete Cosine Transform
DFS	Discrete Fourier Series
DFT	Discrete Fourier Transform
DHT	Discrete Hartley Transform
DjVu	Deja Vu Image Compression
DMS	Discrete Memoryless Source
DPCM	Differential Pulse Code Modulation
dpi	Dots Per Inch
DSL	Digital Subscriber Line
DST	Discrete Sine Transform
DVD	Digital Versatile Disc
DWT	Discrete Wavelet Transform
EBCOT	Embedded Block Coding with Optimized Truncation
EZW	Embedded image coding using Zerotrees of Wavelet coefficients

FELICS	Fast, Efficient Lossless Image Compression System
FFT	Fast Fourier Transform
FLIR	Forward Looking InfraRed
GA	Genetic Algorithm
GIF	Graphics Interchange Format
HC	Huffman Coding
HT	Haar Transform
HTML	HyperText Markup Language
HVS	Human Visual System
IJG	Independent JPEG Group
IP	Internet Protocol
ISDN	Integrated Services Digital Network
ISRG	Improved Seeded Region Growing
ITU	International Telecommunications Union
JBIG	Joint Bilevel Image Group
JBIG2	Joint Bilevel Image Group 2
JPC	JPEG code-stream
JPEG	Joint Photographic Experts Group
JPEG2000	Joint Photographic Experts Group 2000
KLT	Karhunen-Loeve Transform
LAN	Local Area Network
LBG	Linde-Buzo-Gray
LoG	Laplacian of Gaussian
LRCP	Layer-Resolution-Component-Position
LSB	Least Significant Bit
LSP	Linear Successive Projections
LUT	Look-Up Table
LVQ	Learning Vector Quantization
LZ	Lempel-Ziv
LZW	Lempel-Ziv-Welch
MeSH	Medical Subject Heading
MGAR	Multi-level Gaussian Auto-Regressive
MLP	Multi Layer Perceptrons
MRC	Mixed Raster Content
MS	Markov or Markovian Sources
MSB	Most Significant Bit
MSE	Mean Squared Error
MSICT	Multi-Segment Image Coding and Transmission
MTF	Modulation Transfer Function
NLSP	Non-Linear Successive Projections
OCR	Optical Character Recognition
OTF	Optical Transfer Function
PCA	Principal Component Analysis
PCM	Pulse Code Modulation
PCRL	Position-Component-Resolution-Layer

PLA	Page Layout Analysis
PNG	Portable Network Graphics
POCS	Projections Onto Convex Sets
PPM	Packed Packet Headers, Main Header
PPT	Packed Packet Headers, Tile-Part Header
PSF	Point Spread Function
PSNR	Peak Signal to Noise Ratio
PSTN	Public Switched Telecommunications Network
QoS	Quality of Service
RAG	Region Adjacency Graph
RCA	Robust Competitive Agglomeration
RLCP	Resolution-Layer-Component-Position
RLE	Run-Length Encoding
RMSE	Root Mean Squared Error
ROI	Region Of Interest
RPCL	Resolution-Position-Component-Layer
sCIELAB	Spatial CIELab color system
SFC	Shannon-Fano Coding
SNR	Signal to Noise Ration
SOFM	Self-Organizing Feature Maps
SOP	Start Of Packet
SOT	Start Of Tile
SPIHT	Set Partitioning In Hierarchical Trees
STG	Scene Transition Graph
SVD	Singular Value Decomposition
TCP	Transport Control Protocol
TSVQ	Tree-Structured Vector Quantization
URL	Universal Resource Locator
VLC	Variable-Length Code
VLI	Variable-Length Integer
VLSI	Very Large Scale Integration
VQ	Vector Quantizer (P. 309)
VQ	Vector Quantization (P. 70, 302, 309)
WHT	Walsh-Hadamard Transform
WSS	Wide-Sense Stationary
WT	Wavelet Transform

List of Figures

List of Tables

Chapter 1
Vision and Color Theory

—I was born before the Age of the ages, in a Place without place, in a Time without time. In a strange way, however, I feel that I existed prior to my birth...My presence counts the eternal.

Light

1.1 Introduction

Twenty four centuries ago, Plato presented in the history of the *prisoners in the cave* an analogy to human perception. The outline of his theory is that our perception of the world around us consists merely of reflections of the world within us, as captured by our senses and translated by any synthetic procedure performed inside our brains. This theory seems to be basically undeniable today and could be accepted by most. In modern philosophy, psychology and all sciences that deal with vision and visual perception (in medicine and in robotics or mechanics), this reflection is now a basis for developing new theories and methods upon which to build vision theories, or to develop artificial (machine) vision, or to design algorithms to adapt systems to our visual perception and limitations. One of the main philosophical questions that still hold today concerns the insight that while the picture we perceive is the result of processes of our brain and created within it, we are able to distinguish this picture from ourselves and perceive its content as an external surrounding world (Plato 2002; Lehar 2016).

Similarly, problems of concern to psychologists, biologists, medical researchers and engineers, as well as those involved in the cognitive sciences, have to do with modeling of human vision to better understand the mechanisms involved in order to implement them in machine vision systems and in visual information processing applications. The important point here is to keep in mind that the final arbiter, the ultimate judge, is always the human observer, who is to benchmark all developed visual information management systems. Thus, processes such as the comparison between different images, or images produced by processing an initial image, along with concepts such as the image quality are so subjective that it is almost impossible to

© Springer Nature Singapore Pte Ltd. 2017
G. Pavlidis, *Mixed Raster Content*, Signals and Communication Technology,
DOI 10.1007/978-981-10-2830-4_1

establish a frame of reference beyond a subjective human observer. And this, perhaps because the problem of modeling the vision and understanding the root causes for which something looks 'nicer' than something else is quite complicated and difficult. If one accepts the theory of Plato that our perception shows us only a pale imitation, 'a shadow' of the real world (the 'ideal'), then we understand how difficult it is to develop theories and methods for the assessment (even the very concept) of quality in such a world.

It is true that the human eye and its operation has raised many debates and has caused the creation of several studies and publications for over two millennia. The key however, to better understand the main concepts in vision is considered to have been found in 1604 by Kepler, in his *theory of refraction through spherical lenses* (Fig. 1.1). By applying his theory to the eyes, Kepler showed that the eye has a clear function: to focus an inverted image on the retina, the surface on which light is projected (Kepler 1604). The retina, though, is not a passive image sensor, such as films. Instead, the retina is actively transforming images using tens of thousands of cells operating in series and in parallel. It is there that the first transformation of the image takes place, which is none other than digitization (sampling), since there is only a finite number of photosensitive cells, the *photoreceptors*, located at distinct locations that are capable of producing responses of a limited range. These photoreceptors, the so called *cones* and *rods*, have different sensitivity and response to light conditions. They have the ability to change their electrical activity whenever light shines upon them, and to capture and react even to a single photon at every moment. The remarkable in the performance of these sensors is that *it is sufficient for just ten photons to be captured by a cone to create the sensation and perception of light.*

An interesting collection of sources for color science was published in 1970 by MacAdam in his *Sources of Color Science* (MacAdam 1970), where he selected, edited and presented significant sources from Plato and Aristotle to Newton, Palmer, Young, Grassmann, Maxwell, Helmholtz, and more, covering a history of science developments on light and vision between c.380 B.C. and the 1950s.

Since digital image processing (both segmentation and compression that are of concern in this treatise) is evaluated by the human visual system, it seems reasonable to dedicate a chapter to the introduction of this system and the corresponding color theory.

1.2 Light

The development of the quantum theory of light is clearly beyond the scope of this treatise, but the exposition of its basic principles, which impact on the understanding of human vision, is necessary for any introduction to computer vision and image processing.

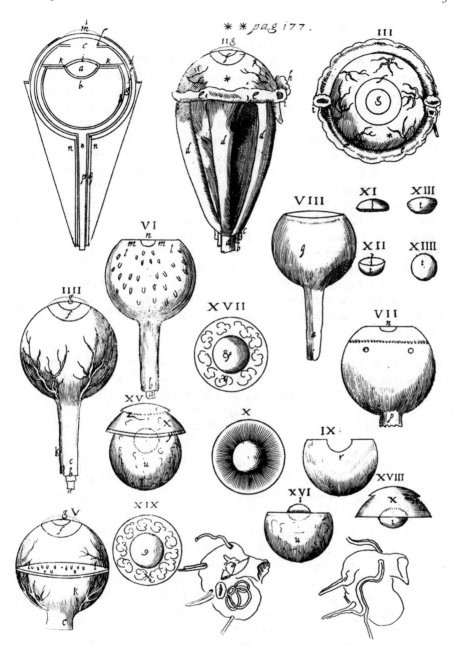

Fig. 1.1 Illustration of the structure of the eyes from Kepler's "Astronomiae Pars Optica"

The understanding and definition of the nature of light has a long history, one that was written by the most prominent scientists throughout the ages and peoples of this world. Newton was among the scientists to have tried to systematically define what light is. In 1675 he proposed that light is a continuous flow of particles, which travel in straight lines. Each particle is called a quantum and every light quantum is a photon. Thus, the intensity of light is measured by the number of those photons. Newton was supporting a particle theory of light. Then a number of scientists supported a wave theory of light, stating that light is a wave and has no mass. Among those scientists are Robert Hooke and Christiaan Huygens in the 17th century, Leonhard Euler and Thomas Young in the 18th century, Augustin-Jean Fresnel and Simon Denis Poisson in the 19th century. With the work of Michael Faraday, James Clerk Maxwell and Heinrich Hertz also in the 19th century there was a shift towards an electromagnetic theory of light. Einstein in the early 20th century, in his special relativity proposed a dual nature of light, with which he managed to solve the paradox of the constant velocity of light by redefining space and time as variable entities and by connecting mass to energy with his famous equation. This theory made possible to solve the paradox of the photoelectric phenomenon. At the same time, Max Planck studied the black body radiation and proposed that the light composes of packets or quanta, the photos, and with other scientists set the basis for quantum mechanics and quantum electrodynamics.

Today we accept a *quantum theory of light*. It is the author's personal impression that a very poetic and yet entirely true (to our current knowledge) view of the nature of light has been imprinted by the esteemed physicist professor Giorgos Grammatikakis in his excellent book *The Autobiography of Light* (Grammatikakis 2005), which is quoted here in the original Greek language. Light presents itself,[1]

«Γεννήθηκα πριν από αιώνες αιώνων, σε έναν Χώρο όπου δεν υπήρχε χώρος και σε έναν Χρόνο όπου δεν υπήρχε χρόνος. Με έναν περίεργο ωστόσο τρόπο, αισθάνομαι ότι προϋπήρχα της γενέσεώς μου. Κι ενώ από τότε όλα έχουν αλλάξει, εγώ αισθάνομαι ότι τίποτα δεν έχει αλλάξει. Η παρουσία μου μετρά το αιώνιο.»

In practice (and for the purposes of this treatise), light defines (although in a limiting sense) a part of the electromagnetic radiation spectrum that is visible to the humans by being detectable by the human visual system. This visible part of the electromagnetic spectrum, in the scale of wavelengths, roughly spans a range between 380 nm (corresponding to what is sensed as a violet color light) to 760 nm (corresponding to what is sensed as a red color light).

One among the very important properties of light propagation is its *refraction*, which occurs when light passes through different media, causing a change in its speed, which is expressed by a bending of its direction. Short wavelengths are refracted more

---

[1] In a somewhat free translation, this passage would read:
"I was born before the Age of the ages, in a Place without place, in a Time without time. In a strange way, however, I feel that I existed prior to my birth. Although since then everything has changed, I feel that nothing has changed. My presence counts the eternal."

Fig. 1.2 The rainbow of colors produced by a prism as a result of refraction

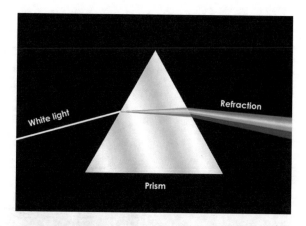

Fig. 1.3 Representation of the effect of the chromatic aberration of a lens

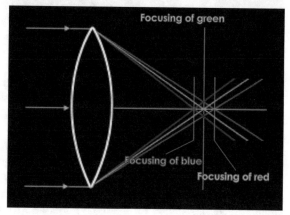

than longer wavelengths. This is why a 'light spectrum' becomes visible in the sky (as the all-loved rainbow), or when white light passes through a prism causing it to produce a rainbow of all colors (Fig. 1.2). The same property causes the occurrence of the *chromatic aberration of lenses*, which is the inability of a lens to focus different wavelengths of light at the same focal point, causing the blue color light to be focused in smaller distances from the lens than green or red color lights (Fig. 1.3).

1.3 The Human Visual System

In order for basic color theory concepts to become apparent it is important to have at least a brief understanding of the Human Visual System (HVS) and its basic structure and functionalities. One very important component of the HVS is the input sensory system, which consists of the eyes. Humans as many other beings on Earth facilitate *a pair of eyes* to be able to visually perceive the world on the basis of two main axes:

Fig. 1.4 Vision and color
perception

the geometry and the spectral manifestation of the world. This means no other than
to understand a scene and to be able to respond to challenges. And even though one
eye would suffice for *color vision*, two eyes are certainly needed to support *depth
perception*, the geometry of the scene.

The eyes are an excellent example of adaptation in evolution and, as suggested
by many research works like the seminal work of Barlow in 1982 (Barlow 1982),
compose a system *no-more and yet no-less than what is required*. In the following
paragraphs, the basic anatomic structure of the human eye is briefly presented, in an
attempt to set the ground for the more important (for the purposes of this treatise)
theory of visual perception and the 'technical' characteristics of visual perception
that are being heavily exploited in image processing applications.

According to Medical Subject Heading (MeSH), *vision* is the sensation of light.
Color perception is defined as the visual identification of each particular color hue
or its absence (Medical Subject Heading National Library of Medicine World of
Ophthalmology 2016). *Visual perception* is defined as the selection, organization and
interpretation of visual stimuli, based on the previous experience of the observer.
If one wanted to give a simple definition for color one would say that *color is
the subjective sensation that occurs when electromagnetic radiation of wavelengths
roughly between 400 and 700 nm gets captured by the eyes* (Fig. 1.4).

A representation of the anatomy of the human eye is shown in Fig. 1.5. It is nearly
a sphere with a 24 mm average diameter. Three membranes enclose the eye, namely,
the *cornea* and the *sclera* on the outside, the *choroid* in the middle and the *retina* on
the inside (Gonzalez and Woods 1992).

The *cornea*, with a diameter of about 11.5 mm and a thickness of about 0.6 mm, it is
perfectly transparent, without blood vessels but with many nerve endings (Goldstein
2006; Cassin and Solomon 1990). It is a touch tissue to cover and protect the lens. To

Fig. 1.5 Diagram of the human eye

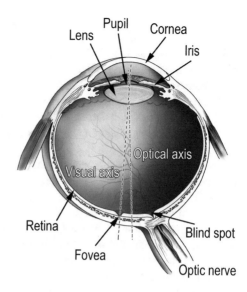

the left and right of the endings of the cornea is the *sclera*, from the Greek σχληρός- 'skleros' (meaning hard), the white part of the outer eye, a very tough fibrous layer, which is completely opaque (Cassin and Solomon 1990).

The *choroid* is a heavily pigmented layer just inside the sclera and on top of it. It purpose is to support the nutrition of the eye and further reduce the backscattering of light within the eye (Gonzalez and Woods 1992; Cassin and Solomon 1990). In front and below the cornea and above the *iris* is the *anterior chamber*, which is a chamber of about 3 mm filled with an aqueous humour (Cassin and Solomon 1990).

The *iris*, from the Greek mythological secondary goddess 'Ιρις related to atmospheric phenomena and particularly the rainbow, is a thin, circular structure in the eye, responsible for controlling the diameter and size of the pupil and thus the amount of light entering the eye. The iris defines the eye color. In an analogy to photographic systems, the iris is equivalent to the diaphragm that serves as the aperture stop (Cassin and Solomon 1990).

The *pupil*, is positioned perpendicular to the light entry direction and surrounded by the iris. It is a actually a hole located in the center of the iris to allow light to enter. It appears black because light rays entering the eye are either absorbed inside the eye directly or absorbed after diffuse reflections within the eye (Cassin and Solomon 1990). The pupil aided by the iris acts like the aperture to the lens, controlling the amount of light entering the eye. In the presence of bright light, light sensitive cells in the retina send signals to the circular iris muscle, which contracts to reduces the size of the pupil (pupillary light reflex). In sight of objects of interest or in the absence of bright light the pupil dilates. The pupil diameter varies from 2–4 mm in bright light to 3–8 mm in the dark, but this might depend on the age (Winn et al. 1994).

The *lens*, or *crystalline lens* is a transparent biconvex structure that refracts light to be focused on the retina. By changing shape the lens functions, as a zoom lens in cameras, to change the focal length of the eye to focus at various distances. The lens is typically circular of 10 mm diameter, contains no blood vessels or nerves and has a light yellow color. It absorbs around 8 % of the incoming light and blocks most ultraviolet light in the wavelength range of 300–400 nm, with shorter wavelengths being blocked by the cornea. It also absorbs infrared light. High intensity ultraviolet or infrared light can cause damage to the lens (Gonzalez and Woods 1992; Mainster 2006). It should be noted that all of the ocular organs have a slight yellowish color to compensate for the phenomenon of chromatic aberration, and transmit optical waves with lengths in the range 300–1600 nm.

The interior of the eye contains a transparent gelatinous colorless substance, the *vitreous humour*, which is a clear gel that fills the space between the lens and the retina. Its main function is to exert pressure on the retina to keep it in place against the choroid and thus to maintain the spherical shape of the eye and the ability to correctly focus the images. It has a composition similar to that of the cornea, contains no blood vessels and almost 99 % of its volume is water, while its main components are phagocytes to remove cellular debris from the visual field (Cassin and Solomon 1990). The *retina*, from the Latin *rete* (meaning 'net'), is the third and inner membrane of the eye, covering most of the inner surface. It is substantially a thin transparent tissue that consists of five cell types (photoreceptors—cones and rods, bipolar cells, horizontal cells, amalgam cells and ganglia) with a function similar to a film or the digital sensor in the cameras. Light goes through all levels of the retina until it reaches the photosensitive elements of the photoreceptors. This initiates a cascade of chemical and electrical events that ultimately trigger nerve impulses that are sent to the brain through the optic nerve (Fig. 1.6).

The photoreceptors, the neurons directly sensitive to light, are positioned in the pigment epithelium, which is rich in blood supply for supplying oxygen to the retina.

Fig. 1.6 Section of the retina with the main functional cells

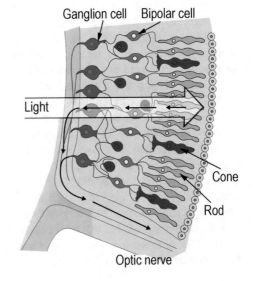

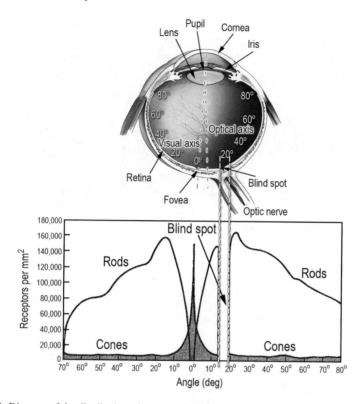

Fig. 1.7   Diagram of the distribution of cones and rods in the retina of the eye

In the retina there is also another type of photoreceptor, the *photosensitive ganglion cells*, which are rare cells but with the important function to provide feedback for reflexive responses to bright daylight (iris contraction) (Gonzalez and Woods 1992; Cassin and Solomon 1990; Wandell 1995). The area of the retina directly perpendicular to the light entrance is the *fovea*, which is responsible for the central vision, the part of the vision with the highest resolution and color sensitivity. It only contains cones, which are thinner and longer in this area and create a denser concentration. *Cones* are the cone-shaped photoreceptive cells that make color vision possible and are sensitive to normal light conditions, whereas, *rods* are rod-shaped cells that are sensitive to low light conditions but cannot distinguish color.

The *distribution of cones and rods* in the retina is illustrated in Fig. 1.7; at the central point (0°) the fovea is shown, where cones are in a very high concentration and moving away in both directions their density drastically decreases with an increase in their size. On the other hand, there are no rods in fovea but their distribution increases dramatically in the off-center periphery of the retina. At roughly 20° angle there is the blind spot, which is a position without photoreceptors since it is the spot optic nerve fibers converge to transmit the signals to the brain. The cons are responsible for the *central vision* or *photopic vision* during the daytime and are around 7 million

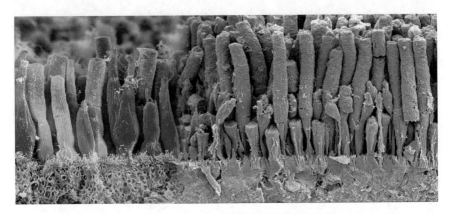

Fig. 1.8 Micro-photos of rods and cones in the human retina

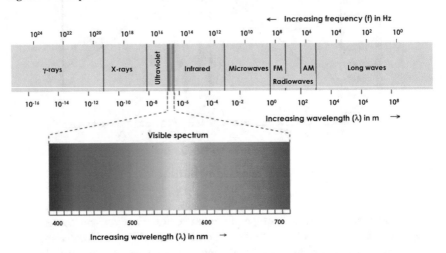

Fig. 1.9 The visual spectrum within the total known electromagnetic spectrum

in the retina, mostly in the fovea. The rods are responsible for the *peripheral vision* or *scotopic vision* in dim-light conditions and are around 100 million in the retina (Gonzalez and Woods 1992; Montag 2016). Figure 1.8 presents two superimposed microscope images of the cones and rods in the human retina (the pseudo-coloring indicates the two distinct types of cells).

Color only makes sense as referring to the human visual system and, therefore, there cannot be a discussion about color on another framework for the scope of this book. Figure 1.9 shows a typical illustration of the 'topology' of the electromagnetic spectrum and the position of the visible spectrum in humans. As mentioned in the previous paragraphs, the human eye has three types of cones that serve as color-sensitive sensors, since they are able to respond to different wavelengths of light. In respect to their spectral sensitivity and according to the wavelength range they absorb,

Fig. 1.10 Pseudo-colored representation of the distribution on L, M, S cones in the fovea

cones are divided into S-cones, which are sensitive to short wavelengths with a peak sensitivity in the violet range (ca. 440 nm), M-cones, which are sensitive to middle wavelengths with a peak sensitivity in the green range (ca. 540 nm) and L-cones, which are sensitive to long wavelengths with a peak sensitivity in the yellow-green range (ca. 570 nm) (Durand and Matusik 2001; Westland et al. 2012).

Figure 1.10 gives an enlarged pseudo-colored representation of the so-called mosaic of cones in the fovea. The colors correspond to the different kinds of cones: red for L, green for M and blue cones for S cones (Cassin and Solomon 1990). It should be noted that the distribution and shape of the mosaic vary (sometimes dramatically) from human to human, producing though a fairly stable image sensing outcome. This distribution is still under study and the mosaic image represents just a surface corresponding to a visual angle[2] of 1° (the central region of the fovea). The spectral sensitivity of the three types of cones that are responsible for the perception of color is shown in Fig. 1.11 (Stockman et al. 1993).

As pointed out by Yellott (1990), it was the 1850s when H. Müller first recognized that vision begins with a spatial sampling of the incoming image on the retina, by combining psychophysics and anatomy. He showed that these discrete photoreceptive elements of the retina must be the rods and cones (Brindley 1960). Müller initially argued that to visually distinguish two points it was sufficient for them simply to stimulate two different cones. On the other hand, C. Bergmann in 1858, supported that a continuous line would be indistinguishable from a row of dots (Bergmann 1858). Both Bergmann and later Helmholtz (1860), supported the three-cones view, where two stimulated cones are separated by an unstimulated, in order to resolve two points. In this way, the estimation of the limits of visual acuity was consistent with the

---

[2]Directions of rays of light are specified by their deviations from any of the ocular axes in terms of degrees (° or deg.), minutes (' or min.) and seconds (", or sec.) of an angle or arc.

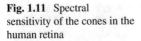

Fig. 1.11 Spectral sensitivity of the cones in the human retina

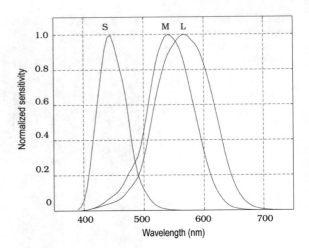

physical dimensions and spacing of the cones on the retina, giving a 60 cycles/degree frequency limit, which will be discussed in following paragraphs. Nowadays, it is commonly accepted that the optical quality of the human eye depends upon the spacing of cones at the central region of the fovea (Snyder and Miller 1977). As the retina discretizes the continuous distribution of light hitting the photoreceptor mosaic, it is expected to be susceptible to sampling artifacts or aliasing, and thus, there exist spatial resolution limits to be considered (Williams 1985) as suggested also by the previous works cited.

A number of interesting and important points are being summarized in the following paragraphs, following the distribution diagram presented in Fig. 1.7 and the problematic introduced in the previous paragraph:

- The ratio between the various cones was determined by estimates from measurements in a number of people of different ages, and amounts roughly to an estimate of (Montag 2016)

$$\text{S-cones} = 7\,\%, \quad \frac{\text{L-cones}}{\text{M-cones}} = 1.5 \tag{1.1}$$

- The central region of about 1.5 mm diameter, corresponding to an angle of $0.34°$, is characterized by the lack of S-cones. Throughout the retina, the ratio of $L+M$ to S-cones is about (Montag 2016)

$$\frac{(L+M)\text{-}cones}{S\text{-}cones} = \frac{100}{1} \tag{1.2}$$

- As shown in Fig. 1.7 the distribution of cones in this central region of the fovea is around 150,000 cells per mm^2 (specifically, Osterberg estimated it to $147.340\,mm^{-2}$ Osterberg 1935). Since that part of the fovea is a circle of about 1.5 mm diameter, the resolution can be estimated (for simplicity a square fovea of 1.5 mm side is considered) as

$$150,000 \times 1.5^2 = 337,000 \text{ elements}$$

$$\sqrt{150,000} \equiv 387.3 \text{ elements/mm per direction} \qquad (1.3)$$

$$\sqrt{150,000} \times 2.54 \approx 983.7 \text{ elements/inch}$$

which is about 980 (dpi) in the terminology used in displays and printers (Gonzalez and Woods 1992; Montag 2016).

- The linear distances on the retina can be estimated from the tangent or sine of the visual angle, given an average focal length of 16.7 mm. For small angles, say for a minimum of $0.1° = \pi/1800$ rad, this is expressed as

$$d = f \cdot \sin(\phi) = 16.7 \text{ mm} \times \sin(\frac{\pi}{180}) = 0.29\frac{\text{mm}}{\text{deg}} = 4.8\frac{\mu\text{m}}{\text{min}} \qquad (1.4)$$

f being the focal length and ϕ the angle. Thus the distance that is resolvable corresponding to the distance between consecutive cones in the fovea, which is estimated as 2–2.5 μm (Miller estimated it to be 3 μm on a hexagonal arrangement Miller 1979) corresponds to an angular separation of $0.5'$ or $30''$. The central fovea, measuring roughly a diameter of 0.29 mm, thus, corresponds to $1°$ visual angle. For comparison, this $1°$ is twice the angular diameter of the sun or the full moon, or a human head at 12–15 m distance (Montag 2016; Greger and Windhorst 1996).

- Assuming a hexagonal arrangement of the cones, Williams (1985) defined the critical frequency in cycles/deg above which aliasing occurs, namely the Nyquist limit[3] as $f_N = \left(\sqrt{3}s\right)^{-1}$, where s is the center-to-center spacing for foveal cones in deg. According to this formula and assuming a $s = 3$ μm cone spacing in the central fovea, where 0.29 mm correspond to $1°$ visual angle, yields that the minimum foveal cone spacing is about 0.62 arc-min.[4] Applying the formula the Nyquist frequency can be attained that (Williams 1985).

$$f_N = \left(\sqrt{3} \times \frac{0.62}{60}\right)^{-1} = \frac{60}{1.732 \times 0.62} = 55.874 \approx 56 \text{ cycles/deg} \qquad (1.5)$$

In literature can be also found other estimates that more or less converge to the same estimate (Williams 1985; Cassin and Solomon 1990; Montag 2016).

- It is noteworthy that the mosaic of the retina along with the overall processing allows for a super-resolution, expressed by an increased power to distinguish misalignments, called *hyperacuity*. Hyperacuity enables humans to distinguish

---

[3] As known from *digital signal processing*, the *Nyquist frequency* is substantially the frequency at which *aliasing* starts to appear. That is, a standard grating of a signal $\cos(2\pi(N/2 + f))$ above the Nyquist frequency cannot be distinguished from a signal $\cos(2\pi(N/2 - f))$ that is an equal amount below that frequency. The Nyquist frequency is practically applied as a lower limit of sampling in numerous engineering application, as the frequency twice the maximum frequency in the samples.

[4] In an ideal model, cone spacing (center-to-center spacing) should be considered as being larger than an individual cone diameter since there is an infinitesimal but still non-zero space between cone cells.

non-aligned objects to within 5 s of arc, which represents a fraction of the width of a cone, which empowers the detection of a misalignment of light sources from a distance of about 63 km (Montag 2016).

- Although the manner in which nerve impulses propagate through the visual system is not fully understood, it is known that the optic nerve contains around 800,000 nerve fibers. Considering there are more than 100,000,000 photoreceptors in the retina, it becomes obvious that interconnections between receptors and fibers is many-to-one (Hecht 1924). Apparently, the cone responses go through a second stage of color coding forming three distinct post-reciprocal channels (Westland et al. 2012),

  1. the $(L + M)$ *luminance channel*, made of additive inputs from the L and M cones
  2. the $(L - M)$ *chromatic channel*, made of opponent inputs from the L and M cones
  3. the $S - (L + M)$ *chromatic channel*, made of opponent inputs from the S and $(L + M)$ cones

- Based on the available data for the eye resolution—foveal cone spacing of about 0.6 arc-min, which according to Nyquist amounts to a resolution of

$$\frac{1}{\frac{0.6}{2}} = \frac{1}{0.3} \text{ arc-min}^{-1} \tag{1.6}$$

Assuming a field of view of 120° by 120°, the number of pixels required to cover the scene with equivalent acuity would be for each of the dimensions,

$$\text{viewfield} \times \text{ resolution} =$$
$$= 120° \times 60 \ \frac{\text{arc-min}}{\text{deg.}} \times \frac{1}{0.3} \text{ arc-min}^{-1} = 24,000 \tag{1.7}$$

equivalent pixels in one dimension.
For a square field of view this amounts to

$$24,000 \times 24,000 = 576,000,000 \text{ pixels or } 576 \text{ MPixels} \tag{1.8}$$

It should be noted that *this is largely an oversimplification*, since the overall estimation is based only on foveal parameters, which account actually only for 1 arc-deg of the central fovea. The density of cones and the visual acuity are significantly reduced with increasing angles (from the optical center). On the other hand, the eye works like a video camera, and moves rapidly to cover bigger and bigger fields of view, constantly focusing on details of interest. Nonetheless, the presence of the second eye perplexes even more the estimations and the analogies.

- At low light levels, the human eye integrates up to about 15 s (Blackwell 1946). During the day, the eye is much less sensitive, over 600 times less (Knowles Middleton 1952). With an ability to function in bright light and still be able to function

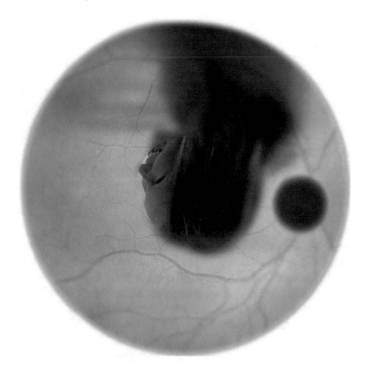

Fig. 1.12 Representation of the formation of an image on the retina

in faint starlight, the human eye has a range of more than $10,000,000:1$. In a single view the eye provides a $10,000$ range in contrast detection, which depends on the overall brightness of the scene, since the eye basically functions a sensor of contrast, and not an absolute sensor like the one in digital cameras.

A synthetic representation of how an image is formatted within the eye and on the retina is shown in Fig. 1.12, which takes into account the basic 'technical' characteristics and response of the eye, as described in the previous paragraphs. The real-world image, which is projected and blended on a photo of the inner eye, is obviously inverted, sharply focuses in a limited circular region around the optical center –the fovea– and gradually blurs out of focus towards the perimeter. The dark spot on the right is the blind spot, where no photoreceptors exist; the dark fuzzy lines are blood vessels.

All signals from the photoreceptors are gathered by the *optic nerve*, which follows a specific route towards the *visual cortex*, which is the part of the cerebral cortex responsible for the processing of the visual information that covers a large area at the back of the brain, as shown in Fig. 1.13. The left-half of the visual field of each eye is led to the left visual cortex whereas the right-half is led to the corresponding right visual cortex. It should be stressed that, since the images in the retina are inverted, the true left visual field is processed in the right visual cortex and vice versa.

Fig. 1.13 The route of the optical information from the eyes to perception

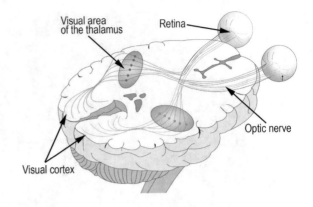

As discussed in the previous paragraphs, the visual system focuses on discriminating between different brightness levels and it is specifically efficient in detecting differences of brightness and alignments. The range of light intensity levels to which the HVS can adapt is enormous (on the order of 10^{10}). Experimental evidence suggests that this subjective brightness is a logarithmic function of the light intensity incident on the eye. As reported in (Gonzalez and Woods 1992) this characteristic is depicted in an intensity-brightness graph in Fig. 1.14. In this graph, the long solid curve represents the range of intensities to which the visual system can adapt. In photopic vision the range is about 10^6 (6 log levels). The transition from scotopic to photopic is gradual roughly over the range from 0.001 to 0.1 mL (-3 to -1 mL in the log scale). Of course, as pointed out in previous paragraphs, the visual system

Fig. 1.14 Intensity of light and perceived brightness

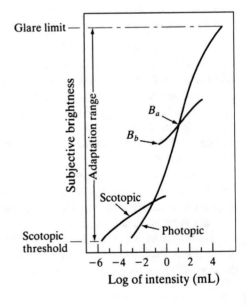

cannot operate on such an enormous dynamic range simultaneously, but rather it changes its overall sensitivity to adapt to the lighting conditions. This is known as the *brightness adaptation*. An example of the manifestation of brightness adaptation is shown in Fig. 1.14, where a brightness level B_a has been selected and the short curve intersecting the long one on B_a represents the range of perceived brightness when the eye adapts to this level. On the lower limit of the small curve, B_b indicates that below this limit all stimuli are perceived as indistinguishable blacks. The upper limit of the small curve is not strictly restricted, but if extended too far it will actually require for a new higher adaptation level B_a.

In addition to the ability of the eye to adapt to a specific intensity level, its ability to discriminate between changes in brightness at any adaptation level is also of great interest. If I is the intensity to which the eye is adapted and ΔI the increase in that intensity, which is just noticeable, then the ratio $\Delta I / I$ may be considered a measure of the discriminative power of the eye. This relation evolved interestingly into what is today known as the Weber-Fechner law (or simply the Weber's law) (Hecht 1924). According to this law, this ratio is constant within a wide range of intensities. Weber in 1834 (Weber 1834) discovered that one could discriminate between two weights if they differed by 1 or 2 parts in 30. Similarly, one could visually distinguish two lines if they differed by 1 part in 100, regardless of the absolute magnitude of the lines. The Weber's law in its initial form is expressed as $\Delta I = K_w \cdot I$, for some constant K_w, the *Weber fraction*, which is an equation of a line with slope K_w and zero intercept. The most used form of the law is by the ratio

$$K_w = \frac{\Delta I}{I} \tag{1.9}$$

which is easily transformed to a linear formula of the form $log(\Delta I) = log(I) + log(K_w)$.

The utility of Weber's law is in assessing the sensitivity of the HVS. In this usage, small values for K_w represent a high sensitivity (since little change is needed to produce a noticeable outcome) and larger values represent lower or even poor sensitivities. As deeply elaborated in (Hecht 1924) *this law only holds roughly within specific limits* of the mesopic-photopic vision, and rises greatly in the far-scotopic and less in far-photopic vision, which means that the law breaks down at the edges of detectable intensities, specifically towards the dark-end, where discrimination becomes very difficult to impossible. This is naturally supported by the findings of the significant work by Koenig and Brodhum (1889) who specifically defined the law's 'picture' in great precision, which has been produced according to those measurements in Fig. 1.15.

As Hecht pointed out (Hecht 1924), Fechner in 1858 (Fechner 1858), using the available astronomical data of that time, described a relation between the magnitude M of a star and its photometric intensity I,

$$M = k \cdot logI + C \tag{1.10}$$

Fig. 1.15 Weber's fraction
expressed for the whole
range of light intensities by
Koenig and Brodhun's data

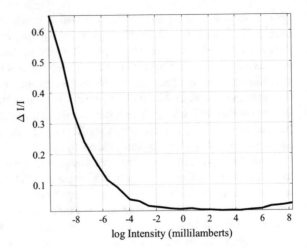

Fechner using his relation and the findings of Weber, developed the idea of a *constant fractional relation between two intensities that produce a threshold difference in perceived brightness*, naming this constant fractional relation the *Weber's law*. Assuming that the difference threshold represented a unit change in perception, he rewrote Weber's law as

$$S = k\frac{\Delta I}{I} \tag{1.11}$$

which by integration gave $S = k\log I + C$, S being the sensation (perceived brightness), an expression that is directly analogous to his calculation of the star magnitude-intensity relation, and which Fechner named as *the psychophysical law*. This law actually linearly connects the perceived brightness to the logarithm of the light intensity (like $S \triangleq f[log(I)]$). So by Fechner's psychophysical lay, two lights will be just discriminable if the response they generate differ by a constant amount ΔS.

Elaborating on this law we get:

$$
\begin{aligned}
\text{Fechner's law} \leftarrow \quad \Delta S &= log(I + \Delta I) - log(I) \Leftrightarrow \\
\Delta S &= log\left(\frac{I + \Delta I}{I}\right) \Leftrightarrow \\
e^{\Delta S} &= \frac{I + \Delta I}{I} \xlongequal{\Delta S = c} K \Leftrightarrow \\
K &= \frac{I + \Delta I}{I} \Leftrightarrow \\
K \cdot I &= I + \Delta I \Leftrightarrow \\
(K - 1)I &= \Delta I \xlongequal{K_w = K - 1} K_w I \Leftrightarrow \\
\Delta I &= K_w I \quad \rightarrow \quad \text{Weber's law}
\end{aligned} \tag{1.12}
$$

which states that if Fechner's psychophysical law holds, then Weber's law naturally emerges.

A classic experiment to determine the capacity of the HVS for brightness discrimination consists of having an observer look at a flat uniformly illuminated area large enough to occupy the entire field of view. This surface is shone from the back by a light source with a controlled varying intensity. This light source takes the form of a short duration flash that appears as a circle in the center of the field. If the increase in the illumination is bright enough, the observer gives a positive response for a perceived change. The ratio of the change in illumination over the background illumination, when the change is perceivable 50 % of the time, is called the *Weber ratio* $\Delta I/I$ (Gonzalez and Woods 1992). If the background (ambient) illumination is constant and the intensity of the flash light source varies from non-detectable to easily perceived, the typical observer can discern a total of a number of different intensity changes. This roughly relates to the number of intensities the observer may recognize at any one point in a monochrome image, which, in a way, counts the number of levels of intensity that can be discriminated. Again, due to the continuous movement of the eye the observer is in a constant brightness adaptation and the results of the simple light source experiment cannot be easily integrated to more general cases (Gonzalez and Woods 1992). The mapping of the Weber ratio over a wide range of light intensities and background levels ($\Delta I/I = f(I; I_0)$, I_0 being the parameter of the various background levels) is also referenced as the *contrast sensitivity* of the visual system. Since the differential of the logarithm of intensity is

$$d\,[log(I)] = \frac{dI}{I} \qquad\qquad (1.13)$$

it turns out that equal changes in the logarithm of the intensity of a light can be related to equal just noticeable differences over the region of intensities for which the Weber ratio is constant. This is the reason some processing algorithms operate on the logarithm of the intensity of images (Pratt 1991).

An interesting phenomenon relates to the fact that the visual system tends to under/over-shoot around the boundary of regions of different intensities, as illustrated in images like those depicted in Figs. 1.16 and 1.17. Specifically, Fig. 1.16 shows a typical representative example of this phenomenon, where, although the intensity within each of the stripes is constant, an observer would perceive a varying brightness pattern near the boundaries. The graph of graylevel image presents the profile of the gray levels to denote that each stripe is of constant brightness. In addition, this graph also includes a representation of the perceived brightness, which exhibits a logarithmic behavior near the stripes' boundaries.[5]

The stripes image is called a *Mach band pattern*, named after E. Mach, who first described the phenomenon in 1865. It is clearly evident that in boundaries of abrupt changes higher intensities seem to get into the dark region and vice versa.

---

[5]This representation is based on the graph provided by Cornsweet (1970).

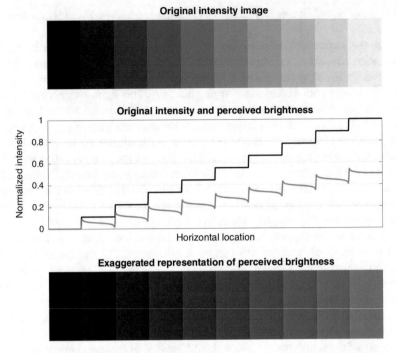

Fig. 1.16 Representation of the difference between intensity and perceived brightness in Mach bands

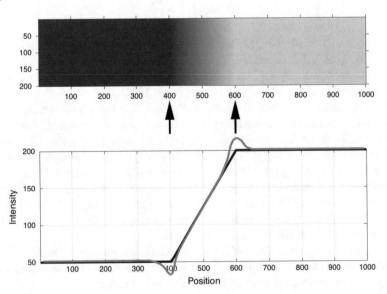

Fig. 1.17 Two intensity levels joined by a ramp transition and the two phantom perceived brightnesses

Fig. 1.18 Illustration of the simultaneous contrast, with small squares of the same intensity

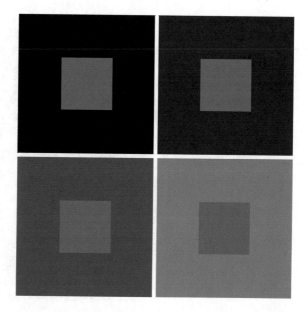

Furthermore, another interesting phenomenon, called *simultaneous contrast*, is related to the fact that a region's perceived brightness does not depend simply on its intensity but the overall relevant intensity context, as Fig. 1.18 demonstrates. In this figure, all the center squares have the exact same intensity, however they appear to become darker as the background gets lighter and vice versa (Gonzalez and Woods 1992).

In principle, in a linear optical system any spatially varying light source is expected to produce an equivalently varying response. By varying the spatial frequency of the input and recording the response it is possible to derive the Optical Transfer Function (OTF) of the system. Given $\omega_x = 2\pi/T_x$ and $\omega_y = 2\pi/T_y$ angular spatial frequencies with T_x, T_y the periods in the two coordinate directions, $\mathcal{H}(\omega_x, \omega_y)$ represents the OTF of the two dimensional linear system. Then the frequency spectra of the input ($\mathcal{I}_i(x, y)$) and output ($\mathcal{I}_o(x, y)$) intensity distributions can be defined as:

$$\mathcal{I}_i(\omega_x, \omega_y) = \iint_{-\infty}^{\infty} \mathcal{I}_i(x, y) e^{-i(\omega_x x + \omega_y y)} \, dx \, dy$$

$$\mathcal{I}_o(\omega_x, \omega_y) = \iint_{-\infty}^{\infty} \mathcal{I}_o(x, y) e^{-i(\omega_x x + \omega_y y)} \, dx \, dy \tag{1.14}$$

And the input and output spectra are related by

$$\mathcal{I}_o(\omega_x, \omega_y) = \mathcal{H}(\omega_x, \omega_y) \mathcal{I}_i(\omega_x, \omega_y) \tag{1.15}$$

Fig. 1.19 Modulated sine wave grating for MTF estimation

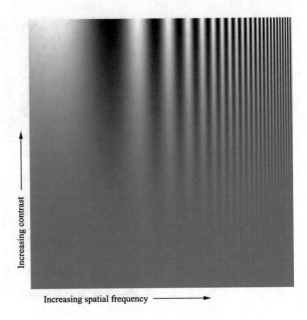

Increasing contrast

Increasing spatial frequency

Taking the inverse Fourier transform the output can be obtained as

$$\mathcal{I}_o(x, y) = \frac{1}{4\pi^2} \iint_{-\infty}^{\infty} \mathcal{I}_o(\omega_x, \omega_y) e^{-i(\omega_x x + \omega_y y)} \, d\omega_x d\omega_y \qquad (1.16)$$

The ratio of the magnitudes of the input and output frequency spectra,

$$\frac{|\mathcal{I}_o(\omega_x, \omega_y)|}{|\mathcal{I}_i(\omega_x, \omega_y)|} = |\mathcal{H}(\omega_x, \omega_y)| \qquad (1.17)$$

is called the Modulation Transfer Function (MTF) of the optical system. An indication of the form of the MTF can be obtained by observing the composite sine wave grating of Fig. 1.19. In this grating, there is a horizontally increasing frequency and a vertically increasing contrast. The envelope of the visible bars is actually a representation of what is called the *contrast sensitivity function*, an inverted image of the relative sensitivity relating to the spatial frequency (Pratt 1991).

The implication of the non-linear response of the eyes as portrayed in the previous paragraphs is graphically depicted in the normalized images of Fig. 1.20; in this graph it is shown that by using the 'linearized' CIELAB color space representation of the perceived brightness (L-channel, bottom image–denoted as L*a*b in the figure) then the corresponding RGB color space luminance representation of a surface that appears to be of the same brightness would be as shown in the above image. The above image is an approximation of a *gamma corrected* linear luminance scale using a gamma value of $\gamma = 2.44$ (*gamma correction* is simply the non-linear mapping

Fig. 1.20 Illustration of the full gamut of perceived brightness and the corresponding light intensities

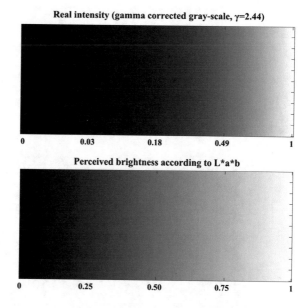

$L = L^\gamma$). As it appears in this graph, the brightness perceived to be at the middle of the L*a*b scale corresponds to a value of 0.18 (or 18 %) in the above scale. Thus a printed surface of a 18 % gray would appear to the human eye as being a middle gray illuminant (equally not a white and not a black). The industry has long been using this specific gray to represent the *middle gray* value in photography for calibration purposes, although the 18 % middle gray appearance is a *curiosity* of the human eye and not of the electronic systems that capture images (practically, the industry today is using various other gray values for the same calibration purposes).

It should be noted that the visual perception, expressing the attachment of a meaning to the content of an image, is a heavily non-linear and rather complex multi-dimensional process. It has been modeled by various researchers either in terms of the monochrome or the color version, the monoscopic and stereoscopic aspects, the connection to memory and other brain centers relating to cognition. It has been many ages ago (like in Plato's parable of the *prisoners in the cave*) since it has been identified that *visual perception is much more than seeing*. The perceived meaning is a result of heavy processing on visual stimuli in the eyes, being filtered by multiple cells and connected with various brain centers onto the visual cortex to form a complete representation that is meaningful to the individual (and only). Although there have been proposed models of vision (for example Gonzalez and Woods 1992; Pratt 1991; Jain 1988; Cormack 2000), visual perception at its full extend is still elusive.

Many efforts tried to identify either what visual perception is or what are its characteristics by experimentation, mainly to pinpoint limitations and basic characteristics. There efforts provided with examples of limitations like the ones known as *optical illusions*, which are evident manifestations of various aspects of the visual

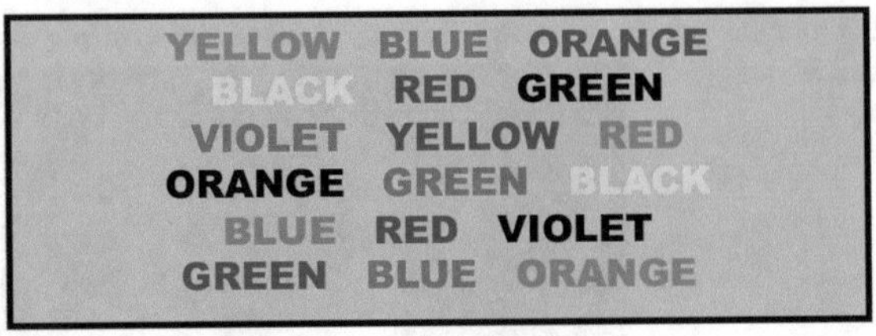

Fig. 1.21 Words and colors experiment

Fig. 1.22 Rotating snakes —illusory rotations (image courtesy of Akiyoshi Kitaoka)

perception. One interesting experiment is the one connecting words and their colors (Fig. 1.21), in which the observer is asked to state the color of a work; after some correct responses the observer finds it rather difficult to continue. Other famous optical illusions include the illusion of motion in a static image depending on the arrangement of colors and intensities in the image (Fig. 1.22), the search of the black dots experiment (the Hermann grid), where the observer is asked to count the number of black dots in junctions of horizontal and vertical lines (Fig. 1.23), ambiguous illusions where the content varies depending on the interpretation of dark of the light regions in the images (Fig. 1.24), illusory distortions in size, length or curvature of the geometry in the image (Fig. 1.25). An interesting aspect of the visual perception is illustrated also in Fig. 1.26, where an English piece of text is presented in a rather scrambled way; as suggested in the text that is easily read, in order to be able to read a text only the first and last letters of the words should be in the correct place.[6]

---

[6]An analysis on this matter was given in 2003 by Matt Davis in the webpage https://www.mrc-cbu. cam.ac.uk/personal/matt.davis/Cmabrigde/.

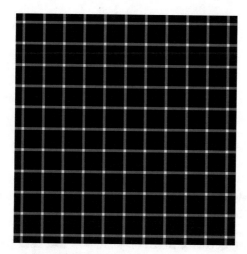

Fig. 1.23 Phantom *black dots* (based on 'Scintillating grid illusion' by Lingelback 1994)

Fig. 1.24 Ambiguity in the interpretation of the image content

Fig. 1.25 Distortions in size, length, curvature; *left to right* 'cafe wall illusion', 'Primrose's field', 'A Bulge' (the latter two images courtesy of Akiyoshi Kitaoka)

Fig. 1.26 Image content
recognition

Aoccdrnig to a rscheearch at Cmabrigde Uinervtisy, it
deosn't mttaer in whht oredr the ltteers in a wrod are, the
olny iprmoatnt tihng is taht the frist and lsat ltteer be in the
rghit pclae. The rset can be a taotl mses and you can sitll
raed it wouthit porbelm. Tihs is bcuseae the huamn mnid
deos not raed ervey lteter by istlef, but the wrod as a wlohe.

1.4 Color Management

The explanation of color vision have puzzled many scientists in the past beginning
with Newton and Maxwell, but the model being used today originates from Thomas
Young's theory around 1800 (Young 1802). Young proposed the trichromatic color
vision model, in which the human eye uses three types of photosensors that respond
to different wavelength bands. Based on the theory that emerged, the color vision
model includes three types of receptors, each of which exhibits a spectral sensitivity
$S_i(\lambda)$, where $i = 1, 2, 3$ and λ the wavelength. In this model, when a light of spectral
energy distribution $C(\lambda)$ excites the receptors, they produce the appropriate signal,

$$s_i = \int_\lambda C(\lambda) S_i(\lambda) \, d\lambda \qquad (1.18)$$

with (λ) varying in the range of the visible spectrum.

In the model proposed by Frei (1974), these signals s_i are subjected to a logarithmic
transfer function and are being combined to produce the outputs,

$$
\begin{aligned}
r_1 &= \log(s_1) && \text{Intensity of light source} \\
r_2 &= \log(s_2) - \log(s_1) = \log(\tfrac{s_2}{s_1}) && \text{Chromaticity of light source \#1} \\
r_3 &= \log(s_3) - \log(s_1) = \log(\tfrac{s_3}{s_1}) && \text{Chromaticity of light source \#2}
\end{aligned}
\qquad (1.19)
$$

Furthermore, these responses r_i pass through linear systems with transfer func-
tions $\mathcal{H}_i(\omega_x, \omega_y)$ to produce the signals to serve as the basis for color perception in
the brain. More detailed information on the modeling of the color vision in humans
can be found in (Cormack 2000).

The three types of cones are misleadingly referred to as red, green and blue. Misleadingly due to the fact that a stimulation of a unique cone sensitive to long, middle
or short wavelength does not necessarily lead to the sensation of a corresponding
red, green or blue color. In addition, as pointed out in the previous paragraphs, the
cones' sensitivity is not maximized in the red, green and blue parts of the spectrum.
The 'red' cones are most sensitive to the yellow-green part and the 'blue' cones
to the violet part of the spectrum. This is why, in order to be exact, one should
refer to them as *long-wavelength-sensitive (L)*, *middle-wavelength sensitive (M)* and
short-wavelength-sensitive (S) photoreceptors, respectively. *The signal each of the
cones generates expresses the rate it absorbs photons regardless of the wavelength,*

Fig. 1.27 Cone
fundamentals (sensitivities)
expressed in normalized
energy units

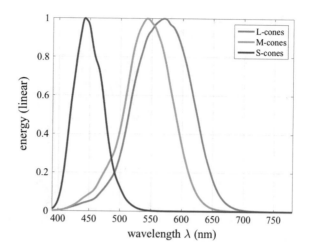

thus, single cones do not transmit information about the wavelengths of the photons
they absorb (no color information is produced). The perception of color depends
upon comparisons of the outputs of the three cone types, each with different spectral
sensitivity. *A normal observer* is able to match any light to a mixture produced by
three fixed-color light sources (like three flashlights of red, green and blue color). The
three functions expressing this matching are known as the Color Matching Functions
(CMF). The *cone fundamentals* can be thought of as CMFs of three imaginary matching lights, each of which exclusively and separately stimulates the one of the three
cone types (Stockman and Sharpe 2000). The cone fundamentals are representations
of the cone sensitivities and are illustrated in Fig. 1.27 based on the measurements
of Stockman and Sharpe[7] (2000).

It is apparent that since the three-receptor model is widely accepted, the minimum amount of sources needed to produce any color is three. Given three primary
sources of light, which are linearly independent with spectral energy distributions
$P_k(\lambda)$, $k = 1, 2, 3$ respectively and supposing that $\int_\lambda P_k(\lambda)d\lambda = 1$, to match a color
$C(\lambda)$ the three primaries are mixed in proportions of β_k, $k = 1, 2, 3$. Then the linear
combination

$$\sum_{k=1}^{3} \beta_k P_k(\lambda) \tag{1.20}$$

should be perceived as the color $C(\lambda)$, that is, the spectral response of the perceived
color will be (with integration limits $[\lambda_{\min}, \lambda_{\max}]$):

---

[7]Data obtained from the *Colour & Vision Research laboratory and database*, Institute of Ophthalmology, University College London, http://www.cvrl.org/cones.htm.

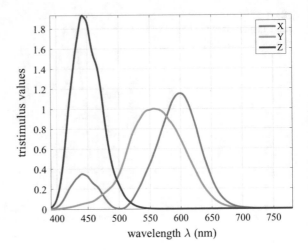

Fig. 1.28 Representation of the color matching functions (CMF)

$$\alpha_i(C) = \int S_i(\lambda)C(\lambda)\,\mathrm{d}\lambda$$

$$\alpha_i(C) = \int \left[\sum_{k=1}^{3} \beta_k P_k(\lambda)\right] S_i(\lambda)\,\mathrm{d}\lambda \qquad (1.21)$$

$$\alpha_i(C) = \sum_{k=1}^{3} \beta_k \int P_k(\lambda)S_i(\lambda)\,\mathrm{d}\lambda \qquad i = 1, 2, 3$$

If one unit of the k-th primary defines the response of the i-th cone as

$$\alpha_{i,k} \triangleq \alpha_i(P_k) = \int S_i(\lambda)P_k(\lambda)\,\mathrm{d}\lambda \qquad i, k = 1, 2, 3 \qquad (1.22)$$

can be derived that

$$\sum_{k=1}^{3} \beta_k P_k(\lambda) = \alpha_i(C) = \int S_i(\lambda)C(\lambda)\,\mathrm{d}\lambda \qquad i = 1, 2, 3 \qquad (1.23)$$

And these are the *color matching equations* or CMF. A typical graph of the CMF is shown in Fig. 1.28 concerning the 2-deg color matching functions as linear transformations of the 2-deg cone fundamentals based on measurements of Stockman and Sharpe[8] (2000) adopted by CIE in 2006 as the current physiologically-relevant fundamental CIE CMF.

If $C(\lambda)$ is an arbitrary color spectral distribution, $P_k(\lambda)$ the primary sources and $S_i(\lambda)$ the spectral sensitivity curves, the quantities β_k can be acquired by the solutions

---

[8]Data obtained from the *Colour & Vision Research laboratory and database*, Institute of Ophthalmology, University College London, http://www.cvrl.org/cmfs.htm.

of the Eq. (1.23). The primary sources are calibrated against a reference white light source with known energy distribution $W(\lambda)$. If w_k denotes the amount the k-th primary required to match the reference white, then the quantities

$$T_k(C) = \frac{\beta_k}{w_k} \qquad k = 1, 2, 3 \tag{1.24}$$

are the *tristimulus values* of color C, with the tristimulus values of the reference white being unity. *The tristimulus values for a color express the relative amount of primaries required to match that color.* The tristimulus values $T_k(\lambda)$ of a unit energy spectral color at wavelength λ give the *spectral matching curves*, which can be obtained by setting $C(\lambda) = \delta(\lambda - \lambda')$ in the CMF equations (1.23), which together with (1.24) yield three simultaneous equations

$$\sum_{k-1}^{3} w_k \alpha_{i,k} T_k(\lambda') = S_i(\lambda') \qquad i = 1, 2, 3 \tag{1.25}$$

for each λ'. Given the spectral tristimulus values $T_k(\lambda)$, the tristimulus values of an arbitrary color $C(\lambda)$ can be calculated as

$$T_k(C) = \int C(\lambda) T_k(\lambda)(d\lambda) \qquad k = 1, 2, 3 \tag{1.26}$$

Since the standard primary sources defined by CIE are

$$\begin{aligned}
P_1 &= \delta(\lambda - \lambda_1), & \lambda_1 &= 700 \text{ nm—red} \\
P_2 &= \delta(\lambda - \lambda_2), & \lambda_2 &= 546.1 \text{ nm—green} \\
P_3 &= \delta(\lambda - \lambda_3), & \lambda_3 &= 435.8 \text{ nm—blue}
\end{aligned} \tag{1.27}$$

using (1.22) can be obtained that $\alpha_{i,k} = S_i(\lambda_k)$, $k = 1, 2, 3$. Since the CIE white source has a flat spectrum $\alpha_i(W) = \int s_i(\lambda) d\lambda$, using these two relations in (1.23) for reference white, it can be derived that

$$\sum_{k=1}^{3} w_k S_i(\lambda_k) = \int S_i(\lambda) \, d\lambda \qquad i = 1, 2, 3 \tag{1.28}$$

which can be solved for w_k provided that $S_i(\lambda_k)$, $1 \le i, k \le 3$ is a nonsingular matrix. Using the spectral sensitivity curves and w_k, (1.25) can be solved for the spectral tristimulus values $T_k(\lambda)$ and obtain their plots as in the typical graph of Fig. 1.29 (Jain 1988).

Testing and quantification of the trichromatic theory of vision involves a process to reproduce any of the monochromatic colors (colors of a single wavelength) as a mixture of three primaries. Commission Internationale d'Eclairage (CIE) in 1931

Fig. 1.29 Spectral matching
tristimulus curves defined in
the CIE system

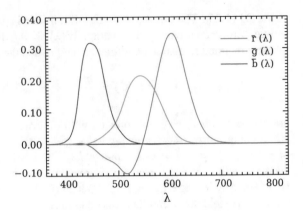

performed these color matching tests using the aforementioned primaries ($\lambda_1 =$ 700 nm—red, $\lambda_2 = 546.1$ nm—green, $\lambda_3 = 435.8$ nm—blue) and standardized the RGB representation the is still being used as the typical additive color space; Fig. 1.29 is actually a representation of the results attained by these experiments.

In 1954, Grassman (1954) concluded in eight (8) axioms to govern trichromatic color matching, which still serve as the basis for color measurements.

1. *Any color can be matched by a mixture of at most three colored lights.*
2. *A color match at one luminance level holds over a wide range of luminance levels.*
3. *A mixture of colored lights cannot be resolved by the human eye to its components.*
   They eye cannot tell the wavelength of a color.
4. *The luminance of a color mixture is equal to the sum of the luminance of its components.*
5. *Law of addition: if color C_1 matches color C_2 and color C_3 matches color C_4, then color C_1 mixed with color C_3 matches color C_2 mixed with color C_4.*

$$\left. \begin{array}{c} [C_1] \equiv [C_2] \\ [C_3] \equiv [C_4] \end{array} \right\} \Rightarrow \alpha[C_1] + \beta[C_3] \equiv \alpha[C_2] + \beta[C_4] \qquad (1.29)$$

6. *Law of subtraction: if a mixture of C_1 and C_2 matches a mixture of C_3 and C_4, and if C_2 matches C_4, then C_1 matches C_3.*

$$\left. \begin{array}{c} \alpha[C_1] + \beta[C_2] \equiv \alpha[C_3] + \beta[C_4] \\ [C_2] \equiv [C_4] \end{array} \right\} \Rightarrow [C_1] \equiv [C_3] \qquad (1.30)$$

7. *Transitive law: if C_1 matches C_2 and C_2 matches C_3, then C_1 matches C_3.*

$$\left. \begin{array}{c} [C_1] \equiv [C_2] \\ [C_2] \equiv [C_3] \end{array} \right\} \Rightarrow [C_1] \equiv [C_3] \qquad (1.31)$$

8. *Color matching: three types of color matches can be defined for any given color C.*

- Direct match by a mixture: $\alpha[C] \equiv \alpha_1[C_1] + \alpha_2[C_2] + \alpha_3[C_3]$
- Indirect match #1: $\alpha[C] + \alpha_1[C_1] \equiv \alpha_2[C_2] + \alpha_3[C_3]$
- Indirect match #2: $\alpha[C] + \alpha_1[C_1] + \alpha_2[C_2] \equiv \alpha_3[C_3]$

Let us consider an example of mixing colors, say red, green and yellow:

$$\begin{aligned}
C_r(\lambda) &= \delta(\lambda - \lambda_r) \quad \lambda_r = 620 \text{ nm—red} \\
C_g(\lambda) &= \delta(\lambda - \lambda_g) \quad \lambda_g = 490 \text{ nm—green} \\
C_y(\lambda) &= \delta(\lambda - \lambda_y) \quad \lambda_y = 560 \text{ nm—yellow}
\end{aligned} \tag{1.32}$$

According to the color sensitivity curves (Fig. 1.27), in this case the responses of the L and M cones to $C_y(\lambda) = \delta(\lambda - \lambda_y)$ are the same as the responses to the mixed color $C_{rg}(\lambda) = C_r(\lambda) + C_g(\lambda) = \delta(\lambda - \lambda_r) + \delta(\lambda - \lambda_g)$, as from (1.23)

$$\int S_i(\lambda) C_y(\lambda) d\lambda = \int S_i(\lambda) C_{rg}(\lambda) d\lambda \qquad (i = L, M)$$

This means that the colors $C_y(\lambda)$ and $C_{rg}(\lambda)$ cannot be distinguished and, actually, *red and green produce yellow.*

Apparently based on the tristimulus color theory, any color $C(\lambda)$ can be matched by mixing the proper amounts of three primaries:

$$C(\lambda) = \sum_{k=1}^{3} \beta_k(C) P_k(\lambda)$$

Since in the tristimulus model the intensities of two incoherent color light sources are added linearly to result new colors (in a typical linear combination, or weighted sum)

$$C_1 + C_2 = \sum_{k=1}^{3} [\beta_k(C_1) + \beta_k(C_2)] P_k(\lambda)$$

it is obvious that a three-dimensional (3D) color space is defined by a set of three primary colors to represent all colors. Each 3D point in this space represents a color as a vector of three components and each mixing of two colors is represented by the vector sum of the vectors of the mixed colors. *Any three linearly independent colors*[9] can become those primaries $P_k(\lambda)$, $(k = 1, 2, 3)$ that represent a basis of the 3D color space. In this setting, any color C can be specified by its color space coordinates $\{\beta_1(C), \beta_2(C), \beta_3(C)\}$, in a similar manner as a point in a 3D Cartesian coordinate system is represented by it typical coordinates $\{x, y, z\}$.

The 3D color space formed by taking the L, M and S cone response intensities as bases is graphically depicted in Fig. 1.30. In this specific color space, any color is represented by a vector initiated from the origin along a direction determined by the

---

[9]Linearly independent meaning that none of the three primaries can be written as the linear combination of the other two: $\sum_k \beta_k P_k(\lambda) \neq 0$, unless $\beta_k = 0$, $\forall k$.

Fig. 1.30 Graphical
representation of the LMS
color space

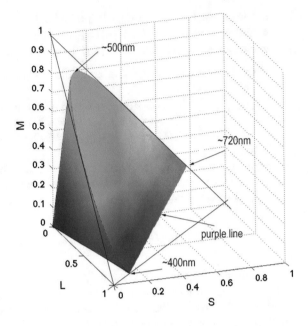

proportions of the LMS responses. All monochromatic colors in the visible range
form a curved surface like a bent fan (Fig. 1.30) that spans from very close to the
L-axis to very close to the S-axis, forming a convex surface. Truncating this curved
surface by a specific intensity results in an edge that is formerly called *the spectrum
locus*. The so called *purple line* or *the line of purples* is a line connecting the two
end points of the graph. Since any color can be represented by a linear combination
of all the spectral colors, it is represented in the LMS color space as a typical vector
summation along the surface, thus all perceived colors must lie within the conic solid
defined by the surface, the origin and the purple line. Taking a cross section on this
solid for a specific intensity defines the *chromaticity diagram* of all perceived colors
at the particular intensity, and this is usually displayed as a projection on a 2D plane.
In order to be able to match any color, CIE developed a set of three hypothetical
primaries X, Y, and Z, which by a linear combination can produce the color being
matched. This new XYZ color space (coordinate system) includes in its positive
octant the conical solid describing the LMS coordinate system (Fig. 1.30). The tristimulus spectral matching curves for these XYZ primaries are shown in Fig. 1.31
and are always positive for the entire range of visible wavelengths. According to
this representation, the chromaticity values or coordinates of any perceived color are
defined by the contribution of those three hypothetical primaries, normalized by the
total energy $X + Y + Z$,

$$x = \frac{X}{X + Y + Z}, \quad y = \frac{Y}{X + Y + Z}, \quad z = \frac{Z}{X + Y + Z} \quad (1.33)$$

Fig. 1.31 The CIE color sensitivity normalization curves

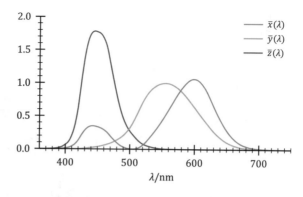

Fig. 1.32 The CIE-xy chromaticity diagram

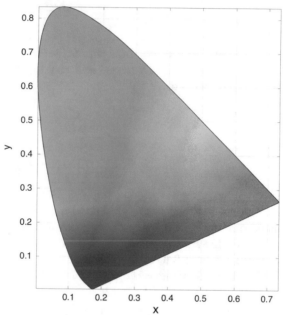

Given that $x + y + z = 1$ it is a common practice to use only two of these coordinates (x and y by convention) in addition to one of the primaries, and specifically the Y, since for the CIE 1931 standard observer definition, this primary corresponds to the luminance.

A typical representation of the CIE chromaticity diagram is shown in Fig. 1.32. It should be stressed though that this is just an illustration since the gamut of the printing system used for the production of this book is smaller than the chromaticity gamut of all possible colors and, thus, it cannot be considered accurate from a colorimetric point of view.

Fig. 1.33 Maxwell's RGB
chromaticity triangle

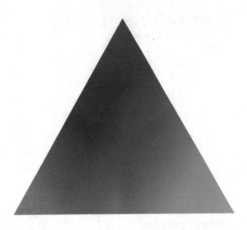

In a similar way the well known *Maxwell's RGB triangle* can be represented
as shown in Fig. 1.33.[10] Of course it is easy to go back to the primaries from the
chromaticity values as,

$$X = (\frac{x}{y})Y$$

$$Y = \frac{y}{y} \tag{1.34}$$

$$Z = (\frac{1-x-y}{y})Y$$

Between 1940 and 1980 a great number of color spaces stemming from transformations of the XYZ color space were proposed as uniform color spaces. It was in
1976 that CIE agreed upon two transformations that led to the CIELAB and CIELUV,
although numerous successful color spaces have already been used (Westland et al.
2012). The CIELAB transformation from XYZ is given by

$$L^* = 116f\left(\frac{Y}{Y_n}\right) - 16$$

$$a^* = 500\left[f\left(\frac{X}{X_n}\right) - f\left(\frac{Y}{Y_n}\right)\right] \tag{1.35}$$

$$b^* = 200\left[f\left(\frac{Y}{Y_n}\right) - f\left(\frac{Z}{Z_n}\right)\right]$$

---

[10]Figures 1.32, 1.33 and 1.34 were plotted by the *Colour toolbox* that accompanies the second edition
of "*Computational Colour Science using MATLAB*" (Westland et al. 2012), available for download
at   http://www.mathworks.com/matlabcentral/fileexchange/40640-computational-colour-science-
using-matlab-2e.

with

$$f(k) = \begin{cases} k^{\frac{1}{3}} & \text{if } k > (\delta)^3 \\ k\frac{1}{3\delta^2} + \frac{2\delta}{3} & \text{otherwise} \end{cases}$$

being a finite-slope approximation to the cube root with $\delta = 6/29$. In many cases the transformation is presented in a simpler and more practical manner as

$$L^* = 25 \left(100\frac{Y}{Y_n} \right)^{\frac{1}{3}} - 16$$

$$a^* = 500 \left[\left(\frac{X}{X_n} \right)^{\frac{1}{3}} - \left(\frac{Y}{Y_n} \right)^{\frac{1}{3}} \right]$$

$$b^* = 200 \left[\left(\frac{Y}{Y_n} \right)^{\frac{1}{3}} - \left(\frac{Z}{Z_n} \right)^{\frac{1}{3}} \right]$$

(1.36)

where X_n, Y_n, Z_n are the tristimulus values for the reference white (or the *neutral points*). In this color coordinate system (or color space) L^* represents the lightness in a $[0, 100]$ percentage range that roughly measures equal amounts of lightness perceptibility, a^* the red-green opponency and b^* the blue-yellow opponency.

In addition, if the tristimulus values for the reference white are known, it is possible to invert Eq. (1.35) as

$$Y = \begin{cases} Y_n \left[\frac{(L^*+16)}{116} \right]^3 & \text{if } L^* > 7.9996 \\ Y_n \frac{L^*}{903.3} & \text{otherwise} \end{cases}$$

$$X = \begin{cases} X_n \left(\frac{a^*}{500} + f_y \right)^3 & \text{if } \left(\frac{a^*}{500} + f_y \right)^3 > 0.008856 \\ X_n \left(\frac{a^*}{500} + f_y - \frac{16}{116} \right) \frac{1}{7.787} & \text{otherwise} \end{cases}$$

(1.37)

$$Z = \begin{cases} Z_n \left(f_y - \frac{b^*}{200} \right)^3 & \text{if } \left(f_y - \frac{b^*}{200} \right)^3 > 0.008856 \\ Z_n \left(f_y - \frac{b^*}{200} - \frac{16}{116} \right) \frac{1}{7.787} & \text{otherwise} \end{cases}$$

where

$$f_y = \begin{cases} \left(\frac{Y}{Y_n} \right)^{1/3} & \text{if } \frac{Y}{Y_n} > 0.008856 \\ 7.787\frac{Y}{Y_n} + \frac{16}{116} & \text{otherwise} \end{cases}$$

Equivalently, the CIELUV color space is defined according to the equations that transform from the XYZ as

Fig. 1.34 2D $a^* - b^*$
CIELAB color space
representation

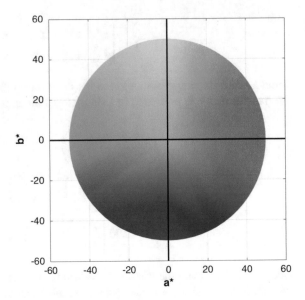

$$L^* = \begin{cases} 116\left(\frac{Y}{Y_n}\right)^{\frac{1}{3}} - 16 & \frac{Y}{Y_n} > 0.008856 \\ 903.3\left(\frac{Y}{Y_n}\right) & \end{cases}$$

$$u^* = 13L^*\left(u' - u_n\right)$$

$$v^* = 13L^*\left(v' - v_n\right)$$

(1.38)

where the subscript n refers to the neutral point and u', v' are the coordinates of the uniform chromaticity space CIE 1976 UCS, which transforms from the XYZ as

$$u' = \frac{4X}{X + 15Y + 3Z}$$

$$v' = \frac{9Y}{X + 15Y + 3Z}$$

It should be noted that CIELUV was proposed to support the representation of self-luminous colors, whereas CIELAB has been proposed as a standard to represent surface colors. In the recent years CIELAB has become the color space exclusively used by most the applications for color representation, appearance and matching. A representation of the $a^* - b^*$ plane of CIELAB is illustrated in Fig. 1.34, while Fig. 1.35 shows the complete color space defined by CIELAB and CIELUV side by side for comparison.[11]

---

[11] Figures 1.35, 1.36 and 1.43 were generated using the *Colorspace Transformations Toolkit* created by Pascal Getreuer (Jan. 2011 edition), available for download at http://www.mathworks.com/matlabcentral/fileexchange/28790-colorspace-transformations.

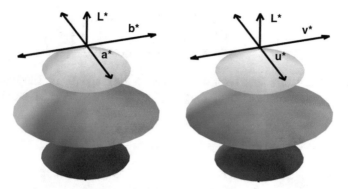

Fig. 1.35 Representation of the CIELAB and CIELUV color spaces

Over years of research and development, along CIELAB and CIELUV a number of other color spaces have been devised in order to cover specific needs and applications. All additive color spaces have been three-dimensional, whereas the well known CMYK color space is a four-dimensional coordinate system specifically defined for printing (subtractive representation of colors). RGB-based color spaces have been influenced by the CIE-RGB primaries definition, namely, R for 700 nm, G for 546.1 nm and B for 435.8 nm corresponding wavelengths, as mentioned in previous paragraphs. Color television systems were the first to define RGB color spaces, with NTSC and PAL dominating the early applications. In these technologies the *YUV and YIQ color spaces* were initially defined, which consisted of a luminosity and two chromaticity channels, in a way to ensure a backwards compatibility with the old black and white television receivers. These transformations in a matrix form are defined (Pratt 1991) as

$$
\begin{pmatrix} Y \\ U \\ V \end{pmatrix} = \begin{pmatrix} 0.299 & 0.587 & 0.114 \\ 0.147 & -0.289 & 0.437 \\ 0.615 & -0.515 & -0.100 \end{pmatrix} \begin{pmatrix} R \\ G \\ B \end{pmatrix},
$$

$$
\begin{pmatrix} Y \\ I \\ Q \end{pmatrix} = \begin{pmatrix} 0.299 & 0.587 & 0.114 \\ 0.596 & -0.274 & -0.322 \\ 0.211 & -0.523 & 0.312 \end{pmatrix} \begin{pmatrix} R \\ G \\ B \end{pmatrix}
$$

(1.39)

In 1996 Hewlett-Packard and Microsoft in an attempt to define a general scheme for the Microsoft Windows OS, HP products, CRT monitors and the Internet, proposed the *sRGB color space*, a color space compatible with the ITU-R BT.709-3 (Rec.709), which is the set of primaries agreed since 1990 for high definition television (HDTV). The transformation from XYZ to sRGB requires an intermediate transformation into linear RGB as

$$\begin{pmatrix} R \\ G \\ B \end{pmatrix} = \begin{pmatrix} 3.2406 & -1.5372 & -0.4986 \\ -0.9689 & 1.8758 & 0.0415 \\ 0.0557 & -0.2040 & 1.0570 \end{pmatrix} \begin{pmatrix} X \\ Y \\ Z \end{pmatrix} \tag{1.40}$$

where the XYZ values have been scaled to be in the range [0, 1]. The RGB values calculated by this formula are in the range [0, 1]. Then the sRGB values are calculated as follows

$$sR = \begin{cases} 12.92R & R \le 0.0031308 \\ (1+0.055)R^{\frac{1}{2.4}} - 0.055 & \text{otherwise} \end{cases}$$

$$sG = \begin{cases} 12.92G & G \le 0.0031308 \\ (1+0.055)G^{\frac{1}{2.4}} - 0.055 & \text{otherwise} \end{cases} \tag{1.41}$$

$$sB = \begin{cases} 12.92B & B \le 0.0031308 \\ (1+0.055)B^{\frac{1}{2.4}} - 0.055 & \text{otherwise} \end{cases}$$

The inverse transform is as follows

$$R = \begin{cases} \frac{sR}{12.92} & sR \le 0.004045 \\ \left(\frac{sR+0.055}{1+0.055}\right)^{2.4} & \text{otherwise} \end{cases}$$

$$G = \begin{cases} \frac{sG}{12.92} & sG \le 0.004045 \\ \left(\frac{sG+0.055}{1+0.055}\right)^{2.4} & \text{otherwise} \end{cases} \tag{1.42}$$

$$B = \begin{cases} \frac{sB}{12.92} & sB \le 0.004045 \\ \left(\frac{sB+0.055}{1+0.055}\right)^{2.4} & \text{otherwise} \end{cases}$$

and then

$$\begin{pmatrix} X \\ Y \\ Z \end{pmatrix} = \begin{pmatrix} 0.4124 & 0.3576 & 0.1805 \\ 0.2126 & 0.7152 & 0.0722 \\ 0.0193 & 0.1192 & 0.9505 \end{pmatrix} \begin{pmatrix} R \\ G \\ B \end{pmatrix} \tag{1.43}$$

An augmentation of sRGB was developed by Adobe in 1998, named *AdobeRGB*, which includes a considerable amount of the colors that can be reproduced by a typical CMYK printing system. Although significantly useful in printing it is very unlikely that most display devices could exploit the total amount of information it conveys, as there are limitations imposed by the specifics of the technology. In addition, AdobeRGB recommends to use images with bit-depths greater than 8 bits per pixel and per channel, which also exceeds the capabilities of many systems.

Figure 1.36 shows a representation of the typical linear RGB color space. Figures 1.37, 1.38, 1.39, 1.40, 1.41 and 1.42 show some color space representations of typical test images in the three dimensional RGB color space. Clearly, natural images exhibit a formation of random sparse blobs within the RGB color cube, whereas the artificial and mixed images exhibit various geometric concentrations and constructs in the three dimensional color space.

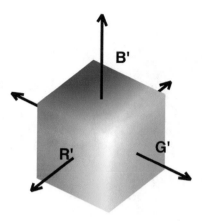

Fig. 1.36 Representation of the linear RGB color space

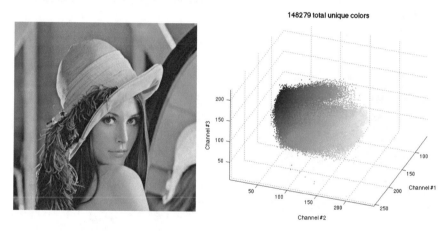

Fig. 1.37 Continuous-tone color image 'lena' (*left*) and the corresponding set of RGB colors (*right*)

A large family of color spaces map the space into a luminance and two chrominance coordinates, basically, accepting the ITU-R Recommendation BT.601-4 for the definition of luminance as

$$Y_{601} = 0.299R + 0.587G + 0.114B$$

(with the RGB values normalized to unity), and changing the representation of chrominance according to the application (Brainard 2003; Poynton 1997). Such color systems include, apart from the aforementioned YUV, YIQ, the widely used YP_bP_r and YC_bC_r.

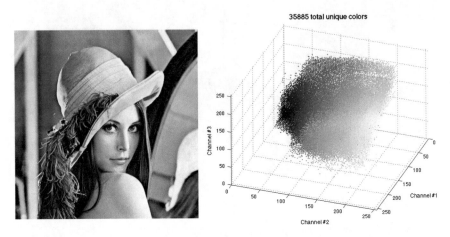

Fig. 1.38 Continuous-tone color image 'lena' after histogram equalization (*left*) and the corresponding set of RGB colors (*right*)

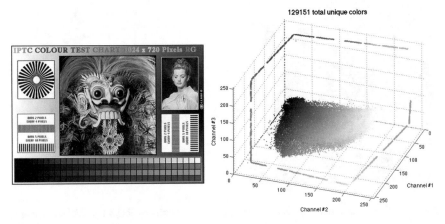

Fig. 1.39 Mixed content color image 'IPCT_test' (*left*) and the corresponding set of RGB colors (*right*)

Specifically, YP_bP_r is defined by the transformation from RGB as

$$\begin{pmatrix} Y \\ P_b \\ P_r \end{pmatrix} = \begin{pmatrix} 0.2990000 & 0.587000 & 0.114000 \\ -0.1687367 & -0.331264 & 0.500000 \\ 0.5000000 & -0.418688 & -0.081312 \end{pmatrix} \begin{pmatrix} R \\ G \\ B \end{pmatrix} \qquad (1.44)$$

with RGB and Y values in the range [0, 1] and the Pb, Pr values in [−0.5, 0.5]. Similarly, YC_bC_r is defined by the transformation from RGB as

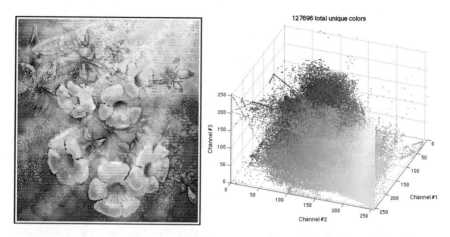

Fig. 1.40 Highly dithered artificial color image 'clegg' (*left*) and the corresponding set of RGB colors (*right*)

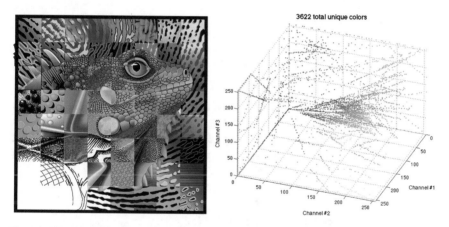

Fig. 1.41 Artificial color image 'frymire' (*left*) and the corresponding set of RGB colors (*right*)

$$
\begin{pmatrix} Y \\ C_b \\ C_r \end{pmatrix} = \begin{pmatrix} 65.481 & 128.553 & 24.966 \\ -37.797 & -74.203 & 112.000 \\ 112.000 & -93.786 & -18.214 \end{pmatrix} \begin{pmatrix} R \\ G \\ B \end{pmatrix} + \begin{pmatrix} 16 \\ 128 \\ 128 \end{pmatrix} \quad (1.45)
$$

with RGB values in the range [0, 1], the Y values in [16, 235] and the C_b, C_r values in [16, 240].

In addition, color spaces like *HSL (Hue-Saturation-Lightness)* or *HSV (Hue-Saturation-Value)* and *HSI (Hue-Saturation-Intensity)* adopt an alternative definition of lightness and define the color hue and purity. HSV is a rather intuitive color space, representing any color by its hue –as the angle on the HSV color wheel– its saturation –as the color's vibrance– and its brightness. HSV relates to sRGB by

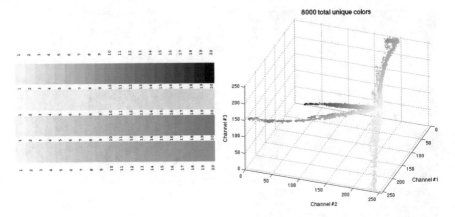

Fig. 1.42 Artificial test color image 'color_shades' (*left*) and the corresponding set of RGB colors (*right*)

$$H = \text{hexagonal hue angle} \qquad\qquad H \in [0, 360)$$
$$V = \max(R, G, B) \qquad\qquad\qquad V \in [0, 1]$$
$$S = \frac{\max(R, G, B) - \min(R, G, B)}{V} \qquad S \in [0, 1] \qquad (1.46)$$

The hue angle H is computed on a hexagon and the color space is represented geo-metrically by a hexagonal cone, as shown in Fig. 1.43-left.

HSL has the same definition for hue as HSV, but somehow different saturation and lightness components, such that all colors tend to white as lightness increases. HSL relates to sRGB by the following

$$H = \text{hexagonal hue angle} \qquad\qquad H \in [0, 360)$$
$$L = \frac{\max(R, G, B) + \min(R, G, B)}{2} \qquad L \in [0, 1]$$
$$S = \frac{\max(R, G, B) - \min(R, G, B)}{1 - |2L - 1|} \qquad S \in [0, 1] \qquad (1.47)$$

This color space is represented by a double hexagonal cone, as shown in Fig. 1.43-right.

Another popular color space in this set is HSI that relates to sRGB by

$$H = \text{polar hue angle} \qquad\qquad H \in [0, 360)$$
$$I = \frac{R + G + B}{3} \qquad\qquad\qquad L \in [0, 1]$$
$$S = 1 - \frac{\min(R, G, B)}{I} \qquad\qquad S \in [0, 1] \qquad (1.48)$$

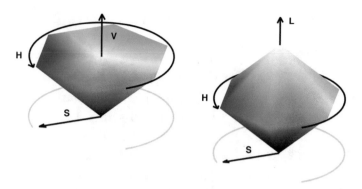

Fig. 1.43 Representation of the HSV (*left*) and the HSL (*right*) color spaces

Apparently, since H is computed as a polar hue angle, unlike HSV and HSL, it is represented on a circle rather than a hexagon. In all these representations the values of the RGB are normalized and the quantity

$$C = \max(R, G, B) - \min(R, G, B) \tag{1.49}$$

is sometimes referenced as the *chroma*.

Returning back to RGB from these color spaces is somehow complicated. Particularly, transformation from HSV to RGB, given the values are as defined in (1.46), requires a number of computations and a set of comparisons to produce intermediate values that are finally combined to reconstruct the original RGB values, as shown in (1.50).

$$C = V \times S$$

$$H' = \frac{H}{60°}$$

$$X = C(1 - |H' \mod 2 - 1|)$$

$$(R_1, G_1, B_1) = \begin{cases} (0, 0, 0) & \text{if } H \text{ is undefined} \\ (C, X, 0) & \text{if } 0 \leq H' < 1 \\ (X, C, 0) & \text{if } 1 \leq H' < 2 \\ (0, C, X) & \text{if } 2 \leq H' < 3 \\ (0, X, C) & \text{if } 3 \leq H' < 4 \\ (X, 0, C) & \text{if } 4 \leq H' < 5 \\ (C, 0, X) & \text{if } 5 \leq H' < 6 \end{cases} \tag{1.50}$$

$$m = V - C$$

$$(R, G, B) = (R_1 + m, G_1 + m, B_1 + m)$$

$$C = (1 - |2L - 1|) \times S$$

$$H' = \frac{H}{60°}$$

$$X = C(1 - |H' \mod 2 - 1|)$$

$$(R_1, G_1, B_1) = \begin{cases} (0, 0, 0) & \text{if } H \text{ is undefined} \\ (C, X, 0) & \text{if } 0 \le H' < 1 \\ (X, C, 0) & \text{if } 1 \le H' < 2 \\ (0, C, X) & \text{if } 2 \le H' < 3 \\ (0, X, C) & \text{if } 3 \le H' < 4 \\ (X, 0, C) & \text{if } 4 \le H' < 5 \\ (C, 0, X) & \text{if } 5 \le H' < 6 \end{cases} \qquad (1.51)$$

$$m = L - \frac{1}{2}C$$

$$(R, G, B) = (R_1 + m, G_1 + m, B_1 + m)$$

Similarly, returning to RGB from HSL is equivalent, with the only difference lying in the computations relating to *chroma* as shown in (1.51). In both cases, intermediate values for RGB components are being estimated before the final reconstruction of the original RGB values.

Modern CIELab color system (CIELAB) extensions try to linearize the correlation between a perceived and the displayed color (eg. the Opponent color space $O_1O_2O_3$ and the Spatial CIELab color system (sCIELAB)) (Brainard 2003; Poynton 1997; Zhang and Wandell 1997). Most often, the images are in the RGB format or are to be displayed in this format, while at intermediate stages it is required to convert from one color system to another.

It is worth noting that it is not so simple to transform to-and-from all the color spaces, since the systems designed to better model the HVS (such as the CIE systems and their spatial extensions—sCIELAB, Opponent Color Space) require the use of parameters highly dependent on ambient lighting conditions and the nature and characteristics of the light source or the imaging medium. It should also be stressed that the choice of the color description system for a digital image can play an important role in a compression process, since after converting to an appropriate color space a decorrelation of the color channels can be achieved, which reduces significantly the redundancies between the channels, resulting in better compression results and reconstruction quality (both subjective and objective).

Figures 1.44, 1.45, 1.46, 1.47, 1.48 and 1.49 show different color representations of a single digital color image. The image is originally in the RGB color space as shown in Fig. 1.44. These representations include three components, each of which is an intensity image (presented in grayscale). YUV, YIQ, YC_bC_r, CIELAB representations include a luminance component and two opponent chromaticity components that express the blue-yellow and red-green opponencies. All chromaticity

Fig. 1.44   A test color image and its RGB components

Fig. 1.45   The HSV representation of the test image

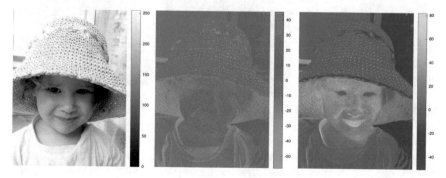

Fig. 1.46   The YUV representation of the test image

Fig. 1.47 The YIQ representation of the test image

Fig. 1.48 The YC$_b$C$_r$ representation of the test image

Fig. 1.49 The CIELAB representation of the test image

components are pseudo-colored to provide a general representation of the opponency that is embedded. In the HSV representation, H is pseudo-colored to show the actual hue represented in each pixel location.

References

Barlow, H. B. (1982). What causes trichromacy? A theoretical analysis using comb-filtered spectra. *Vision Research, 22,* 635–644.

Bergmann, C. (1858). Anatomisches und physiologisches über die netzhaut des auges. *Zeitschrift für rationelle Medicine, 2,* 83–108.

Blackwell, H. R. (1946). Contrast thresholds of the human eye. *Journal of the Optical Society of America, 36*(11), 624–643.

Brainard, D. H. (2003). Color appearance and color difference specification. In *The Science of Color* (2nd ed., pp. 191–216). Optical Society of America.

Brindley, G. S. (1960). *Physiology of the retina and visual pathway.* Baltimore, MD; London: Edward Arnold Publishers Ltd, ASIN; B001ISFUZS.

Cassin, B., & Solomon, S. (1990). *Dictionary of eye terminology.* Gainesville: Triad Publishing Co.

Cormack, L. K. (2000). *Computational models of early human vision* (pp. 271–288). San Diego: Academic Press.

Cornsweet, T. N. (1970). *Visual perception.* Academic Press Inc. ISBN 978-0121897505.

Durand, F., & Matusik, W. (2001). *Lecture notes on computer graphics [6.837].*

Fechner, G. T. (1858). Ueber ein' wichtiges psychophysisches Grundgesetz und dessen Beziehung zur Schaetzung der Sterngroessen. *Abhandl, k. saechs. Ges. Wissensch., Math.-Phys. Kl., IV*(31).

Frei, W. (1974). *A new model of color vision and some practical limitations.* Technical report. University of Southern California, Image Processing Institute, USCEE Report. 530.

Goldstein, E. B. (2006). *Sensation & perception.* Canada: Wadsworth Publishing Co Inc.

Gonzalez, R. C., & Woods, R. E. (1992). *Digital image processing* (3 ed.). Prentice Hall. ISBN 978-0201508031.

Grammatikakis, G. (2005). *The autobiography of light.* Creta University Press. ISBN 9789605242077.

Grassmann, H. G. (1954). Theory of compound colours. *Philosophic Magazine, 4*(7), 254–264.

Greger, R., & Windhorst, U. (1996). *Comprehensive human physiology: From cellular mechanisms to integration* (Vol. 1). Springer, ASIN: B01F9QNK0G.

Hecht, S. (1924). The visual discrimination of intensity and the Weber-Fechner law. *The Journal of General Physiology, 7,* 235–267.

Helmholtz, H. (1860). *Handbuch der Physiologichen Optik* (Vol. II). New York: English edition republished by Dover Publications 1962.

Jain, K. A. (1988). *Fundamentals of digital image processing.* New Jersey: Prentice-Hall.

Kepler, J. (1604). *Optics, paralipomena to witelo and the optical part of astronomy* (English translation by William H. Donahue). Santa Fe: Green Lion Press.

Knowles Middleton, W. E. (1952). *Vision through the atmosphere.* University of Toronto Press, ASIN: B0000CIGQF.

Koenig, A., & Brodhun, E. (1889). Experimentelle Untersuchungen fiber die psychophyslsche Fundamentalformel in Bezug auf den Gesichtssinn. *Sitzungsber. k. Akad. Wissensch., 641.*

Lehar, S. (2016). *Plato's cave and the nature of visuospatial perception.*

MacAdam, D. L. (1970). *Sources of color science.* Cambridge, Massachusetts and London, England: The MIT Press.

Mainster, M. A. (2006). Violet and blue light blocking intraocular lenses: photoprotection versus photoreception. *British Journal of Ophthalmology, 90*(6), 784–792.

Medical Subject Heading National Library of Medicine World of Ophthalmology, MeSH. (2016). *Vision.*

Miller, W. H. (1979). Ocular optical filtering, Chap. (Vol. 3, pp. 70–135). Berlin: Springer.

Montag, E. D. (2016). *Rods & cones - online course material.*

Osterberg, G. (1935). Topography of the layer of rods and cones in the human retina. *Acta Opthalmologica (Supplement), 6,* 1–103.

Plato. c.380 B.C. (2002). *Politeia* (pp. 503–511). 2002 edn. Ekdosis Polis. ISBN 978-960-8132-71-9. Chap. Z.

Poynton, C. (1997). *Frequently asked questions about color.*

Pratt, W. (1991). *Digital image processing* (2 ed.). Wiley-Interscience Publication. ISBN 0-471-85766-1.

Snyder, A. W., & Miller, W. H. (1977). Photoreceptor diameter and spacing for highest resolving power. *Journal of the Optical Society of America, 67,* 696–698.

Stockman, A., MacLeod, D. I. A., & Johnson, N. E. (1993). Spectral sensitivities of the human sensitive cones. *Journal of the Optical Society of America, 10,* 2491–2521.

Stockman, A., & Sharpe, L. T. (2000). Spectral sensitivities of the middle- and long-wavelength sensitive cones derived from measurements in observers of known genotype. *Vision Research, 40,* 1711–1737.

Wandell, B. (1995). *Foundations of vision: behaviour, neuroscience and computation.* USA: Sinauer Associates Inc.

Weber, E. H. (1834). *De pulsu, resorptione, auditu et tactu annotationes anatomicae et physiologicae.* Leipsic/Nabu Press (2011). ISBN 978-1247277530.

Westland, S., Ripamonti, C., & Cheung, V. (2012). *Computational colour science using Matlab* (2 ed.). Wiley-Blackwell. ISBN 978-0470665695.

Williams, D. R. (1985). Aliasing in human foveal vision. *Vision Research, 25*(2), 195–205.

Winn, B., Whitaker, D., Elliott, D. B., & Phillips, N. J. (1994). Factors affecting light-adapted pupil size in normal human subjects. *Investigative Ophthalmology & Visual Science, 35*(March), 1132–1137.

Yellott, J. I. (1990). The photoreceptor mosaic as an image sampling device. *In:Advances in photoreception: proceedings of a symposium on frontiers of visual science.* DC: National Research Council (US) Committee on Vision. National Academies Press (US).

Young, T. (1802). On the theory of light and colors. *Philosophical Transactions of the Royal Society of London, 92,* 20–71.

Zhang, X., & Wandell, B. A. (1997). A spatial extension of CIELAB for digital color image reproduction. *Journal of the Society for Information Display, 5,* 61–63.

Chapter 2
Data Coding and Image Compression

—The fundamental problem of communication is that of reproducing at one point either exactly or approximately a message selected at another point.

Shannon, Claude

2.1 Introduction

The need for compression of images becomes apparent when one counts the number of bits needed to represent the information content within each image. Let us consider, for example, the storage space required by the following types of images:

- A color image of compact size, say 512×512 pixels at 8 bpp and 3 color channels requires about 6.3×10^6 bits (768 KB).
- A digitized at 12 μm color negative photo of 24×36 mm (35 mm), 3000×2000 pixels at 8 bpp, requires 144×10^6 bits (about 17.5 MB).
- A digitized at 70 μm radiogram of 11×17 inch, 5000×6000 pixels at 12 bpp, requires 360×10^6 bits (about 44 MB).
- A multispectral LANDSAT image of 6000×6000 pixels per spectral band at 8 bpp for 6 spectral bands, requires 1.73×10^9 bits (about 211 MB).

These examples, and those mentioned in the Preface indicate the need for a method to tackle with this massive amount of data. In the following paragraphs an introduction into the realm of data compression is given and approaches for image compression are presented.

© Springer Nature Singapore Pte Ltd. 2017
G. Pavlidis, *Mixed Raster Content*, Signals and Communication Technology,
DOI 10.1007/978-981-10-2830-4_2

2.2   Quantification of Information

Since *Information Theory* can provide answers to two fundamental questions, "what is the ultimate data compression" (Entropy), and "what is the ultimate transmission rate of communication" (channel capacity) (Cover and Thomas 2006), questions which are also fundamental in the explorations in this treatise, it is important to include an introduction on these subjects. A need that naturally arises is to model and quantify 'information'. Fortunately, this issue has been addressed back one century ago. Hartley in his seminal paper in 1928 (Hartley 1928) realized that

> *...as commonly used, information is a very elastic term, and it will first be necessary to set up for it a more specific meaning...*

so his first care was about measuring information. Information communication was the context in this work. Hartley noticed that during the communication of a message, the precision of the information depends upon what other sequences might have been chosen. He realized that whenever new information became available towards completing the communicated message, ambiguity lowered, by excluding any other possibilities for the communicated meaning. Furthermore, by realizing that the symbols that comprise the information, in terms of quality and quantity, constitute an integral part of a process to measure the information, he first tried to eliminate undesirable side effects of this insight that are imposed by non-physical (like psychological) quantities. For example, the number of symbols available to two persons that speak different languages to communicate is negligible as compared to the available number of symbols to two persons speaking the same language. Hartley's approach was to propose that any physical system that transmits information should be indifferent to the content and interpretation and should focus on the ability of the system to distinguish the transmitted symbols.

Hartley assumed a system able to transmit s symbols by doing a number of n possible selections (arrangements of s symbols into sequences of n members). He deduced that the amount of information should be proportional to the number of selections n

$$H = Kn \tag{2.1}$$

where K a constant that depends on the number of symbols s available at each selection. By comparing two such systems that produce the same amount of possible sequences he found that

$$\left. \begin{array}{l} s_1^{n_1} = s_2^{n_2} \\ H = K_1 n_1 = K_2 n_2 \end{array} \right\} \Rightarrow \frac{K_1}{\log s_1} = \frac{K_2}{\log s_2} \tag{2.2}$$

where $s_k^{n_k}$ the number of possible distinguishable sequences of system k. Apparently this relation holds only if

$$K = K_0 \log s \tag{2.3}$$

with K_0 being the same for all systems, which may be omitted by making the base of the logarithm arbitrary. Then connecting (2.3) with (2.1),

$$H = n \log s \Rightarrow H = \log s^n \qquad (2.4)$$

This is equivalent to stating that *a practical measure of information is the logarithm of the number of possible symbol sequences.* If for example $n = 1$ (that is there is a single selection available) then the amount of information is equal to the logarithm of the number of symbols. Hartley, by introducing a logarithmic measure of information for communication (the logarithm of the alphabet size), had hacked his ways into the world of information and defined the first means to quantify it.

Shannon took on the work by Hartley and build a complete *Mathematical Theory of Communication* (Shannon 1948). Shannon approved Hartley's choice of a logarithmic measure for various reasons and pointed out that the choice of the logarithmic base corresponds to the choice of a unit for measuring information. So, if the base is two (2) then information is measured in *binary digits, or bits* (suggested by J.W. Tukey). Shannon distinguished the systems into two major categories of study, the *Discrete Noiseless Systems* and the *Discrete Systems with Noise*. He defined the *Discrete Channel* as

> *the system in which a sequence of choices from a finite set of elementary symbols can be transmitted from one point to another, where each symbol is assumed to have a certain duration in time*

Shannon then proceeded to define the *Discrete Source* of information using mathematical formalism, by realizing that such a source would produce a message symbol by symbol according to certain probabilities depending on the particular symbols and preceding symbol choices. Apparently what this formulation describes is a *stochastic process*. Shannon turned it also the other way around and stated that

> *...any stochastic process that produces a discrete sequence of symbols chosen from a finite set may be considered a discrete source.*

This was a crucial step in defining a mathematical description of a source of information. Building on this, Shannon defined the various series of approximations, like the zero-order, first-order, etc., in symbols and symbol sequences, bringing the notions of *Markov processes*[1] into the new information theory. Shannon further restricted the domain of interest into those Markov processes which exhibit a means of statistical homogeneity. The special case of processes that correspond to this domain are the *ergodic processes*, which exhibit the property that any collection of random samples from an ergodic process must represent the average statistical properties of the entire process. Subsequently, *an ergodic source* produces sequences with the same statistical properties.

---

[1] Shannon used the older spelling 'Markoff' for the famous Russian mathematician Andrei Andreyevich Markov.

After representing the discrete information sources as ergodic Markov sources, Shannon searched for a quantity to measure the amount of information produced by such sources, which can also be the rate at which information is produced by these sources. Supposing a set of possible events with corresponding probabilities $p_1, p_2, ..., p_n$ he searched for a measure $H(p_1, p_2, ..., p_n)$ with the following properties:

- H should be continuous in the p_i.
- If all the p_i are equal, $p_i = \frac{1}{n}$, then H should be a monotonic increasing function of n.
- If a choice be broken down into two successive choices, the original H should be the weighted sum of the individual values of H.

By elaborating on these desired properties, Shannon concluded that this measure is

$$H = -K \sum_{i=1}^{n} p_i \log p_i \tag{2.5}$$

Shannon realized that quantities of the form

$$H = - \sum_{i=1}^{n} p_i \log p_i \tag{2.6}$$

play a central role in quantifying information and thus the name *entropy* was coined to describe them, as they are directly related to the same definition of entropy in statistical mechanics. According to this definition, the entropy of two random variables with probabilities p and $q = 1 - p$ is $H = -(p \log p + q \log q)$ which gives rise to the famous bell-shaped graph shown in Fig. 2.1.

This quantity, the entropy H, has some interesting properties that further support its usability as an information measure:

1. It is a positive value except for one single case, where a single event is certain, thus the entropy collapses to zero,

$$H = \begin{cases} = 0, & p_i = 0, \ \forall i, \text{ except for one } p_j = 1 \\ > 0, & \text{otherwise} \end{cases} \tag{2.7}$$

2. The entropy is maximized for equally probable events, thus the most uncertain situation is successfully captured,

$$H = H_{max} = \log n, \qquad p_i = \frac{1}{n} \tag{2.8}$$

3. The entropy of a joint event of events x and y with $p(i, j)$ probability of joint occurrence of i for the first and j for the second is,

Fig. 2.1 Entropy of two random variables with probabilities p and $(1 - p)$

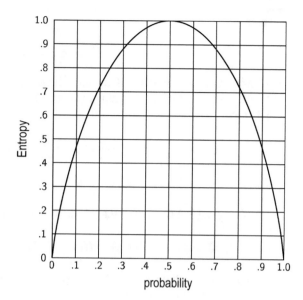

$$H(x, y) = -\sum_{i,j} p(i, j) \log p(i, j)$$

$$\text{while} \begin{cases} H(x) = -\sum_{i,j} p(i, j) \log \sum_j p(i, j) \\ H(y) = -\sum_{i,j} p(i, j) \log \sum_i p(i, j) \end{cases} \tag{2.9}$$

and it is easily shown that,

$$H(x, y) \leq H(x) + H(y) \tag{2.10}$$

with equality holding for the case of the events being independent ($p(i, j) = p(i)p(j)$).

4. Any equalization of the probabilities $p_i, p_2, ..., p_n$ or averaging of the form,

$$p_{i'} = \sum_j a_{ij} p_j, \quad \sum_i a_{ij} = \sum_j a_{ij} = 1, \ \forall a_{ij} \geq 0 \tag{2.11}$$

increases the entropy H.

5. Supposing two random events as in the 3rd case, the *conditional entropy* of y, $H_x(y)$ (modern notation for conditional entropy using the symbolism $H(y|x)$, but the original Shannon's notation is used here), is defined as the average of the entropy of y for each value of x, weighted according to the probability of getting that particular x,

$$H_x(y) = -\sum_{i,j} p(i, j) \log p_i(j) \tag{2.12}$$

where

$$p_i(j) = \frac{p(i,j)}{\sum_j p(i,j)} \tag{2.13}$$

is the conditional probability that y has the value j. Apparently, (2.12) quantifies the amount of uncertainty regarding y when x is known, on the average. Combining (2.12) with (2.13),

$$
\begin{aligned}
H_x(y) &= -\sum_{i,j} p(i,j) \, \log p(i,j) + \sum_{i,j} p(i,j) \, \log \sum_j p(i,j) \\
&= H(x,y) - H(x) \iff H(x,y) = H(x) + H_x(y)
\end{aligned}
\tag{2.14}
$$

6. Combining the 3rd (2.10) and the 5th (2.14) properties,

$$H(x) + H(y) \geq H(x,y) = H(x) + H_x(y) \iff H(y) \geq H_x(y) \tag{2.15}$$

which indicates that the knowledge of x decreases the uncertainty of y, unless the events x and y are independent.

The theory of entropy expanded to describe the uncertainty in an information source, which produced two very significant theorems limiting the data rates that can be attained in a compression system. By defining,

$$G_N = -\frac{1}{N} \sum_i p(B_i) \, \log p(B_i) \tag{2.16}$$

$p(B_i)$ being the probability of a sequence B_i of source symbols, and the summation defined over all sequences containing N symbols. This G_N is a monotonic decreasing function of N and

$$\lim_{N \to \infty} G_N = H \tag{2.17}$$

This describes the zero-order case, where there is no dependence on previously produced symbols. In all other cases, where the probability of a sequence B_i being followed by a symbol s_j is denoted by $p(B_i, s_j)$, where also the conditional probability of s_j after B_i is $p_{B_i}(s_j) = \frac{p(B_i, s_j)}{p(B_i)}$,

$$F_N = -\sum_{i,j} p(B_i, s_j) \, \log p_{B_i}(s_j) \tag{2.18}$$

summing over all blocks B_i of $N-1$ symbols and over all symbols s_j. Then F_N is also a monotonic decreasing function of N, and by combining (2.16) with (2.18) holds that,

$$F_N = NG_N - (N-1)G_{N-1}$$

$$G_N = \frac{1}{N} \sum_{n=1}^{N} F_N \qquad (2.19)$$

$$F_N \le G_N, \quad \lim_{N \to \infty} F_N = H$$

These very important results show that a series of approximations to H can be obtained by considering only the statistical structure of the sequences over the N symbols, and F_N is a better approximation, in fact being the N-th order approximation to the information source. F_N is the *conditional entropy* of the next symbol when knowing the $N-1$ preceding symbols. Shannon defined the ratio of the entropy of a source to the maximum value as the *relative entropy* of the source, identifying this as *the maximum compression possible*, and defined as *redundancy* the one minus the relative entropy.

At the time Shannon published this theory it was thought to be impossible to send information at a positive rate with negligible probability of error. Shannon proved that the probability of error could be made nearly zero for all communication rates below channel capacity. Shannon's *fundamental theorem for a noiseless channel*, today referenced as *the noiseless coding theorem*, stated that it is impossible to encode the output of a source in such a way as to transmit at an average rate less than or equal to $\frac{C}{H}$, where C the channel capacity (bits per second) and H the entropy of the source (bits per sample). In this theory Shannon defined the upper limit of any possible compression (the entropy H) for a random processes such as music, speech or images, by identifying that in such processes there is an irreducible complexity below which the signal cannot be compressed. Shannon argued that if the entropy of the source is less than the capacity of the channel, asymptotically error-free communication can be achieved (Cover and Thomas 2006).

Some typical examples of source entropy computations are as follows (bit-rates are reported typically in Bits Per Sample (bps)). Given $X = \{a, b, c, d, e\}$ with all symbols equiprobable ($p(a) = \ldots = p(e) = 1/5$), then the entropy of X would be,

$$H(X) = -5 \times \frac{1}{5} \log_2 \frac{1}{5} = 2.32 \, \text{bps} \qquad (2.20)$$

and a typical message produced by such a source could be (Shannon 1948),

B D C B C E C C C A D C B D D A A E C E E A
A B B D A E E C A C E E B A E E C B C E A D

When the probabilities of the symbols change, say, $p(a) = 1/2.5$, $p(b) = p(e) = 1/10$, $p(c) = p(d) = 1/5$, then the entropy of X becomes

$$H(X) = -\frac{1}{2.5} \log_2 \frac{1}{2.5} - \frac{1}{10} \log_2 \frac{1}{10} - \frac{1}{5} \log_2 \frac{1}{5} - \qquad (2.21)$$

$$- \frac{1}{5} \log_2 \frac{1}{5} - \frac{1}{10} \log_2 \frac{1}{10} = 2.12 \, \text{bps}$$

and a typical message produced by such a source could be (Shannon 1948),

A A A C D C B D C E A A D A D A C E D A
E A D C A B E D A D D C E C A A A A A D

Results change dramatically when joint and conditional probabilities are at work,

$$X = \begin{cases} a & p(a) = 9/27 \\ b & p(b) = 16/27 \\ c & p(c) = 2/27 \end{cases}$$

$$p_i(j) = \begin{pmatrix} p_a(a) = 0 & p_a(b) = 4/5 & p_a(c) = 1/5 \\ p_b(a) = 1/2 & p_b(b) = 1/2 & p_b(c) = 0 \\ p_c(a) = 1/2 & p_c(b) = 2/5 & p_c(c) = 1/10 \end{pmatrix} \tag{2.22}$$

$$p(i,j) = \begin{pmatrix} p(a,a) = 0 & p(a,b) = 4/15 & p(a,c) = 1/15 \\ p(b,a) = 8/27 & p(b,b) = 8/27 & p(b,c) = 0 \\ p(c,a) = 1/27 & p(c,b) = 4/135 & p(c,c) = 1/135 \end{pmatrix}$$

a typical message produced by such a source would be (Shannon 1948),

A B B A B A B A B A B A B A B A B B B A B B
B B B A B A B A B A B A B A B B B A C A C A
B B A B B B B A B B B A B A C B B B B A B A

This scales up considerably when moving from a source of symbols to a source of more complicated output, say words. Furthermore, by imposing a second-order word approximation and using the total English alphabet, Shannon was able to produce sequences such as (Shannon 1948),

THE HEAD AND IN FRONTAL ATTACK ON AN ENGLISH WRITER THAT THE CHARACTER OF THIS POINT IS THEREFORE ANOTHER METHOD FOR THE LET-TERS THAT THE TIME OF WHO EVER TOLD THE PROBLEM FOR AN UNEX-PECTED

which represents a sentence that could, actually, make some sense, even though Shannon's primary concern had to do with the decoupling of the message and the meaning.

When Shannon's attention switched to the case of channels with noise, then the focus had to be shifted from the compact representation of information to the reliable communication. Building on entropy as the measure of uncertainty, Shannon found it reasonable to use *the conditional entropy of the message, knowing the received signal*, as a measure of the missing information, which represents the amount of information needed at the received to correct the corrupted message,

$$R = R(x) - R_y(x) \tag{2.23}$$

R being the transmission rate, $R(x)$ the rate that corresponds to the $H(x)$ entropy of the source and $R_y(x)$ the rate that corresponds to the $H_y(x)$ conditional entropy of the source given the message received at the receiver. In a typical example, supposing two possible symbols 0 and 1, and a transmission at a rate of 1000 symbols per second with probabilities $p_0 = p_1 = 1/2$, and assuming that during transmission the noise alters 1 in 100 received symbols, if a 0 is received the *a posteriori* probability that a 0 was transmitted is 0.99, and that a 1 was transmitted is 0.01. Hence,

$$H_y(x) = -[0.99 \log_2 0.99 + 0.01 \log_2 0.01] = 0.081 \text{ bits per symbol} \quad (2.24)$$

which corresponds to $R_y(x) = 81$ bits per second in this case. Then the system is transmitting at a rate $R = R(x) - R_y(x) = 1000 - 81 = 919$ bits per second. In the extreme case where a 0 is equally likely to be received as a 0 or 1 and similarly for 1, the a posteriori probabilities are $1/2, 1/2$ and

$$H_y(x) = [0.5 \log_2 0.5 + 0.5 \log_2 0.5] = 1 \text{ bits per symbol} \quad (2.25)$$

which amounts to $R_y(x) = 1000$ bits per second. The rate of transmission is then $R = R(x) - R_y(x) = 1000 - 1000 = 0$ as expected. On this intuition, Shannon built his definition of the *capacity of a channel* as $C = max(H(x) - H_y(x))$ (with the 'max' in respect to all possible information sources used as input to the channel) and stated his *fundamental theorem for a discrete channel with noise*, today referenced also as *the noisy coding theorem*, which defines the limits for the rate of information that can be transmitted over a noisy channel keeping the error at a specific range. More formally, If the rate of the source is less than or equal to the capacity of the channel there exists a coding system such that the output of the source can be transmitted over the channel with an arbitrarily small frequency of errors.

A major achievement of the theory set out by Hartley and Shannon was the fundamental modeling of information as a probabilistic process, which can be measured in a manner that agrees with intuition. According to this modeling, a random event x with a probability of occurrence $p(x)$ is said to convey an amount of information content defined as (Gonzalez and Woods 1992)

$$I(x) = \log \frac{1}{p(x)} = -\log p(x) \quad (2.26)$$

where $I(x)$ is the *self-information* of x. Apparently, the self-information of an event is inversely related to the probability of its occurrence. So, for example, in the case of a certain event ($p(x) = 1$) there is no self-information ($I(x) = 0$) to attribute to this event. Intuitively, there is no meaning in transmitting a message about this event since its occurrence is certain. In any other case, (2.26) states that *the more surprising the event is, the more the information content it conveys*.

Other scientists built upon those theories or took their own path into the realm of information theory, such as Fisher who defined *the notion of sufficient statistic* (Fisher 1922) ($T(X)$ is sufficient relative to $\{f_\theta(x)\} \Leftrightarrow$ if $I(\theta; X) = I(\theta; T(X))$ for all distributions on θ), Lehmann and Scheffe who introduced the minimal sufficient statistic (Lehmann and Scheffe 1950), Kullback and Leibler who defined *the rela-*

tive entropy (or Kullback–Leibler distance, information divergence, cross entropy) (Kullback and Leibler 1951) ($D(p||q) = \sum_x p(x) \log \frac{p(x)}{q(x)}$), and Fano who proved his *Fano's inequality*, giving a lower bound on the error probability of any decoder (Fano 1952) ($P_e = Pr\hat{X}(Y) \neq X \Rightarrow H(P_e) + P_e \log |\mathcal{X}| \geq H(X|Y)$, $\mathcal{X}$ denoting the support of X).

Any process that produces information can be considered as a source of symbol sequences, which are pre-selected from a finite set. The very text of this treatise, for example, is written using a source that includes all the American Standard Code for Information Interchange (ASCII) symbols (basically, Unicode symbols). Similarly, a computer performs calculations using binary data, which can be considered as sequences of symbols created by a source with a binary alphabet (0 and 1). Thus an n-bpp digital image is created by a source with an alphabet of 2^n symbols (which represents all possible values). The classification of terms of the sequence generated by a source-image can be based on the values of adjacent pixels resulting in a one-dimensional scanning (1-D raster scan), or may be based on values resulting from two-dimensional image regions (2-D blocks of pixels). The development of models for different input images makes it easier to measure the information carried by the symbol sequences (Rabbani and Jones 1991a). Let us consider a couple of simple examples from the image processing domain to see how the very basic theory of entropy applies in two-dimensional signals. Suppose there is a simple graylevel (or grayscale) image shown in Fig. 2.2.

This image comprises of the following pixel values

$$\begin{pmatrix} 40 & 40 & 40 & 80 & 80 & 160 & 200 & 200 \\ 40 & 40 & 40 & 80 & 80 & 160 & 200 & 200 \\ 40 & 40 & 40 & 80 & 80 & 160 & 200 & 200 \\ 40 & 40 & 40 & 80 & 80 & 160 & 200 & 200 \\ 40 & 40 & 40 & 80 & 80 & 160 & 200 & 200 \\ 40 & 40 & 40 & 80 & 80 & 160 & 200 & 200 \\ 40 & 40 & 40 & 80 & 80 & 160 & 200 & 200 \\ 40 & 40 & 40 & 80 & 80 & 160 & 200 & 200 \end{pmatrix}$$

Suppose also that the image is represented with an 8 bpp resolution, which corresponds to a source producing 256 (2^8) different integer values ranging in [0, 255]. If all these values are equiprobable then this particular image is one of the $2^{8 \times 8 \times 8} =$

Fig. 2.2 Simple graylevel image

$2^512 \approx 1.34 \times 10^{154}$ equally probable 8×8 images that can be produced by the specific source. In this image, four symbols (gray values) are present,

$$p(g): \begin{cases} g_1 = 40 & p(40) = 0.375 \quad \text{(counts of '40' divided by all image pixels)} \\ g_2 = 80 & p(80) = 0.250 \\ g_3 = 160 & p(160) = 0.125 \\ g_4 = 200 & p(200) = 0.250 \end{cases}$$

where g the variable for the gray values. If a uniform distribution is assumed (equiprobable gray values) then apparently the source the produces such images is characterized by an entropy of 8 bpp. If the entropy of the specific image is examined then according to the fundamental law of entropy (2.6),

$$H = -\sum_g p(g) \log_2 p(g) \approx 1.9056 \text{ bpp}$$

which amounts to $1.9056 \times 8 \times 8 = 122$ total bits and is the first-order entropy approximation, since all pixel values are supposed to be independent. A second-order approximation may be attained if the image is converted to one line of pixels and all pairs be counted,

40 40 40 80 80 160 200 200 40 40 40 80 80 160 200 200
40 40 40 80 80 160 200 200 40 40 40 80 80 160 200 200
40 40 40 80 80 160 200 200 40 40 40 80 80 160 200 200
40 40 40 80 80 160 200 200 40 40 40 80 80 160 200 200

It is easy to identify the gray level pairs listed in Table 2.1. In this case, the entropy is estimated by (2.6) as $2.75/2 = 1.375$ bpp, with the division by 2 imposed by the fact that pixels are taken in pairs. Apparently, a better estimate of the entropy is achieved, and it is expected that higher-order approximations provide even better estimates for a considerable computational cost, though. It should be noted that if

Table 2.1 Second-order entropy approximation with gray level pairs

Gray level pair	Counts	Probabilities
(40, 40)	16	0.250
(40, 80)	8	0.125
(80, 80)	8	0.125
(80, 160)	8	0.125
(160, 200)	8	0.125
(200, 200)	8	0.125
(200, 40)	8	0.125

the pixels are statistically independent any higher-order approximation collapses to the first-order approximation.

The case changes even more if the image pixels are subtracted to form a new 'difference' image,[2]

$$\begin{pmatrix} 40\ 0\ 0\ 40\ 0\ 80\ 40\ 0 \\ 40\ 0\ 0\ 40\ 0\ 80\ 40\ 0 \\ 40\ 0\ 0\ 40\ 0\ 80\ 40\ 0 \\ 40\ 0\ 0\ 40\ 0\ 80\ 40\ 0 \\ 40\ 0\ 0\ 40\ 0\ 80\ 40\ 0 \\ 40\ 0\ 0\ 40\ 0\ 80\ 40\ 0 \\ 40\ 0\ 0\ 40\ 0\ 80\ 40\ 0 \\ 40\ 0\ 0\ 40\ 0\ 80\ 40\ 0 \end{pmatrix}$$

In this case the symbols are reduced to three (0, 40, 80) with probabilities,

$$p(g): \begin{cases} g_1 = 0 & p(0) = 0.500 \\ g_2 = 40 & p(40) = 0.375 \\ g_3 = 80 & p(80) = 0.125 \end{cases}$$

and the estimated entropy is 1.4056 bpp. Thus, by taking a 'difference' image, an image that encodes the difference of the adjacent pixel values, even with a first-order approximation, a lower entropy is measured, which corresponds to fewer bits to describe the original image data (90 bits in this example).

It should be noted here that the entropy estimate provided by (2.6) can be computed using matrix operations. In case the probabilities of the symbols $p(x_i)$ are arranged in a row vector $\mathbf{P}$ then the computation of entropy can become a matrix multiplication,

$$H = -\mathbf{P}^T \cdot \log \mathbf{P} = \begin{bmatrix} p(x_1) \cdots p(x_n) \end{bmatrix} \cdot \begin{bmatrix} \log p(x_1) \\ \vdots \\ \log p(x_n) \end{bmatrix} \qquad (2.27)$$

where the ^{T} notation denotes the transpose and the logarithm operates on each of the elements of the matrix, producing a row vector of the logarithms of the probabilities.

2.2.1 Discrete Memoryless Sources

The simplest form of an information source is what is referenced as a *Discrete Memoryless Source (DMS)*, wherein the output symbols are statistically independent. A DMS S is characterized by its alphabet $S = \{s_1, s_2, ..., s_n\}$ and the corresponding

---

[2]Each new pixel (to the right of the first) represents the difference of its original value and the value of the previous pixel.

probabilities $P = p(s_1), p(s_2), ..., p(s_n)$. An important concept is the average information provided by this DMS, but before defining it, one should define a way of quantifying information in an event. Quantification of information content should, of course, have some practical consequences; for example, it is reasonable to accept that the emergence of a less probable event (or symbol) contains more information than the appearance of a more probable (expected). It should be also noted that the information content of several independent events, regarded as a new fact, is the sum of the information of independent events (Rabbani and Jones 1991a).

Defining $I(s_i)$ the information received by the appearance of a particular symbol s_i in terms of probability,

$$I(s_i) = \log_k \left(\frac{1}{p(s_i)} \right) \tag{2.28}$$

where the base of the logarithm, k, determines the information unit. If $k = 2$, the information is binary (in the form of bits).

Taking the average value of this quantity for all possible symbols of the DMS, the average information per symbol, $H(S)$, known as *entropy* may be computed as:

$$H(S) = \sum_{i=1}^{n} p(s_i) I(s_i) = - \sum_{i=1}^{n} p(s_i) \log_2 p(s_i) \text{ bps} \tag{2.29}$$

As an example, consider one DMS S producing four symbols, $S = \{A, B, C, D\}$ with probabilities $P(A) = 0.60$, $P(B) = 0.30$, $P(C) = 0.05$, $P(D) = 0.05$. By using (2.29) the source entropy is $H(S) = -[0.6 \log_2 0.6 + 0.3 \log_2 0.3 + 2 \times 0.05 \log_2 0.05] \approx 1.4 \text{ bps}$.

The entropy of a source can be interpreted in two ways. By definition is the average value of information per symbol of the input source. It can also be defined as the average value of information per input source symbol that an observer needs to spend in order to alleviate the uncertainty of the source. For example, an observer is likely to want to recognize a symbol of the unknown source asking simple questions that require YES/NO (or TRUE/FALSE) answers. Each question is equivalent to an information bit. The observer needs to devise smart methods to minimize the average number of questions, which are required for the disclosure of a symbol or even a number of symbols. As intelligent though a method could be, it could never reveal symbols with an average number of binary questions less than the entropy of the source (Rabbani and Jones 1991a).

2.2.2 Extensions of Discrete Memoryless Sources

It is often useful to deal with symbol tables rather than pure symbols. Consider one DMS S with an alphabet of length n. Consider also that the output of the source is grouped in sets of N symbols. Each group is considered to be a new symbol of an extended new source S^N, which has an alphabet size n^N. This new source is called *the N-th extension of S*. Since the sources 'have no memory', the probability of a symbol $\sigma_i = (s_{i1}, s_{i2}, ..., s_{iN})$ in S^N is given (Rabbani and Jones 1991a):

$$p(\sigma_i) = p(s_{i1})p(s_{i2})...p(s_{iN}) \tag{2.30}$$

It can therefore be shown that the entropy per symbol of the extended source S^N is N times the entropy per symbol of the source S:

$$H(S^N) = N \times H(S) \tag{2.31}$$

2.2.3   Markov Sources

The model of a DMS is too restrictive for many applications. In practice, it is observed that there is a previous section of a message to major influence the likelihood of the next symbol (i.e. the source has a 'memory'). Thus, in digital images, the probability of a pixel to get a specific value may depend significantly on the values of the pixels preceding. Sources with this feature are modeled as Markov or Markovian Sources (MS). An mth-order MS is a source, at which the probability of a symbol depends on the previous m symbols (m being a finite number). This probability in this case is a conditional probability expressed as $p(s_i|s_{j_1}, s_{j_2}, ..., s_{j_m})$, where $i, j_p (p = 1, .., m) = 1, 2, ..., n$. So an mth-order MS has n^m states. For an *ergodic MS* there is a unique probability distribution on the set of states, called stationary or equilibrium distribution (Rabbani and Jones 1991a). For the computation of the entropy of an mth-order MS the work is as follows (Rabbani and Jones 1991a):

- First calculate the entropy given that the source is in a particular state $s_{j_1}, s_{j_2}, ..., s_{j_m}$:

$$H(S|s_{j_1}, s_{j_2}, ..., s_{j_m}) = -\sum_{i=1}^{n} p(s_{j_1}, s_{j_2}, ..., s_{j_m}) \log_p (s_{j_1}, s_{j_2}, ..., s_{j_m}) \tag{2.32}$$

- Then sum, for all possible states, the products of the probabilities that the source are in that state with the corresponding entropy:

$$H(S) = \sum_{S^m} p(s_{j_1}, s_{j_2}, ..., s_{j_m}) H(S|s_{j_1}, s_{j_2}, ..., s_{j_m}) \tag{2.33}$$

Figure 2.3 depicts an example of a state diagram of a second order Markov process with binary alphabet $S = 0, 1$. The conditional probabilities are

Fig. 2.3 Example of a states diagram for a Markov process

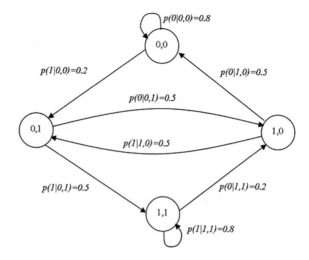

$$p(0|0, 0) = p(1|1, 1) = 0.8$$
$$p(1|0, 0) = p(0|1, 1) = 0.2 \qquad (2.34)$$
$$p(0|0, 1) = p(0|1, 0) = p(1|1, 0) = p(1|0, 1) = 0.5$$

There are four possible states $(0, 0)$, $(0, 1)$, $(1, 0)$ and $(1, 1)$. Because of the symmetry the stationary probability distribution for these states satisfies

$$p(0, 0) = p(1, 1)$$
$$p(0, 1) = p(1, 0) \qquad (2.35)$$

In addition, the source will be in one of the given states at any time. So,

$$p(0, 0) + p(0, 1) + p(1, 0) + p(1, 1) = 1 \qquad (2.36)$$

Apparently, there are two ways to reach state $(0, 0)$,

- either by encountering a 0 symbol while the system is in state $(0, 0)$ with 0.8 probability
- or by encountering a 0 symbol while the system is in state $(1, 0)$ with 0.5 probability

Similarly, for state $(0, 1)$,

$$p(0, 0) = 0.5 \, p(0, 1) + 0.8 \, p(0, 0)$$
$$p(0, 1) = 0.2 \, p(0, 0) + 0.5 \, p(0, 1) \qquad (2.37)$$

The solution of these equations leads to

$$p(0, 0) = p(1, 1) = 5/14$$
$$p(0, 1) = p(1, 0) = 2/14$$
(2.38)

Then the entropy can be calculated as

$$H(S) = -\sum_{2^3} p(s_{i_1}, s_{i_2}, s_i) \log_2(s_i | s_{i_1}, s_{i_2}) = 0.801 \text{ bps}$$
(2.39)

2.2.4 The Theorem of Lossless Coding

According to the theory proposed by Shannon, there is a direct relation between the entropy and the information content. In this approach any source S can be simply considered being *ergodic* with an alphabet of size n, and entropy $H(S)$. Whenever segments of N input symbols from this source are being encoded into binary words, $\forall \delta > 0$, it is possible, by choosing a large enough N, to create a code in such a way that the average number of bits per input symbol $\overline{L}$, will satisfy the inequality (Rabbani and Jones 1991a)

$$H(S) \le \overline{L} < H(S) + \delta$$
(2.40)

This inequality is usually referenced as the *theorem of lossless coding* and expresses the fact that *each source may be losslessly encoded with a code, in which the average number of bits per symbol for any input symbol is close, but never less than the entropy of the source*. But while the theorem assures of the existence of a code that can achieve a value close to the entropy of the source, it does not provide any information on the way of its creation. In practice, variable sized codes with extended source models are being used to achieve the desired performance (Rabbani and Jones 1991a). A code is called *compact* (for a given source), when its average length is less than or equal to the average length of all other codes that satisfy the *prefix codes condition*[3] for the same source and the same alphabet.

2.3 Digital Image Compression

After the necessary introduction to the information theory based essentially on the probability theory, the transition to the study of more complex sources can be done with two-dimensional data: the digital images. The examples presented thus far

---

[3]The prefix condition, which represents a necessary and sufficient condition for the creation of variable length codes, states that no code can be the beginning of another code for the same alphabet.

indicate that the digitization of images leads to a massive amount of data. The amount, though, of bits actually required to describe the information of an image can be much smaller due to the existence of information *redundancy*. The main objective of research in image compression is to reduce the number of bits required for the representation, by removing such redundancies. These redundancies are either of *statistical nature* and can be identified through proper modeling of the source, or of *optical nature* directly connected to the HVS, identified by appropriate modeling of the HVS itself. At the same time, establishing fundamental limits for the performance of each type of compression method to a specific kind of image, through the use of the basic principles of information theory are also considered. Apart from these basic objectives of the relevant research it is necessary to develop various algorithms that can 'fit' in different applications. There are several approaches to image compression, but generally all fall into two basic categories, *lossless* and *lossy* image compression (Fig. 2.4):

- *Lossless compression*—or *reversible coding*, in which the reconstructed image (decompressed) after compression is numerically identical to the original image (at pixel level). Obviously, lossless compression is ideal, since no part of the original information is likely to be lost or distorted. In this way, however, only small compression ratios can be achieved, typically of the order of 2 : 1. This type of compression has to do with the modeling of statistical characteristics of the image.

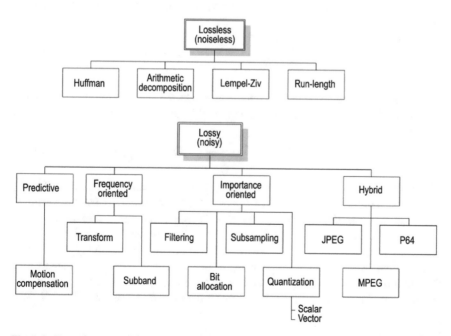

Fig. 2.4 General categorization of compression methods

- *Lossy compression*—or *irreversible coding*, wherein the reconstructed image is degenerate with respect to the original. Lossy compression can achieve significantly higher compression ratios. However, compression is achieved at the cost of increasing distortion, i.e. losing parts of the original data. It is important that the degeneration of the reconstructed image may or may not be noticeable, or it is possible to select the level and nature of distortions that might be acceptable. This type of compression makes heavy use of the concept of HVS modeling and exploitation of redundancies of an optical nature. The term *compression "without noticeable losses"* is often used to describe compression methods with losses that result in images with no noticeable degeneration (under normal or specific viewing conditions). It should be noted that the scope of the term "noticeable loss" in the definition of "compression without noticeable loss" is quite subjective and should be used with caution. Clearly, a method that implements compression with "no noticeable losses" that is specifically designed for display (e.g. for a computer screen of 19 in. and for an observer at a distance of about one meter) could be largely insufficient for other purposes, such as e.g. for printing on paper or on film.

In lossy compression it is allowed (or is tacitly accepted) to produce errors in order to increase the compression ratio. These errors can be quantified by distortion estimators. Usually this distortion is denoted as $D(I, \hat{I})$ and expresses the qualitative difference between an original image $I \equiv i[x, y]$ and a reconstruction $\hat{I} \equiv \hat{i}[x, y]$ from the compressed image.[4]

A widespread distortion estimator is the Mean Squared Error (MSE) defined as:

$$\text{MSE} = \frac{1}{N_1 \cdot N_2} \sum_{x=0}^{N_1-1} \sum_{y=0}^{N_2-1} (i[x, y] - \hat{i}[x, y])^2 \qquad (2.41)$$

In image compression, this estimator is expressed through another equivalent estimator, PSNR, which takes the place of an objective image quality estimator, and is defined as:

$$\text{PSNR} = 10 \log_{10} \frac{(2^B - 1)^2}{\text{MSE}} = 20 \log_{10} \frac{2^B - 1}{\text{RMSE}} \qquad (2.42)$$

where B is the color depth of the image in bpp and Root Mean Squared Error (RMSE) is the square root of the MSE. *PSNR is expressed in dB and is the most widespread mathematical formulation of the qualitative difference between two images.*

Along with the assessment of the distortion, there is also the assessment of the efficiency of compression. Since the objective of compression is the representation of an original image by a string consisting of bits (usually referenced as a bitstream), the aim is to maintain the length of this string as short as possible. In the general case an image of dimensions $N_1 \times N_2$ with a color depth of B bpp requires $N_1 \times N_2 \times B$

---

[4]This is a typical symbolism for digital images. Every image consists of a sequence of two-dimensional samples. The parameters x and y represent these two dimensions: $0 \leq x < N_1, 0 \leq y < N_2, N_1$ and N_2 being the image dimensions in the horizontal and vertical directions respectively.

Table 2.2 Typical compression performance values

Lossless	High quality lossy	Medium quality lossy	Low quality lossy
(B-3) bpp	1 bpp	0.5 bpp	0.25 bpp

bits for a complete representation. After compressing this image, the compression ratio may be defined as:

$$\text{compression ratio} = \frac{N_1 N_2 B}{\|c\|} \qquad (2.43)$$

where $\|c\|$ is the length of the final bitstream c.

Equally, the data rate (or bit-rate) in bpp is defined as (Taubman and Marcellin 2002b):

$$\text{bit} - \text{rate} = \frac{\|c\|}{N_1 N_2} \qquad (2.44)$$

For the case of lossy compression, the compression rate (bit-rate) is an important performance measure, particularly since the least significant bits of large color depth images can be removed without significant (noticeable) optical distortion. The measure, therefore, of the average number of bits spent for each sample of the image is usually the most important measure of compression efficiency, regardless of the accuracy with which these samples were represented initially. Table 2.2 presents indicative values of compression rate for natural images (photographs of the real world) in bpp, both for the case of lossless and lossy compression, adopting a convention that the encoded lossy images are being viewed in typical monitor resolution of 90 pixels per inch (or 22 pixels per mm) (Taubman and Marcellin 2002b). It should be noted that the values are indicative for each color channel (i.e. directly applicable to images of gray tones) and can show considerable variations depending on the image content.

2.3.1 Exploiting Redundancy for Compression

Without compression, an image is represented, generally, by a large amount of bits. Compression is only possible if some of those bits can be regarded as redundant or in some cases irrelevant. Two main categories of redundancy can be distinguished:

- spatial or statistical redundancy, and
- spectral redundancy or color correlation

In parallel, various application specifics render some of the information content of images irrelevant. This is a very important aspect relating to a potentially large amount of data able to be discarded, but it will not be discussed here, as it is application-dependent.

2.3.1.1   Spatial and Statistical Redundancy

Consider the simple case of a binary image $N_1 \times N_2$ consisting of white and black pixels, and suppose each pixel is treated as an integer of B bits long. The image will then be represented by $N_1 \times N_2 \times B$ bits. If an encoding-decoding system knows a priori that the image is binary, it is then able to use only one bit for each pixel and thereby achieve a compression ratio of $B : 1$. But in the general case, the system may just assume that some subset of combinations of pixels exhibits a higher incidence rate than others. So if it is known that the next sample belongs with high probability to the set {0, 1} then it is expected to spend a little more than 1 bit for the actual value of the sample. This is indicated by the theory of entropy, which shows the smallest amount of data expected to be consumed for the representation of a random variable (image samples in this case). This is the overall of the a priori knowledge for the input image regarded as a random variable.

Additionally, the system may assume that a small pixel neighborhood is very likely to exhibit small and smooth variation. Making this assumption, the system can benefit from previous pixel values to reduce the number of bits required to represent the next. The assumption is based on probability theory. This relative a priori knowledge can be described by the joint probability of two random variables, and thus the expected number of bits are already known through conditional entropy.

So what is readily understood is that if in a given image the number of pixels needed to extract robust statistics to predict new pixel values, then the compression rate is expected to be high.

2.3.1.2   Spectral Redundancy

Considering any digital color image as a union of numerous point spread functions (on each pixel) on the three dimensions color dimensions R, G and B, leads to an obvious recognition of inter-pixel color dependencies. That is, color variations in small regions of natural color images are expected to be limited or at least gradual. The consequence of this is that color images do not exhibit abrupt color changes, since there is usually a significant amount of color correlation. The obvious way to tackle with such correlation would be to apply a transformation to another color space that is decorrelated, minimizing the amount of redundancy in the color space of the image. Theoretically, the optimum (in a least squares sense) transformation in this case would be the Karhunen-Loeve Transform (KLT), or otherwise to apply the Principal Component Analysis (PCA) on the color space. This typically means that the covariance matrix of the image color pixels be computed, the eigenvalues of the covariance matrix be extracted and the corresponding eigenvectors be identified. The three eigenvectors, which are orthogonal, define a newcolor space for the image which has the least possible color redundancy (from a statistical point of view). Figure 2.5 shows how the principal components of a color image would look like after the application of PCA on the image color space.

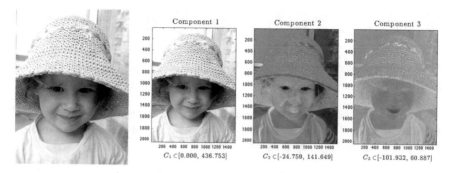

Fig. 2.5 KLT color space representation of a color image

In addition, it is known, according to many studies, that a human observer is significantly less sensitive to sudden changes in chromaticity than in luminance. This feature of the HVS is usually modeled by adopting a linear and sometimes a non-linear transformation of the original RGB representation of an image into a luminance-chromaticity representation (typically followed by a downsampling of the chromaticity channels). An example of this transformation is the transformation of RGB to YC_bC_r (Taubman and Marcellin 2002b):

$$\begin{pmatrix} Y \\ C_b \\ C_r \end{pmatrix} = \begin{pmatrix} 0.299 & 0.587 & 0.1140 \\ -0.169 & -0.331 & 0.5000 \\ 0.500 & -0.419 & -0.0813 \end{pmatrix} \begin{pmatrix} R \\ G \\ B \end{pmatrix} \tag{2.45}$$

Usually, after the application of the transformation, subsampling of the chromaticity channels (C_b, C_r) takes place either in the horizontal, or in the vertical direction or both. Important research in the field of modeling of the HVS (as already reported in the previous chapter), has made other similar transformations possible, which lead to different color systems. In those systems (such as CIELAB and sCIELAB) a linearization of the distances between colors as perceived by the HVS is targeted. In any case, conversion to another color system and sub-sampling of the channels with less visual importance leads to discarding of a significant amount of samples, without imposing a significant (noticeable) visual distortion, by just exploiting the characteristics (limitations) of the HVS.

2.3.2 Structure of a Basic Compression System

An abstraction of a compression-decompression system is shown in Fig. 2.6 as two mappings M and $\overline{M^{-1}}$ respectively. To achieve lossless compression, $\overline{M^{-1}} = M^{-1}$ should hold. In lossy compression, mapping M is not reversible, so the notation $\overline{M^{-1}}$ is used to indicate that the decompression system is an approximate inverse.

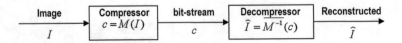

Fig. 2.6 Compression as a mapping

The compression system may be considered as one large Look-Up Table (LUT) with $2^{N_1 N_2 B}$ values. In the general case of designing an encoder M of fixed-length code words, an obvious way is to use the relation

$$c = M(I) = \arg \min_{c'} D(I, \overline{M^{-1}}(c')) \tag{2.46}$$

where D with a measure of the distortion and $\|c\|$ the fixed coding length imposed by the particular encoder, for which the decoder can be regarded as a LUT with $2^{\|c\|}$ values.

This condition is sufficient to maintain the LUT in a small number of elements both to the compressor and the decompressor. Thus, the compressor produces the string whose reconstruction by the decompression will be the most equivalent to the original image, based on the measure of distortion estimation. This approach results in that the mapping $\overline{M^{-1}}$ is the reverse of M:

$$M(\overline{M^{-1}}(c)) = c \tag{2.47}$$

Such a compression system is immune to distortion due to a repetitive compression-decompression.

This is basically the idea behind Vector Quantization (VQ), the generality of which is very attractive. However, the exponential increase in the size of the LUT with the increase in complexity, renders it an impractical solution (except for small images with few samples). It is therefore necessary to implement a separation of the various complementary structures within the mappings M and M^{-1}. The basic structure of a general compression scheme, as it stands following this distinction, and is the basis for almost all today's compression systems, is presented in Fig. 2.7. The first step is to transform the input into a new set of samples, particularly suitable for compression. In the second step, the new samples are quantized. During the final third step, the quantized samples are encoded to form the final compression bitstream.

In the following paragraphs, a description of the key parts of a structured compression system is presented. In this presentation it is considered that any color space transformation has already occurred on the input image data.

2.3.2.1 Transformation

The term *image transform* typically refers to a class of unitary matrices used for the representation of images. Similarly to the one-dimensional case, where a signal may

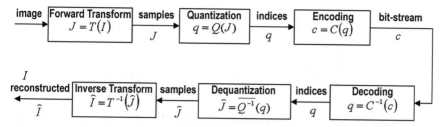

Fig. 2.7 Elements of a structured compression system

be represented by a series of *(orthogonal) basis functions*, images are represented by a series of two-dimensional basis functions or *basis images*. In the one-dimensional case, a continuous function may be expressed as a series expansion using orthonormal (orthogonal unitary) coordinate vectors, in order to produce a set of coefficients, which may be used to represent the original function,

$$\mathbf{J} = \mathbf{TI} \Leftrightarrow J(u) = \sum_{x=0}^{N-1} T(u,x)I(x), \quad 0 \le u \le N-1 \tag{2.48}$$

where $\mathbf{T}$ is the orthonormal transformation matrix and $\mathbf{I}$ the input data matrix. Since the transformation is orthonormal it holds that $\mathbf{T}^{-1} = \mathbf{T}^{*T}$, and the inversion of the transformation is,

$$\mathbf{I} = \mathbf{T}^{-1}\mathbf{J} \Leftrightarrow I(x) = \sum_{u=0}^{N-1} J(u)T^*(u,x), \quad 0 \le x \le N-1 \tag{2.49}$$

Apparently, Eq. (2.49) can be viewed as a series expansion of $I(x)$. In this representation, the columns of $\mathbf{T}^{-1}$ are the basis vectors of $\mathbf{I}$. It is also apparent in this expansion that the elements $J(u)$ are the coefficients that are needed to scale the basis vectors in order to reconstruct the original data (Jain 1988).

In image processing, the signals are two-dimensional and the above approach scales up to the following pair of transformations,

$$J(u,v) = \sum_{x=0}^{N-1}\sum_{y=0}^{N-1} T_{u,v}(x,y)I(x,y), \quad 0 \le u,v \le N-1$$

$$\tag{2.50}$$

$$I(x,y) = \sum_{u=0}^{N-1}\sum_{v=0}^{N-1} J(u,v)T_{u,v}^*(x,y), \quad 0 \le x,y \le N-1$$

where $T_{u,v}(x,y)$ is a set of orthonormal basis functions satisfying the *orthogonality* and the *completeness* properties,

$$\sum_{x=0}^{N-1}\sum_{y=0}^{N-1} T_{u,v}(x, y) T_{u',v'}^*(x, y) = \delta(u - u', v - v')$$

$$\sum_{u=0}^{N-1}\sum_{v=0}^{N-1} T_{u,v}(x, y) T_{u,v}^*(x', y') = \delta(x - x', y - y')$$

(2.51)

'δ' denoting the impulse function. The $J(u, v)$ elements are the *transform coefficient* and the set of all transform coefficients $\mathbf{J} \triangleq \{J(u, v)\}$ is the *transformed image* (Jain 1988). In this context, the transformation matrix $\mathbf{T}$ is also called the *transformation kernel* and a two-dimensional transformation is said to be *separable* if its kernel can be written as a multiplication of two kernels acting on the two spatial dimensions of the data,

$$T_{u,v}(x, y) = T_C(u, x) \cdot T_R(v, y) \tag{2.52}$$

where the c, R indices denote the column- and row-wise one-dimensional operation respectively. Thus an image transform can be written as,

$$J(u, v) = \sum_{y=0}^{N-1}\left[\sum_{x=0}^{N-1} T_C(u, x)I(x, y)\right]T_R(v, y) \tag{2.53}$$

With all these representations converted into matrix form,

$$\mathbf{J} = \mathbf{T_C} \cdot \mathbf{I} \cdot \mathbf{T_R^T} \tag{2.54}$$

while the inverse transforms follow the same formalism (Pratt 1991).

In image compression, the *transformation* (or simply *transform*) is responsible for the conversion of input samples to a format that allows for easier and more efficient quantization and encoding. A transformation captures inherent statistical dependencies among samples in the original data, so that the transformed sample would include only small local correlation. In the ideal case, the samples after the application of the transformation must be statistically independent. Additionally, the transformation must separate the information into parts that can be characterized as 'important' or not, so that the redundant (less or non-important) information be quantized with higher quantization level or even be eliminated. Specifically, in the unitary transformation $\mathbf{J} = \mathbf{T} \cdot \mathbf{I}$ holds that,

$$\|\mathbf{J}\|^2 \triangleq \mathbf{J}^{*T}\mathbf{J} = \mathbf{I}^{*T}\mathbf{T}^{*T}\mathbf{T}^T\mathbf{T}\mathbf{I} = \mathbf{I}^{*T}\mathbf{I} \triangleq \|\mathbf{I}\|^2 \Rightarrow \|\mathbf{J}\|^2 = \|\mathbf{I}\|^2 \tag{2.55}$$

which states that the *signal energy is preserved*. In addition, this equation shows that a unitary transformation is a *simple rotation in the N-dimensional vector space*, or the components of $\mathbf{J}$ are the *projections* of $\mathbf{I}$ on the new basis (Jain 1988).

In addition, since in unitary transforms the energy is preserved, if the transformation concentrates the energy of the signal to a small amount of coefficients would significantly aid in the compression. The fact is that unitary transforms actually do pack a large fraction of the energy into few coefficients. If $\boldsymbol{\mu}_I$ and $\boldsymbol{\Sigma}_I$ represent the mean value and the covariance of an input data vector $\mathbf{I}$ then the corresponding quantities for the transform coefficients are,

$$
\begin{aligned}
\boldsymbol{\mu}_J &\triangleq E[\mathbf{J}] = E[\mathbf{TI}] = \mathbf{T}E[\mathbf{I}] = \mathbf{T}\boldsymbol{\mu}_I \\
\boldsymbol{\Sigma}_J &= E\left[(\mathbf{J} - \boldsymbol{\mu}_J)(\mathbf{J} - \boldsymbol{\mu}_J)^{*T}\right] \\
&= \mathbf{T}\left(E\left[(\mathbf{I} - \boldsymbol{\mu}_I)(\mathbf{I} - \boldsymbol{\mu}_I)^{*T}\right]\right)\mathbf{T}^{*T} = \mathbf{T}\boldsymbol{\Sigma}_I\mathbf{T}^{*T}
\end{aligned}
\tag{2.56}
$$

In the diagonal of $\boldsymbol{\Sigma}_I$ lie the variances of the transform coefficients,

$$
\sigma_J^2(k) = [\boldsymbol{\Sigma}_J]_{k,k} = \left[\mathbf{T}\boldsymbol{\Sigma}_I\mathbf{T}^{*T}\right]_{k,k}
\tag{2.57}
$$

and since $\mathbf{T}$ is unitary,

$$
\sum_{k=0}^{N-1} |\mu_J(k)|^2 = \boldsymbol{\mu}_J^{*T}\boldsymbol{\mu}_J = \boldsymbol{\mu}_I^{*T}\mathbf{T}^{*T}\mathbf{T}\boldsymbol{\mu}_I = \sum_{n=0}^{N-1} |\mu_I(n)|^2
\tag{2.58}
$$

$$
\sum_{k=0}^{N-1} \sigma_J^2(k) = Tr\left[\mathbf{T}\boldsymbol{\Sigma}_I\mathbf{T}^{*T}\right] = Tr[\boldsymbol{\Sigma}_I] = \sum_{n=0}^{N-1} \sigma_I^2(n)
\tag{2.59}
$$

so,

$$
\sum_{k=0}^{N-1} E[|J(k)|^2] = \sum_{n=0}^{N-1} E[|I(n)|^2]
\tag{2.60}
$$

and the average energy $E[|J(k)|^2]$ of the transform coefficients $J(k)$ tends to be unevenly distributed, even though the energy in the input might be evenly distributed. In addition, the off-diagonal elements of the covariance matrix $\boldsymbol{\Sigma}_J$ tend to become small compared to the diagonal elements, which indicates that *the transform coefficients tend to be uncorrelated* (Jain 1988).

Last but certainly not least, the eigenvalues and the determinant of unitary transform matrices have unity magnitude and the entropy of the input signal is preserved during such a transformation, which states that *unitary transformations preserve the information content of a signal* (Jain 1988).

Known transformations that satisfy these requirements are either those which usually act on parts of the images (block transforms) as the *Karhunen-Loeve transform*, the *Fourier transform* and the *cosine transform*, or global transforms applied into spectral regions (sub-band transforms), as the *Wavelet transform*. In subsequent paragraphs a brief and comprehensive description of these transformations as applied in (2D data) image compression is included.

In order to adopt a more practical description and notation in the context of image compression, the operators of transformations are being defined in relation to their Point Spread Function (PSF). The PSF of an operator T is the result of the application of the operator on a point source, and for the case of 2-D are

$$T \left[\text{point source}\right] = \text{point spread function, or}$$
$$T \left[\delta(x - \alpha, y - \beta)\right] = h(x, \alpha, y, \beta) \tag{2.61}$$

where $\delta(x - \alpha, y - \beta)$ is an impulse function that expresses a point source of intensity 1 at point (α, β). The physical meaning of function $h(c, \alpha, y, \beta)$ is that it expresses how the input value in position (x, y) affects the output value at position (α, β).

When the operator is linear and the point source is c times more intense, then the result of the operator is c times larger, thus is

$$T[c \cdot \delta(x - \alpha, y - \beta)] = c \cdot h(x, \alpha, y, \beta) \tag{2.62}$$

Considering an image as a collection of such point sources each of different intensity, one may, ultimately, represent the image as the sum of all these point sources. In this case, the result of applying an operator characterized by a PSF $h(x, \alpha, y, \beta)$ to a square $N \times N$ image $I(x, y)$ can be written as

$$J(\alpha, \beta) = \sum_{x=0}^{N-1} \sum_{y=0}^{N-1} I(x, y) h(x, \alpha, y, \beta) \tag{2.63}$$

where the value of (α, β) is a linear combination of the values at all image locations (x, y) weighted by the PSF of the transformation.

Apparently, the problem in image compression is to define a transformation $h(x, \alpha, y, \beta)$ such that the input image can be losslessly represented with fewer bits. Additionally, if the effect expressed by the PSF of a linear transformation is independent of the actual position but depends on the relative position of pixels that affect and are affected, then the PSF is *translation invariant*,

$$h(c, \alpha, y, \beta) = h(x - \alpha, y - \beta) \tag{2.64}$$

and then (2.63) expresses a convolution

$$J(\alpha, \beta) = \sum_{x=0}^{N-1} \sum_{y=0}^{N-1} I(x, y) h(x - \alpha, y - \beta) \tag{2.65}$$

Further, if the columns are influenced independently of the rows of the image, the PSF is *separable*,

$$h(c, \alpha, y, \beta) = h_c(x, \alpha) \cdot h_r(y, \beta) \tag{2.66}$$

where the index c is used for the columns, while the index r for the rows, and thus (2.63) is written as the sequential application of two one-dimensional transformations,

$$J(\alpha, \beta) = \sum_{x=0}^{N-1} h_c(x, \alpha) \sum_{y=0}^{N-1} I(x, y) h_r(y, \beta) \tag{2.67}$$

And if the PSF is *translation invariant and separable*, then (2.63) is written as a sequence of two one-dimensional convolutions,

$$J(\alpha, \beta) = \sum_{x=0}^{N-1} h_c(x - \alpha) \sum_{y=0}^{N-1} I(x, y) h_r(y - \beta) \tag{2.68}$$

To transform these equations into matrix form, (2.67) can be expressed as

$$\mathbf{J} = \mathbf{h}_c^T \cdot \mathbf{I} \cdot \mathbf{h}_r \tag{2.69}$$

The transformation in this equation is *unitary* if matrices $\mathbf{h}_c$ and $\mathbf{h}_r$ are unitary. The concept of unitary is similar to that of the *orthogonal* except that it relates to complex-valued quantities (of unit magnitude).

Reversing (2.69), the image $\mathbf{I}$ can be recovered as

$$\mathbf{I} = \mathbf{h}_c \cdot \mathbf{J} \cdot \mathbf{h}_r^T \tag{2.70}$$

which in the form of elements is written as

$$I(x, y) = \sum_{\alpha=0}^{N-1} \sum_{\beta=0}^{N-1} J(\alpha, \beta) \mathbf{h}_r(y, \beta) \mathbf{h}_c(x, \alpha) \tag{2.71}$$

where for constants α and β, $\mathbf{h}_c$ and $\mathbf{h}_r$ are vectors. Their product is of $N \times N$ dimension. Taking all possible combinations of columns of $\mathbf{h}_r$ and rows of $\mathbf{h}_c$ the *basis matrices* are produced. Thus, Eq. (2.71) is the expression of the original image with respect to these basis matrices. Finally, the data obtained from the transformation of the original image are the coefficients with which each of the basis images (matrices) must be multiplied before being summed with the others to reconstruct the entire image.

Karhunen-Loeve Transform

Supposing a Gaussian random variable $\mathbf{I} = (I_1, I_2, ..., I_k)$, the covariance $\sigma_{i,j}$ between I_i and I_j is defined as

$$\sigma_{i,j} = E\left[(I_i - \mu_i)(I_j - \mu_j)\right] \tag{2.72}$$

where μ_i, μ_j, are the mean values of I_i and I_j respectively. The $k \times k$ covariance matrix of the random variable $\mathbf{I}$ is

$$\Sigma_I = [\sigma_{i,j}]_{i,j=1}^k \tag{2.73}$$

If $\mathbf{A}$ is the $k \times k$ orthogonal matrix whose rows are the eigenvectors of Σ_I, then the Karhunen-Loeve Transform (KLT) of $\mathbf{I}$ is defined as

$$\mathbf{Y}^T = \mathbf{A} \cdot \mathbf{I}^T \tag{2.74}$$

Vector $\mathbf{Y}$ is a Gaussian variable with statistically independent elements.

In practice, the question is to find the set of basis images, which are optimal for a family of images with similar statistical characteristics. To answer this question, the assumption that images are *ergodic processes* is made, namely:

• the mean value over the total family of images equals to the spatial mean in each of the images (Fig. 2.8)
• the covariance matrix over the total family of images is equal to the covariance matrix of each of the images

Of course, the ergodicity assumption is correct only in the case of uniform noise. But it can be approximately correct for small portions of an image. The application of the transformation on a graylevel (one-color-channel) image $\mathbf{I}$ of $N \times N$ dimensions involves:

1. Conversion of the image $\mathbf{I}$ to a column vector i of $N^2 \times 1$ dimensions, by putting the columns one below the other
2. Calculation of the covariance matrix Σ that is of size $N^2 \times N^2$
3. Calculation of the eigenvalues of the covariance matrix
4. Calculation of the corresponding eigenvectors
5. Creation of matrix $\mathbf{A}$ of the eigenvectors (as rows) using an ordering imposed by sorting the eigenvalues in a decreasing order
6. Compute the transformed image is: $\hat{\mathbf{I}} = \mathbf{A}(\mathbf{I} - \boldsymbol{\mu})$ where $\boldsymbol{\mu}$ a vector with all elements equal to the mean value of the image

Fig. 2.8 Ergodicity in images

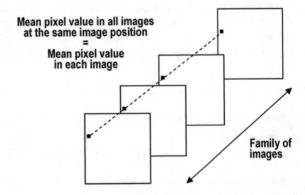

Mean pixel value in all images at the same image position
=
Mean pixel value in each image

Family of images

This transformation is optimal from the energy standpoint; if only a limited number of the KLT transform coefficients are stored, they are expected to hold the most part of the total energy in comparison to all other transformations. Unfortunately, the basis functions of the KLT dependent on the image and require the calculation of the covariance. Thus, a fast KLT algorithm is not possible if the transformation is applied to the entire image (global). Therefore, the KLT is limited to only few applications in image compression (Rabbani and Jones 1991a). Nevertheless, alternative implementations have already been proposed based on the assumed ergodicity in natural digital images and that KLT is equivalent to Singular Value Decomposition (SVD), which decomposes the original data into three matrices (factorization) in the typical form $I = U\Sigma V^T$, representing the new coordinate system to project the original image (columns and rows of matrices U, V) and the amount of scaling to apply (on the diagonal of Σ—the variances). Figure 2.9 shows an example of the KLT on a graylevel image. The original image is shown on the left; the two noisy images on the right represent the U and V^T matrices. The graph below the noisy images is a portion of the graphical representation of the variances in the diagonal of Σ, which illustrates how the most of the energy is concentrated in the first values. The image can then be represented by a limited amount of basis vectors and variances to impose efficient coding and, consequently, significant compression. Figure 2.10 shows the reconstruction that can be achieved by using a limited number of dimensions, by discarding (setting to zero) a number of variances in Σ. Specifically, the first reconstruction is attained by keeping only the first element in Σ containing almost no

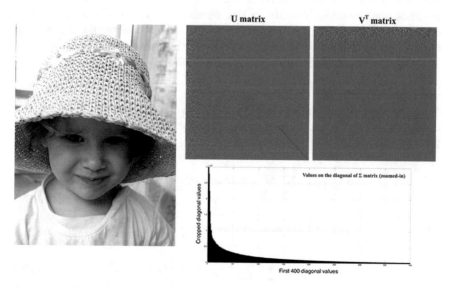

Fig. 2.9 KLT (PCA) on a graylevel image

1 element of Σ - 16.58 dB   7 elements of Σ - 20.87 dB   23 elements of Σ - 23.38 dB   67 elements of Σ - 27.38 dB

118 element of Σ - 30.57 dB   173 elements of Σ - 33.28 dB   290 elements of Σ - 37.67 dB   534 elements of Σ - 43.17 dB

Fig. 2.10 Progressive reconstruction of the graylevel image undergone KLT

valuable information; what is really surprising is that by keeping only about 100 elements in Σ a very good reconstruction can be achieved, like in the case of 118 elements that results in 30.57 dB PSNR, which is fairly a threshold above which images are considered of an acceptable quality. It should be noted that all elements of the diagonal of Σ are 1536 for this example, so the 290 elements are just a small 19 % fraction of the original data producing a significant 37.67 dB quality.

Discrete Fourier Transform

The Discrete Fourier Transform (DFT) is often used in the spectral analysis and filter design. Following the line of unitary transforms, for an $M \times N$ input matrix $\mathbf{I}$ (image), the two-dimensional DFT is defined as

$$J(u, v) = \sum_{x=0}^{M-1}\sum_{y=0}^{N-1} I(x, y)e^{-2\pi i(\frac{ux}{M} + \frac{vy}{N})}, \quad i = \sqrt{-1} \tag{2.75}$$

whereas the inverse transform is defined as:

$$I(x, y) = \frac{1}{M \cdot N}\sum_{u=0}^{M-1}\sum_{v=0}^{N-1} J(u, v)e^{2\pi i(\frac{ux}{M} + \frac{vy}{N})}, \quad i = \sqrt{-1} \tag{2.76}$$

The DFT decomposes an image into its spectral bands. The parameters u, v are called spatial frequencies of the transformation. The kernel of the transformation may be separated as follows

$$e^{-2\pi i(\frac{ux}{M} + \frac{vy}{N})} = e^{-\frac{2\pi iux}{M}} \cdot e^{-\frac{2\pi ivy}{N}} \tag{2.77}$$

so that the two-dimensional transformation is implemented by two successive one-dimensional transformations,

$$J(u, v) = \sum_{x=0}^{M-1} e^{-\frac{2\pi iux}{M}} \sum_{y=0}^{N-1} I(x, y) e^{-\frac{2\pi ivy}{N}}$$

$$= \sum_{x=0}^{M-1} J(x, v) e^{-\frac{2\pi iux}{M}} \tag{2.78}$$

$$\text{where } J(x, v) = \sum_{y=0}^{N-1} I(x, y) e^{-\frac{2\pi ivy}{N}}$$

Extensive study has been performed on fast implementations of DFT (such as the Fast Fourier Transform (FFT)), which typically exhibit a computational complexity of $O(N \log_2 N)$ for a transformation of N-points (Rabbani and Jones 1991a). In general, the transform coefficients generated by the DFT are complex numbers so their storage and handling is problematic. Actually there are $2N^2$ transform coefficients, but because of the symmetry of the conjugate elements, it holds that

$$J(u, v) = J^*(-u + lN, -v + mN) \tag{2.79}$$

for $l, m = 0, 1, 2, \ldots$ and J^* the complex conjugate of J. Thus, almost half of the elements are redundant, which is a disadvantage of the DFT. Another disadvantage of the DFT is the occurrence of false spectral components due to indirect periodicity in the image boundaries. When the DFT is used for compression with high compression ratio, these spurious components can lead to the appearance of sudden changes (*blocking artifacts*) between the parts of the image in which the transformation is applied. To express the transformation in the form of matrices, according to what has been defined in the introductory paragraphs, it suffices to define the functions h_r and h_c

$$h_c(x, u) = e^{-i\frac{2\pi xu}{M}}, \quad h_r(y, v) = e^{-i\frac{2\pi yv}{N}} \tag{2.80}$$

Figures 2.11, 2.12, 2.13, 2.14, and 2.15 show an example of the application of the DFT and the possible reconstructions using a limited amount of the transform coefficients. In this example only the magnitude of the complex coefficients is considered. The reconstruction is begin done by applying a filtering on the transform domain, zeroing out most of the transform coefficients as shown in the figures. It is apparent that the transform successfully compacts the image data into only a few

Fig. 2.11 DFT representation of a graylevel image by applying an FFT

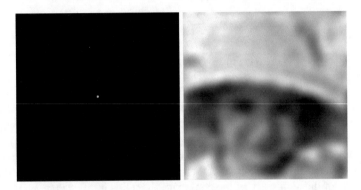

Fig. 2.12 Reconstruction using 0.014% of the DFT coefficients at 19.78 dB PSNR

Fig. 2.13 Reconstruction using 0.34% of the DFT coefficients at 22.70 dB PSNR

Fig. 2.14 Reconstruction using 1.34% of the DFT coefficients at 26.07 dB PSNR

Fig. 2.15 Reconstruction using 10.25% of the DFT coefficients at 33.05 dB PSNR

transform coefficients, achieving an acceptable image quality (a good approximation of the original image data) with a very limited amount of data.

Discrete Cosine Transform

For practical applications in image processing the DFT cannot be the first choice as it poses computational difficulties by producing complex valued coefficients. This issue along with a need for fast and practical implementations of KLT gave rise to real-valued related transformations, such as the Discrete Cosine Transform (DCT). The DCT is defined by the transformation matrix $\mathbf{C} = \{C(u, x)\}$,

$$C(u, x) = \begin{cases} \sqrt{\frac{1}{N}} & u = 0, \ 0 \leq x \leq N-1 \\ \sqrt{\frac{2}{N}} \cos \frac{(2x+1)u\pi}{2N} & 1 \leq u \leq N-1, \ 0 \leq x \leq N-1 \end{cases} \tag{2.81}$$

The two-dimensional DCT of a square $N \times N$ matrix $\mathbf{I}$ is defined as

$$J(u, v) = \sum_{x=0}^{N-1}\sum_{y=0}^{N-1} C(u, x)I(x, y)C(v, y)$$

$$J(u, v) = \sum_{x=0}^{N-1}\sum_{y=0}^{N-1} I(x, y)c(u)c(v)\cos\frac{(2x+1)u\pi}{2N}\cos\frac{(2y+1)v\pi}{2N} \qquad (2.82)$$

$$J(u, v) = c(u)c(v)\sum_{x=0}^{N-1}\sum_{y=0}^{N-1} I(x, y)\cos\frac{(2x+1)u\pi}{2N}\cos\frac{(2y+1)v\pi}{2N}$$

with,

$$c(\xi) = \begin{cases} \sqrt{\frac{1}{N}}, & \xi = 0 \\ \sqrt{\frac{2}{N}}, & \text{otherwise} \end{cases}$$

Alternatively, the forward DCT can be found in the literature defined as,

$$J(u, v) = \frac{2}{N}c(u)c(v)\sum_{x=0}^{N-1}\sum_{y=0}^{N-1} I(x, y)\cos\frac{(2x+1)u\pi}{2N}\cos\frac{(2y+1)v\pi}{2N} \qquad (2.83)$$

with,

$$c(\xi) = \begin{cases} \sqrt{\frac{1}{2}}, & \xi = 0 \\ 1, & \text{otherwise} \end{cases}$$

The inverse two-dimensional DCT is defined as:

$$I(x, y) = \sum_{u=0}^{N-1}\sum_{v=0}^{N-1} c(u)c(v)J(u, v)\cos\frac{(2x+1)u\pi}{2N}\cos\frac{(2y+1)v\pi}{2N} \qquad (2.84)$$

with,

$$c(\xi) = \begin{cases} \sqrt{\frac{1}{N}}, & \xi = 0 \\ \sqrt{\frac{2}{N}}, & \text{otherwise} \end{cases}$$

or by using the alternative formulation,

$$I(x, y) = \frac{2}{N}\sum_{u=0}^{N-1}\sum_{v=0}^{N-1} c(u)c(v)J(u, v)\cos\frac{(2x+1)u\pi}{2N}\cos\frac{(2y+1)v\pi}{2N} \qquad (2.85)$$

with,

$$c(\xi) = \begin{cases} \sqrt{\frac{1}{2}}, & \xi = 0 \\ 1, & \text{otherwise} \end{cases}$$

Fig. 2.16   DCT representation of a graylevel image

Fig. 2.17   Reconstruction using 50 of the DCT coefficients (0.05%) at 20.69 dB PSNR

A practical example of the application of DCT on a graylevel image is shown in Figs. 2.16, 2.17, 2.18, 2.19, and 2.20. The first image shows a representation of the original image and the transform coefficients after the transformation. The following images present gradual reconstructions using an increasing amount of transform coefficients.[5] Apparently, DCT is also good at producing compact representations of images as with only a small fraction of the coefficients it is able to produce a high quality reconstruction, as shown in Fig. 2.20, in which a reconstruction using just 35 % of the coefficients led to an image with 39.8 dBs of quality. It is noted that the black regions in the transform domain denote the coefficient being zeroed out. The diagonal scheme used is in line with the ordering of increasing frequency in two dimensions.

For natural images that usually exhibit a high correlation among neighboring pixels, the performance of the DCT approaches that of KLT. It can be shown that for a source that is modeled as a first-order Markovian, and since the correlation of

---

[5]It should be noted that in this example the transform has been applied once on the total image area, which is not the usual case. Usually, these transforms are being applied in a block-by-block basis, typically of 8 × 8 pixels.

Fig. 2.18 Reconstruction using 200 of the DCT coefficients (0.85%) at 24.88 dB PSNR

Fig. 2.19 Reconstruction using 546 of the DCT coefficients (6.3%) at 31.68 dB PSNR

Fig. 2.20 Reconstruction using 1296 of the DCT coefficients (35.6%) at 39.82 dB PSNR

adjacent pixels tends to unity, the DCT basis functions are similar to those of the KLT. Theoretically, one of the properties of random processes that are stationary in the wide sense (Wide-Sense Stationary (WSS)) is that their Fourier coefficients are uncorrelated. As the number of coefficients increases, DFT and DCT, as well as other similar transformations, in the frequency domain, *diagonalize the covariance matrix of the source*. Therefore, these transformations are asymptotically equivalent to the KLT for WSS sources, up to an extent of a special reordering of the transform coefficients (like the well known *zig-zag ordering* of the DCT coefficients during the application of JPEG compression—as will be described in following paragraphs). Although most of the images cannot be readily modeled as WSS processes, the DCT has proven to be a robust approximation to the KLT for natural images.

Following the definitions in previous paragraphs and in accordance with the definition of the inverse DCT in (2.85), the DCT basis functions (or basis images in this case) are defined as

$$b(u, v, x, y)_{x,y=0,\dots,N-1} = c(u)c(v)\cos\frac{(2x+1)u\pi}{2N}\cos\frac{(2y+1)v\pi}{2N} \quad (2.86)$$

The basis functions of DCT for an image of size 8×8 pixels are shown in Fig. 2.21. Like the DFT, the DCT has a fast implementation with computational complexity $O(N \log N)$ for a transformation of N points. Nevertheless, DCT is more efficient in compression applications, as it does not exhibit the phenomenon of occurrence of spurious spectral components. The characteristic transformation matrices, as defined in (2.70), are calculated for the case of an 8×8 pixels image, are

$$\mathbf{h}_c = \mathbf{h}_r = $$

$$\begin{pmatrix}
0.3536 & 0.3536 & 0.3536 & 0.3536 & 0.3536 & 0.3536 & 0.3536 & 0.3536 \\
0.4904 & 0.4157 & 0.2778 & 0.0975 & -0.0975 & -0.2778 & -0.4157 & -0.4904 \\
0.4619 & 0.1913 & -0.1913 & -0.4619 & -0.4619 & -0.1913 & 0.1913 & 0.4619 \\
0.4157 & -0.0975 & -0.4904 & -0.2778 & 0.2778 & 0.4904 & 0.0975 & -0.4157 \\
0.3536 & -0.3536 & -0.3536 & 0.3536 & 0.3536 & -0.3536 & -0.3536 & 0.3536 \\
0.2778 & -0.4904 & 0.0975 & 0.4157 & -0.4157 & -0.0975 & 0.4904 & -0.2778 \\
0.1913 & -0.4619 & 0.4619 & -0.1913 & -0.1913 & 0.4619 & -0.4619 & 0.1913 \\
0.0975 & -0.2778 & 0.4157 & -0.4904 & 0.4904 & -0.4157 & 0.2778 & -0.0975
\end{pmatrix}$$

$$(2.87)$$

The DFT is a mapping into a Discrete Fourier Series (DFS) of a sequence of finite length and therefore, exhibits an indirect periodicity, which is illustrated in Fig. 2.22a (Rabbani and Jones 1991a). This periodicity is the result of the sampling in the frequency domain. During the reconstruction of the original sequence in this way, discontinuities between the encoded parts emerge. These discontinuities lead to spurious high frequency components, which can cause significant degradation in the efficiency of the transformation. Though these spurious components are not part of the original sequence, they are required for reconstructing the boundaries in the periodic sequence.

Fig. 2.21 The 8 × 8 2-D
DCT basis images

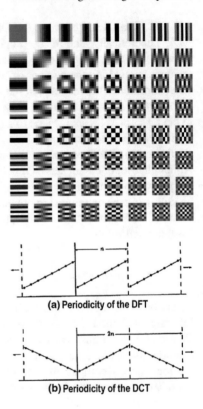

Fig. 2.22 Periodicities of
a DFT and **b** DCT

(a) Periodicity of the DFT

(b) Periodicity of the DCT

Striving for improved efficiency of transformation through the rejection of these components leads to errors in the reconstruction at the region boundaries. In image coding, wherein the image is divided into non-overlapping parts to form two-dimensional sequences, such errors in the reconstruction result in the appearance of intense deformation at the boundaries between adjacent segments. To eliminate these inconsistencies in the boundaries, the original sequence of N points can be extended to a $2N$-point sequence by mirroring the original image at the boundaries. The extended sequence is repeated to form a periodic sequence that is required for the DFS representation. As shown in Fig. 2.22b, this periodic sequence exhibits no such discontinuities in the boundaries and no spurious spectral components in the DCT.

The process of calculating the $2N$-point DFT of the extended sequence (from the N-point original) is identical to the calculation of the DCT of the original sequence. Actually, a DCT can be computed as a $2N$-point FFT. Due to its symmetry DCT exhibits two main advantages over DFT, namely,

- The DCT does not generate spurious spectral components, and thus, the efficiency of coding remains high, while simultaneously, diminishing the *blocking artifacts* at the boundaries of the coding regions.

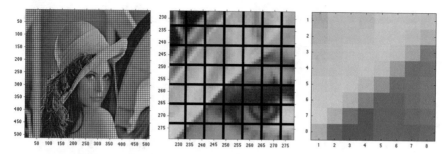

Fig. 2.23 Tiling of test image '*lena*' into 8 × 8 non-overlapping image blocks, zoomed-in portion of the image and a magnified selected image block

- The DCT calculations require only real numbers, while the DFT calculations involve complex numbers.

These advantages rendered DCT the most widely-used compression transform for both static and moving images (video). In overall, DCT can lead to efficient compression when the spatial correlation between neighboring pixels is significant, usually the case in natural images.

Finally, since the transform has been defined in 2-D and the corresponding basis images have been shown let us consider how the transform actually represents a color image. In practice, DCT (like many of the transforms) is being applied on non-overlapping image blocks, which simplifies the computations and make extremely fast parallel implementations possible. Figure 2.23 presents such a tiling on the color image '*lena*', which separates the image into 8 × 8 non-overlapping image blocks. Suppose during a sequential encoding process, the current block is the one in the pixel locations (250–257, 250–257) as shown in the figure. The values of the pixels in this image block, shown only for the green (G) channel,[6] are,

$$G = \begin{bmatrix} 157 & 176 & 177 & 174 & 190 & 186 & 187 & 183 \\ 166 & 177 & 180 & 183 & 187 & 188 & 181 & 167 \\ 177 & 181 & 188 & 189 & 194 & 185 & 160 & 95 \\ 181 & 184 & 189 & 193 & 185 & 143 & 81 & 74 \\ 188 & 192 & 192 & 179 & 133 & 75 & 77 & 76 \\ 193 & 195 & 175 & 108 & 67 & 61 & 74 & 77 \\ 200 & 169 & 93 & 53 & 69 & 57 & 70 & 72 \\ 163 & 74 & 52 & 53 & 59 & 65 & 61 & 65 \end{bmatrix}$$

---

[6]The green channel was selected because it is easy to follow; the upper left triangular region of the image, which is yelowish is expected to exhibit higher values in this channel than those in the rest of the block, which shows up redish.

Application of the transform on this image block and channel results in the matrix,

$$DCT_G = \begin{bmatrix} 1112 & 204 & 5 & 20 & 1 & -2 & -4 & 10 \\ 285 & -112 & -112 & -12 & -23 & -6 & -6 & -3 \\ -60 & -118 & 78 & 40 & 5 & 2 & 3 & 2 \\ 17 & 37 & 48 & -61 & -15 & -12 & -9 & 3 \\ -21 & -8 & -37 & -5 & 35 & 2 & 9 & -3 \\ 2 & 16 & 2 & 24 & -7 & -22 & -4 & 1 \\ -4 & -3 & -8 & -2 & -17 & 16 & 8 & 3 \\ 7 & 2 & 1 & 7 & 0 & 6 & -17 & 7 \end{bmatrix}$$

with the values rounded to the nearest integer for better visualization.

Apparently, the first coefficient (top-left) is significantly higher than all the other (in absolute values). So, what do actually these coefficient values represent? Clearly, they represent the weight (multiplier) of the appropriate basis image so that the weighted summation of all coefficients and basis images would approximate the original image. This is graphically depicted in Fig. 2.24 where above each of the basis images the corresponding weight has been added. This is equivalent to a series expansion, in which the image block is decomposed into,

$$I = \sum_{x=1}^{8} \sum_{y=1}^{8} c(x, y) \cdot b(x, y) = [c_1 \ ... \ c_{x \cdot y}] \begin{bmatrix} b_1 \\ \vdots \\ b_{x \cdot y} \end{bmatrix}$$

Fig. 2.24 DCT domain representation of image

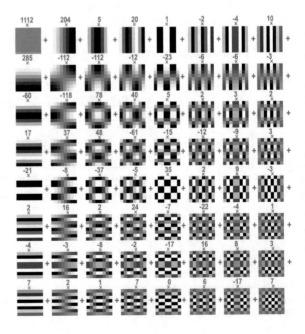

where $c(x, y)$ the coefficient from the *DCT* matrix and $b(x, y)$ the basis image (Fig. 2.21). *Geometrically, this can be viewed as the representation in a linear image space, in which the basis images are the unit vectors of the axes and the coefficients are the coordinates.*

Four entirely different scenarios of reconstruction are being reviewed in the example depicted in Fig. 2.25. Specifically, the first scenario represents the reconstruction possible if only the first row of DCT coefficients matrix are being used. In the second scenario, the main diagonal coefficients have been selected and all others were discarded. In the third scenario the coefficients from the first row and first column of the matrix have been used for the reconstruction. Finally, for the fourth scenario the upper-left triangular part of the matrix has been used for a reconstruction. In all scenarios, Fig. 2.25 presents the original image block data, the reconstructed data using the limited number of coefficients and the difference between the two, with values exaggerated for emphasis. Each reconstruction is titled by the RMSE and the quality in PSNR dBs.

Discrete Sine Transform

The Discrete Sine Transform (DST) is defined similarly to the definition of DCT by the transformation matrix $\mathbf{C} = \{C(u, x)\}$,

$$C(u, x) = \sqrt{\frac{2}{N + 1}} \sin \frac{(x + 1)(u + 1)\pi}{N + 1} \tag{2.88}$$

The two-dimensional DST of a square $N \times N$ matrix $\mathbf{I}$ is defined as

$$J(u, v) = \sum_{x=0}^{N-1} \sum_{y=0}^{N-1} C(u, x) I(x, y) C(v, y)$$

$$J(u, v) = \sum_{x=0}^{N-1} \sum_{y=0}^{N-1} I(x, y) \sqrt{\frac{2}{N + 1}} \sin \frac{(x + 1)(u + 1)\pi}{N + 1} \sqrt{\frac{2}{N + 1}} \sin \frac{(y + 1)(v + 1)\pi}{N + 1}$$

$$J(u, v) = \frac{2}{N + 1} \sum_{x=0}^{N-1} \sum_{y=0}^{N-1} I(x, y) \sin \frac{(x + 1)(u + 1)\pi}{N + 1} \sin \frac{(y + 1)(v + 1)\pi}{N + 1}$$

$$\tag{2.89}$$

The inverse two-dimensional DST is defined as:

$$I(u, v) = \frac{2}{N + 1} \sum_{x=0}^{N-1} \sum_{y=0}^{N-1} J(x, y) \sin \frac{(x + 1)(u + 1)\pi}{N + 1} \sin \frac{(y + 1)(v + 1)\pi}{N + 1} \tag{2.90}$$

A practical example of the application of DST on a graylevel image is shown in Figs. 2.26, 2.27, 2.28, 2.29, and 2.30. The first image shows a representation of the original image and the transform coefficients after the transformation. The following images present gradual reconstructions using an increasing amount of transform

Scenario #1: Using coefficients only on the first row of the matrix
Reconstruction RMSE=35.517, Quality=17.8 dB

Original image data   Reconstructed image data   Scaled difference

Scenario #2: Using coefficients only in the main diagonal of the matrix
Reconstruction RMSE=38.079, Quality=17.3 dB

Original image data   Reconstructed image data   Scaled difference

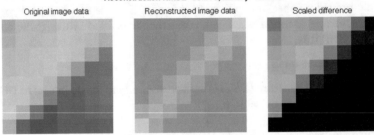

Scenario #3: Using coefficients on the first row and column of the matrix
Reconstruction RMSE=22.162, Quality=21.9 dB

Original image data   Reconstructed image data   Scaled difference

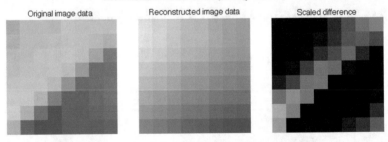

Scenario #4: Using coefficients on the up-left triangular region of the matrix
Reconstruction RMSE=6.242, Quality=32.8 dB

Original image data   Reconstructed image data   Scaled difference

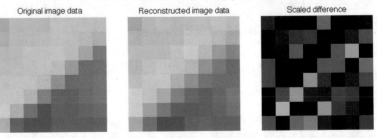

Fig. 2.25 Reconstructions of a block from image '*lena*' from a limited number of DCT coefficients

Fig. 2.26 DST representation of a graylevel image

Fig. 2.27 Reconstruction using 50 of the DST coefficients (0.05%) at 19.95 dB PSNR

Fig. 2.28 Reconstruction using 200 of the DST coefficients (0.85%) at 24.52 dB PSNR

Fig. 2.29 Reconstruction using 546 of the DST coefficients (6.3%) at 31.37 dB PSNR

Fig. 2.30 Reconstruction using 1296 of the DST coefficients (35.6%) at 39.82 dB PSNR

coefficients.[7] Apparently, DST is also good at producing compact representations of images as with only a small fraction of the coefficients it is able to produce a high quality reconstruction, as shown in Fig. 2.30, in which a reconstruction using just 35 % of the coefficients led to an image with 39.8 dBs of quality, very close the quality achieved in DCT. Also in this example, the black regions in the transform domain denote the coefficient being zeroed out using the usual diagonal scheme, in line with the ordering of increasing frequency in two dimensions.

In DST the 2-D basis functions (images) are

$$b(u, v, x, y)_{x,y=0,\ldots,N-1} = \frac{2}{N+1}\sin\frac{(x+1)(u+1)\pi}{N+1}\sin\frac{(y+1)(v+1)\pi}{N+1} \quad (2.91)$$

---

[7]Again, as in the case of the example given for DCT, the transform has been applied once on the total image area, which is not the usual case; in the usual case the transform is being applied in a block-by-block basis, typically of 8×8 pixels.

Fig. 2.31 The 8 × 8 2-D basis functions (images) of DST

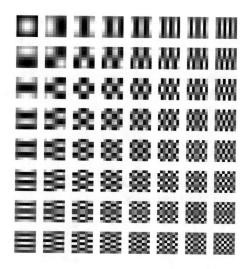

A representation of these basis functions (basis images) appears in Fig. 2.31. The characteristic transformation matrix, as in the case of the DCT, is defined, for the case of an 8 × 8 pixels image, as

$$\mathbf{h}_c = \mathbf{h}_r =$$

$$
\begin{pmatrix}
0.1612 & 0.3030 & 0.4082 & 0.4642 & 0.4642 & 0.4082 & 0.3030 & 0.1612 \\
0.3030 & 0.4642 & 0.4082 & 0.1612 & -0.1612 & -0.4082 & -0.4642 & -0.3030 \\
0.4082 & 0.4082 & 0.0000 & -0.4082 & -0.4082 & 0.0000 & 0.4082 & 0.4082 \\
0.4642 & 0.1612 & -0.4082 & -0.3030 & 0.3030 & 0.4082 & -0.1612 & -0.4642 \\
0.4642 & -0.1612 & -0.4082 & 0.3030 & 0.3030 & -0.4082 & -0.1612 & 0.4642 \\
0.4082 & -0.4082 & 0.0000 & 0.4082 & -0.4082 & 0.0000 & 0.4082 & -0.4082 \\
0.3030 & -0.4642 & 0.4082 & -0.1612 & -0.1612 & 0.4082 & -0.4642 & 0.3030 \\
0.1612 & -0.3030 & 0.4082 & -0.4642 & 0.4642 & -0.4082 & 0.3030 & -0.1612
\end{pmatrix}
$$

$$(2.92)$$

Discrete Hartley Transform

The Discrete Hartley Transform (DHT) has been proposed by an alternative to the Fourier transforms by Bracewell (1983), based on a previous work by Hartley (1942) based on the DFT (2.75) as,

$$\mathbf{J} = Re\{\mathbf{J}_F\} - Im\{\mathbf{J}_F\} \qquad (2.93)$$

where $\mathbf{J}$ the Hartley transformed signal, $\mathbf{J}_F$ the Fourier transformed signal, and $Re\{\}$ and $Im\{\}$ the real and imaginary parts of the transform. Formulating this equation using the notation in (2.75), requires to consider two approaches as pointed out by (Watson and Poirson 1986). These two approaches involve a slight variation in the definition of the basis functions (in the general case of $M \times N$ 2-D signals) as,

$$b(u, v, x, y) = \begin{cases} \cos\left(2\pi\left(\frac{ux}{M} + \frac{vy}{N}\right)\right) \\ \cos\left(\frac{2\pi ux}{M}\right) case\left(\frac{2\pi vy}{N}\right) \end{cases} \tag{2.94}$$

where $\cos(\theta) = \cos(\theta) + \sin(\theta)$, which has been shown to be an orthogonal function. The first definition gives the following definition for the transform,

$$J(u, v) = \sum_{x=0}^{M-1}\sum_{y=0}^{N-1} I(x, y)\cos\left(\frac{2\pi ux}{M} + \frac{2\pi vy}{N}\right) \tag{2.95}$$

The alternative definition of the basis functions gives rise to the following definition of the transform,

$$J(u, v) = \sum_{x=0}^{M-1}\sum_{y=0}^{N-1} I(x, y)\cos\frac{2\pi ux}{M}\cos\frac{2\pi vy}{N} \tag{2.96}$$

The inverse transform is defined as,

$$I(x, y) = \frac{1}{MN}\sum_{u=0}^{M-1}\sum_{v=0}^{N-1} J(u, v)\cos\frac{2\pi ux}{M}\cos\frac{2\pi vy}{N} \tag{2.97}$$

which is typically a separable 2-D transform and was the preferred representation by Hartley. Apparently, the Hartley Transform is equivalent to the typical Fourier Transform and the choice between them is a matter of the application.

A representation of the basis functions (basis images) for an 8×8 transform appears in Fig. 2.32. An example of the application of DHT is shown in Figs. 2.33 and 2.34 for the case of a graylevel image. The coefficients in Fig. 2.33 have been shifted to the centered of the spatial range exactly as in the case of the DFT. The three-dimensional representation of the transform coefficients magnitude shown in Fig. 2.34 reveals how this transform concentrates the image energy near the mean value and practically only in low spatial frequencies, achieving a rather compact representation. This graph shows the pseudo-colored logarithm of the coefficient magnitudes for better representation.

Walsh-Hadamard Transform

There is a family of transformations, which generalize the Fourier transforms, in which the basis functions are non-sinusoidal. The are all based on discretized versions of *Walsh functions*,[8] to formulate the transformation matrices $\mathbf{h}_c$ and $\mathbf{h}_r$. These matrices may be defined in various ways, the simpler being by the use of *Hadamard matrices*,

---

[8] A family of special piecewise constant functions assuming only the two values, ± 1.

Fig. 2.32 The 8 × 8 2-D basis functions (images) of DHT

Fig. 2.33 Example of the application of DHT on a gray level image

$$\mathbf{H}_{2^n} = \sqrt{\frac{1}{2}} \begin{pmatrix} \mathbf{H}_{2^{n-1}} & \mathbf{H}_{2^{n-1}} \\ \mathbf{H}_{2^{n-1}} & -\mathbf{H}_{2^{n-1}} \end{pmatrix}, \quad \mathbf{H}_0 = 1, \quad n > 0 \tag{2.98}$$

where the $\sqrt{\frac{1}{2}}$ normalization factor is sometimes omitted. This definition produces $2^n \times 2^n$ matrices for $n > 0$ with all elements ± 1 and a normalization factor that changes as $1/2^{\frac{n}{2}}$, so as to have the elements defined as

$$(H_{2^n})_{i,j} = \frac{1}{2^{\frac{n}{2}}} (-1)^{i \cdot j} \tag{2.99}$$

with the dot product being the bitwise dot product of the binary representations of i and j. The rows of such a matrix are Walsh functions.

Fig. 2.34 Three-dimensional pseudo-colored representation of the DHT transform coefficients

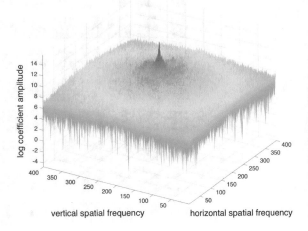

According to this definition,

$$\mathbf{H}_2 = \sqrt{\frac{1}{2}} \begin{pmatrix} 1 & 1 \\ 1 & -1 \end{pmatrix}, \quad \mathbf{H}_4 = \frac{1}{2} \begin{pmatrix} 1 & 1 & 1 & 1 \\ 1 & -1 & 1 & -1 \\ 1 & 1 & -1 & -1 \\ 1 & -1 & -1 & 1 \end{pmatrix} \qquad (2.100)$$

Scaling up to get the typical 8×8 transformation matrix for $n = 3$ produces,

$$\mathbf{h}_c = \mathbf{h}_r = \mathbf{H}_8 = \frac{1}{2\sqrt{2}} \begin{pmatrix} 1 & 1 & 1 & 1 & 1 & 1 & 1 & 1 \\ 1 & -1 & 1 & -1 & 1 & -1 & 1 & -1 \\ 1 & 1 & -1 & -1 & 1 & 1 & -1 & -1 \\ 1 & -1 & -1 & 1 & 1 & -1 & -1 & 1 \\ 1 & 1 & 1 & 1 & -1 & -1 & -1 & -1 \\ 1 & -1 & 1 & -1 & 1 & -1 & 1 & -1 \\ 1 & 1 & -1 & -1 & -1 & -1 & 1 & 1 \\ 1 & -1 & -1 & 1 & -1 & 1 & 1 & -1 \end{pmatrix} \qquad (2.101)$$

A 'curiosity' in the formation of the basis images (the 2-D case) is that a reordering of the computed function should be imposed in order to get the basis images in gradually incrementing spatial frequency in both dimensions. This reordering is imposed on the bit-levels of the coordinates—that is all coordinates are converted to their binary representation and everything is computed on their bits using a right-hand MSB notation—(leaving out the scaling factor) as

$$(H_{2^n})_{u,v,x,y} = (-1)^{\sum_{j=0}^{b-1} \left(g(u_j)x_j + g(v_j)y_j \right)} \quad \text{with} \quad g(\xi_j) = \begin{cases} \xi_0, \\ \xi_{b-1} + \xi_{b-2} \\ \vdots \quad \vdots \\ \xi_0 + \xi_1 \end{cases} \qquad (2.102)$$

where b are the total bitplanes needed to describe the numbers involved and all summations are considered in modulo-2 arithmetic, that is the modulo-2 is taken after each summation. If the binary representations are taken as column vectors then these summations are transformed to dot products as,

$$(H_{2^n})_{u,v,x,y} = (-1)^{(g(\mathbf{u})^T \cdot \mathbf{x} + g(\mathbf{v})^T \cdot \mathbf{y}) \bmod 2} \qquad (2.103)$$

with g defined as in (2.102) and the reordering imposed on the elements of the corresponding vectors.

A representation of the basis images appears in Figs. 2.35 and 2.36 for the 4×4 and 8×8 transform respectively. A practical example of the application of WHT on a graylevel image is shown in Figs. 2.37, 2.38, 2.39, 2.40, 2.41 and 2.42. The first image shows a representation of the original image and the transform coefficients produced by the transformation, whereas the other images present gradual reconstructions using an increasing amount of transform coefficients. As in the previous examples, the black regions in the transform domain denote the coefficient being zeroed out.

Due to the computational simplicity of Walsh-Hadamard Transform (WHT) in comparison to Fourier transforms, since the WHT does not involve multiplications or divisions (only ± 1 factors), this transform is significantly more appealing in applications that have access to low computational resources. It has been used for image compression when the computational power of computer was not strong enough to support images of high dimensions in limited time. NASA made heavy use of the WHT during the 1960s and early 1970s space missions for photo compression, 'real-time' video transmission and error correction over unreliable channels.

Fig. 2.35 The 4×4 2-D basis images of WHT

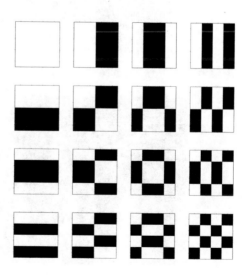

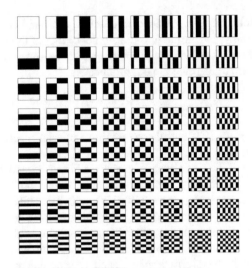

Fig. 2.36 The 8 × 8 2-D basis images of WHT

Fig. 2.37 WHT representation of a graylevel image

Fig. 2.38 Reconstruction using 2% of the WHT coefficients at 23.75 dB PSNR

Fig. 2.39 Reconstruction using 8% of the WHT coefficients at 27.09 dB PSNR

Fig. 2.40 Reconstruction using 18% of the WHT coefficients at 30.27 dB PSNR

Fig. 2.41 Reconstruction using 32% of the WHT coefficients at 32.52 dB PSNR

Fig. 2.42 Reconstruction using 50% of the WHT coefficients at 36.79 dB PSNR

Haar Transform

Alfred Haar in his seminal 1910 *theory of orthogonal function systems* (Haar 1910), defined the so called *orthogonal function system* χ, as the most representative of the class of orthogonal systems as,

$$\chi_0(s) = 1, \quad \forall s \in [0, 1]$$

$$\chi_1(s) = \begin{cases} 1 & 0 \le s < \frac{1}{2} \\ -1 & \frac{1}{2} \le s < 1 \end{cases}$$

$$\chi_2^{(1)}(s) = \begin{cases} \sqrt{2} & 0 \le s < \frac{1}{4} \\ -\sqrt{2} & \frac{1}{4} < s < \frac{1}{2} \\ 0 & \frac{1}{2} < s \le 0 \end{cases} \tag{2.104}$$

$$\chi_2^{(2)}(s) = \begin{cases} 0 & 0 \le s < \frac{1}{2} \\ \sqrt{2} & \frac{1}{2} < s < \frac{3}{4} \\ -\sqrt{2} & \frac{3}{4} < s \le 0 \end{cases}$$

so, by dividing the interval [0, 1] into 2^n equal parts and by denoting these subintervals by $i_n^{(1)}, i_n^{(2)}, \ldots, i_n^{(2^n)}$,

$$\chi_n^{(k)}(s) = \begin{cases} 0 & s \in i_n^{(1)}, i_n^{(2)}, \ldots, i_n^{(2k-2)} \\ \sqrt{2^{n-1}} & s \in i_n^{(2k-1)} \\ -\sqrt{2^{n-1}} & s \in i_n^{(2k)} \\ 0 & s \in i_n^{(2k+1)}, \ldots, i_n^{(2^n)} \end{cases} \tag{2.105}$$

$$k = 1, 2, \ldots, 2^{n-1}$$

At points 0 and 1, each function $\chi_n^{(k)}(s)$ is assigned the value it assumes in the intervals $\left[0, \frac{1}{2^n}\right]$ or $\left[1 - \frac{1}{2^n}, 1\right]$. According to this definition, $\chi_n^{(k)}(s)$ is a piecewise constant function with discontinuities only in the points $\frac{2k-2}{2^n}$, $\frac{2k-1}{2^n}$, $\frac{2k}{2^n}$, where $\chi_n^{(k)}(s)$ assumes (by agreement) a value equal to the arithmetic means of the values in the intervals adjoining at those points. Using this definition, infinitely many function χ form *a complete orthogonal function system*. Haar proved that *this system is orthogonal and complete*.

The Haar Transform (HT) was developed as a family of transformations, in which the characteristic transformation tables $\mathbf{h}_c$ and $\mathbf{h}_r$ are discretized forms of *Haar functions*. The HT is essentially a simple compression process. In one dimension, the HT transforms a vector of two elements $\mathbf{x} = (x_1 \ x_2)^T$ to a vector $\mathbf{y} = (y_1 \ y_2)^T$, as follows

$$\begin{pmatrix} y_1 \\ y_2 \end{pmatrix} = \mathbf{H} \cdot \begin{pmatrix} x_1 \\ x_2 \end{pmatrix} , \text{ where } \mathbf{H} = \frac{1}{\sqrt{2}} \begin{pmatrix} 1 & 1 \\ 1 & -1 \end{pmatrix} \quad (2.106)$$

or, in simple form $\mathbf{y} = \mathbf{H} \cdot \mathbf{x}$, and according to the definition of $\mathbf{H}$, y_1 and y_2 are the sum and the difference of x_1 and x_2, divided by $\sqrt{2}$ for energy conservation. It should be noted that the matrix $\mathbf{H}$ is orthonormal (comprises of orthogonal unit-length vectors), and therefore, $\mathbf{H}^{-1} = \mathbf{H}^T = \mathbf{H}$ (as $\mathbf{H}$ is symmetric, $\mathbf{H}^T = H$). Therefore, the inverse transformation can be solved by (2.106) as follows

$$\mathbf{y} = \mathbf{H} \cdot \mathbf{x} \Leftrightarrow \mathbf{H}^{-1} \cdot \mathbf{y} = \mathbf{H}^{-1} \cdot \mathbf{H} \cdot \mathbf{x} \Leftrightarrow \mathbf{H}^T \cdot \mathbf{y} = \mathbf{I} \cdot \mathbf{x} \Leftrightarrow \mathbf{H} \cdot \mathbf{y} = \mathbf{x}$$

$$\begin{pmatrix} x_1 \\ x_2 \end{pmatrix} = \mathbf{H}^T \cdot \begin{pmatrix} y_1 \\ y_2 \end{pmatrix} , \text{ where } \mathbf{H}^T = \mathbf{H} = \frac{1}{\sqrt{2}} \begin{pmatrix} 1 & 1 \\ 1 & -1 \end{pmatrix} \quad (2.107)$$

In two dimensions, $\mathbf{x}$ and $\mathbf{y}$ are 2×2 matrices. Since the HT is separable it can be applied first to the columns and then the rows of x. The relation which expresses the process is $\mathbf{y} = \mathbf{H} \cdot \mathbf{x} \cdot \mathbf{H}^T$, and the inverse HT becomes $\mathbf{x} = \mathbf{H}^T \cdot \mathbf{y} \cdot \mathbf{H}$. Specifically, what happens during the HT is that given the input matrix $\mathbf{x}$

$$\mathbf{x} = \begin{pmatrix} a & b \\ c & d \end{pmatrix} \quad (2.108)$$

then the output of the transform is

$$\mathbf{y} = \frac{1}{2} \begin{pmatrix} a+b+c+d & a-b+c-d \\ a+b-c-d & a-b-c+d \end{pmatrix} \quad (2.109)$$

Equations (2.108) and (2.109) correspond to the following filtering procedures (which take into account the halving factor):

- Top-Left: $a+b+c+d$ = mean value or a 2D low-pass filter (LL—Low in both dimensions)
- Top-Right: $a-b+c-d$ = average horizontal gradient or horizontal high-pass and vertical low-pass filter (HL)
- Bottom-Left: $a+b-c-d$ = average vertical gradient or horizontal low-pass and vertical high-pass filter (LH)

- Bottom-Right: $a-b-c+d$ = diagonal gradient or 2D high-pass filter (HH—High in both dimensions)

An example of the HT on a graylevel image is shown in Fig. 2.43, where only one step in the overall possible decomposition is being depicted. Apparently, the application of the transform results in a band-based decomposition of the original image. One step of the transform produces four bands, which correspond to a low-pass filtered replica of the original image (upper-left quarter) usually denoted as the LL band, and three high-pass filtered replicas (the three dark quarter images) that capture the horizontal (HL), vertical (LH) and diagonal (HH) high spatial frequencies. The three high frequency bands are displayed in log-scale for better illustration. Figure 2.44 shows the histograms of all these images. Apparently the histogram of the LL band is very similar to that of the original image, whereas the histograms of the high frequency bands are similar to two-sided geometric distributions centered

Fig. 2.43   The result of the application of a 1-step HT on a digital image

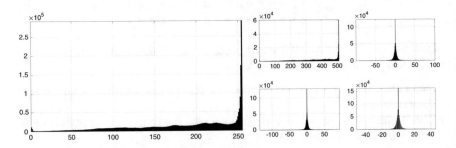

Fig. 2.44   The histograms of the images in Fig. 2.43

Table 2.3 Energy levels per image band after HT

LL (%)	HL (%)	LH (%)	HH (%)
96.5	2.2	0.9	0.4

around zero. The energy is unevenly distributed among the transform bands, as shown in Table 2.3, which lists the relative energy levels in each band normalized to the total energy of the image.

In order to create transform matrices of any size, one has to begin with the Haar functions,

$$h_{p,q}(x) = \frac{1}{\sqrt{N}} \begin{cases} 1 & p = q = 0,\ x \in [0, 1] \\ 2^{\frac{p}{2}} & \frac{q-1}{2^p} \leq x < \frac{q-\frac{1}{2}}{2^p} \\ -2^{\frac{p}{2}} & \frac{q-\frac{1}{2}}{2^p} \leq x < \frac{q}{2^p} \\ 0 & \text{otherwise for } x \in [0, 1] \end{cases} \tag{2.110}$$

where,

$$\begin{aligned} 0 &\leq p \leq n - 1 \\ q &= 0, 1 \quad p = 0 \\ 1 &\leq q \leq 2^p \quad p \neq 0 \\ N &= 2^n \end{aligned} \tag{2.111}$$

By discretizing the values of x at m/N, $m = 0, \ldots, N - 1$ these functions produce the transformation matrices. This way, the characteristic 8×8 transformation matrix, as defined in previous cases, is given for $N = 8$ as

$$\mathbf{h}_c = \mathbf{h}_r = \frac{1}{\sqrt{8}} \begin{pmatrix} 1 & 1 & 1 & 1 & 1 & 1 & 1 & 1 \\ 1 & 1 & 1 & 1 & -1 & -1 & -1 & -1 \\ \sqrt{2} & \sqrt{2} & -\sqrt{2} & -\sqrt{2} & 0 & 0 & 0 & 0 \\ 0 & 0 & 0 & 0 & \sqrt{2} & \sqrt{2} & -\sqrt{2} & -\sqrt{2} \\ 2 & -2 & 0 & 0 & 0 & 0 & 0 & 0 \\ 0 & 0 & 2 & -2 & 0 & 0 & 0 & 0 \\ 0 & 0 & 0 & 0 & 2 & -2 & 0 & 0 \\ 0 & 0 & 0 & 0 & 0 & 0 & 2 & -2 \end{pmatrix} \tag{2.112}$$

Wavelet Transform

Supposing a real or complex function $\psi(t)$ with the following properties:

• the integral of the function equals zero:

$$\int_{-\infty}^{\infty} \psi(t)dt = 0 \tag{2.113}$$

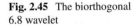

Fig. 2.45 The biorthogonal 6.8 wavelet

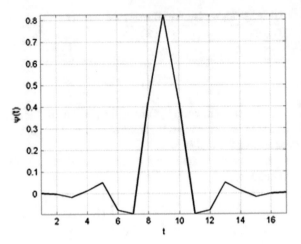

- the function is square integrable (or has finite energy):

$$\int_{-\infty}^{\infty} |\psi(t)|^2 dt < \infty \tag{2.114}$$

- the function satisfies the convention:

$$C \equiv \int_{-\infty}^{\infty} \frac{|\Psi(\omega)|^2}{|\omega|} d\omega \tag{2.115}$$

where $\Psi(\omega)$ the Fourier transform of $\psi(t)$, then the function $\psi(t)$ is a *mother wavelet or simply a wavelet*.

And while the assumption expressed in (2.115) is useful for the formulation of the inverse transformation, the first two conditions are sufficient for the definition of the Continuous-time Wavelet Transform (CWT) and explain the reason why the function is called wavelet (wavelet). The first condition (2.113) contains the information that the function itself is characterized as an oscillation or has a wavy nature. Unlike a continuous sinusoidal function, it is a 'small' wave. The second condition (2.114) ensures that most of the function energy is limited to a finite temporal duration. These two conditions are satisfied relatively easy and there is a multitude of functions that are suitable as mother wavelets (Rao and Bopardikar 1998).

Figure 2.45 illustrates the 'famous' *biorthogonal 6.8* wavelet.[9] This wavelet takes values in a closed region, i.e. it has a finite duration. On the contrary, there is the possibility of infinite duration wavelets, such as the *Morlet wavelet*, as illustrated in Fig. 2.46. This wavelet is created through the modulation of a cosine function by a Gaussian function. While it is of infinite duration, most of the energy is confined to

---

[9]'Famous' due to its acceptance and usage in image compression applications, like in the Joint Photographic Experts Group 2000 (JPEG2000) image compression standard.

Fig. 2.46   The Morlet
wavelet

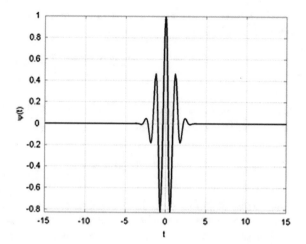

a finite duration. As shown in the graph, more than 99 % of the energy is confined
within the time frame $|t| \leq 2.5$ s (Rao and Bopardikar 1998).

Consider a square integrable function $f(t)$. The CWT of this function relative to
the wavelet is defined (Rao and Bopardikar 1998)

$$W(a, b) \equiv \int_{-\infty}^{\infty} f(t) \frac{1}{\sqrt{|a|}} \psi^* \left(\frac{t - b}{a} \right) dt \qquad (2.116)$$

where a and b are real numbers and $*$ indicates the complex conjugate. Accordingly,
this transformation is a function of two variables. It should be noted that both $f(t)$
and $\psi(t)$ belong to the $L^2(R)$ space of the square integrable functions, which is called
the *set of energy signals*. Defining:

$$\psi_{a,b}(t) \equiv \frac{1}{\sqrt{|a|}} \psi^* \left(\frac{t - b}{a} \right) \qquad (2.117)$$

(2.116) can be written as:

$$W(a, b) \equiv \int_{-\infty}^{\infty} f(t) \psi^*_{a,b}(t) dt \qquad (2.118)$$

It should also be noted that:

$$\psi_{1,0}(t) = \psi(t) \qquad (2.119)$$

The normalization factor $1/\sqrt{|a|}$ guarantees that the energy remains constant for all a and b, so

$$\int_{-\infty}^{\infty} |\psi_{a,b}(t)|^2 dt = \int_{-\infty}^{\infty} |\psi(t)|^2 dt \tag{2.120}$$

for every a and b. For a given value of a, the function $\psi_{a,b}(t)$ expresses the translation (or shifting) of $\psi_{a,0}(t)$ by b on the horizontal time axis. In addition, from

$$\psi_{a,0}(t) \equiv \frac{1}{\sqrt{|a|}} \psi^* \left(\frac{t}{a} \right) \tag{2.121}$$

can be concluded that $\psi_{a,0}(t)$ is an extension in time and scale (magnitude) of $\psi(t)$. Since a determines the size of the time extension or expansion, it is called the expansion or scale variable. When $a > 1$, $\psi(t)$ is expanded on the time axis, and when $0 < a < 1$ it is retracted. Negative values of a result in a time reversal while scaling. Thus, since the CWT is created by scaling and shifting of a function, the wavelet transform is called mother wavelet (Rao and Bopardikar 1998). It should be emphasized that the biggest advantage of Wavelet-based transforms over Fourier-based transforms is that they *provide a better time and frequency domain localization*, obeying, of course, the uncertainty principle, as expressed in time-frequency analysis

$$\Delta t_\psi \Delta \omega_\psi = c_\psi \tag{2.122}$$

where c_ψ is a constant dependent on the wavelet used, Δt_ψ the mean squared duration and $\Delta \omega_\psi$ the mean squared frequency range. It is understood that the smaller the value of the constant c_ψ the more accurate time-frequency analysis can this wavelet achieve (Rao and Bopardikar 1998).

Finally, when the third condition of the mother wavelet, as expressed by Eq. (2.115) is satisfied for $0 < C < \infty$, then the inverse CWT is defined as

$$f(t) = \frac{1}{C} \int_{a=-\infty}^{\infty} \int_{b=-\infty}^{\infty} \frac{1}{|a|^2} W(a,b) \psi_{a,b}(t) dadb \tag{2.123}$$

It should of course be noted that this is a sufficient but not a necessary condition for achieving a mapping of all CWT in $L^2(R)$. To discretize CWT, the following representation is adopted

$$f(t) = \sum_{k=-\infty}^{\infty} \sum_{l=-\infty}^{\infty} d(k,l) 2^{-k/2} \psi(2^{-k}t - 1) \tag{2.124}$$

It is worth noting that unlike Eq. (2.123), in which continuous shifts and scalings are involved, Eq. (2.124) involves discrete values. Scaling takes values of the form

Fig. 2.47 The 'cells' of
time-frequency analysis that
correspond to a dyadic
sampling

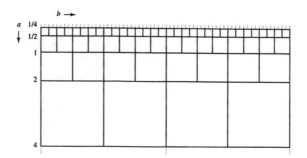

$a = 2^k$ (where k is an integer), whereas for each scaling 2^k, the shift takes values of
the form $b = 2^k l$ (where l is an integer). Thus, the values of $d(k, l)$ relate to values
of the transformation at $a = 2^k$ and $b = 2^k l$, which corresponds to sampling the
coefficients (a, b) on a grid, as shown in the Fig. 2.47. This process is called *dyadic
sampling* because successive discrete values of scale and the corresponding sampling
intervals differ by a factor of two (2) (Rao and Bopardikar 1998).

The two-dimensional sequence $d(k, l)$ is called the Discrete Wavelet Transform
(DWT) of $f(t)$. As is evident, the DWT remains a transformation of a continuous
signal. The discretization refers only to the variables a and b. In this sense, it is
analogous to Fourier series, which represent in a discrete frequency domain (periodic)
a continuous (in time) signal. For this reason, the DWT is referred to as *Continuous-
Time Wavelet Series*.

For the application of Wavelet Transform (WT) in images, a two-dimensional
transformation is required. It turns out, however, that the WT is separable, and therefore, it is possible to sequentially apply two one-dimensional transformations (e.g.,
first on the rows and then the columns of the image), which greatly simplifies the
calculations. Key features of use of the WT in compression are (Rao and Bopardikar
1998):

- it involves the idea of the multi-resolution image representation
- it applies (or at least could apply) to the entire image and, therefore, it does not
  suffer from blocking artifacts (evident in DCT)
- it can be used (by applying integer wavelet filter coefficients) for simultaneous
  lossless and lossy coding, and embedding of both in the same output file
- it can provide decomposition of the image in spectral bands, where each band can
  be quantized according to the importance of the visual content

The one-dimensional transform is implemented by applying two analysis filters
followed by downsampling, as shown in Fig. 2.48. The one-dimensional sample
sequence is filtered by a low pass and a high pass filter, is then sub-sampled by a
factor of 2, which ultimately results in a representation of a low and a high frequency
band. Of course, the inverse transform is the exact reverse procedure (Rabbani and
Joshi 2002; Rabbani and Cruz 2001).

Fig. 2.48 Analysis filter
bank

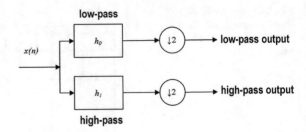

Two methods leading to the same result can be applied to compute the transformation (Rabbani and Joshi 2002; Rabbani and Cruz 2001):

- *convolution*
- *lifting*: an alternative method to compute the transform coefficients using the following three-step process:

1. *separation*: initially the input signal is divided into two sequences, taking samples in even locations (or times) for one sequence and samples in odd locations for the other. This step is commonly called the *lazy wavelet transform*
2. *prediction and update*: assuming that the input signal shows a significant correlation between successive samples, one may deduce with some certainty that an even sample can predict the value of the neighboring odd sample. Then the output is updated with corrections in the prediction that preceded. The process in this step can be expressed as:

$$(s_{odd_{i-1}}, s_{even_{i-1}}) \leftarrow S(s_i) s_{odd_{i-1}} - = P(s_{even_{i-1}}) s_{even_{i-1}} + = U(s_{odd_{i-1}}) \quad (2.125)$$

and graphically represented in the Fig. 2.49, and can be textually description as follows:
  - separation S of the input into even and odd samples
  - prediction P of odd samples from previous even samples
  - update U of predicted value
3. *normalization*: at the final stage, the output of the previous step is normalized to form the final transform coefficients.

Figure 2.50 shows the result of applying the DWT on an image. The two steps are shown, in which the transformation is applied sequentially due to the *separability*. During the first step the DWT is applied to the columns of the original image resulting in an image like the one shown in Fig. 2.50a. At the second step, the transformation is applied to the rows of the image produced by the first step resulting in the image shown in Fig. 2.50b. It should be noted that the values of the parts of the images that reflect the high frequency content (the 'noisy' gray regions) have been normalized to allow a better representation.

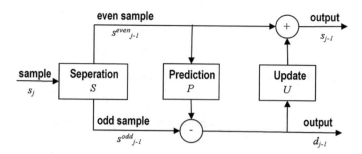

Fig. 2.49 Representation of the lifting method for WT

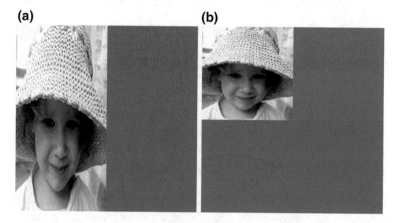

Fig. 2.50 DWT as a separable image transformation: **a** the transformation on the columns of the image and **b** the transformation on the rows of the image after the first step

Statistics in the Transform Domain

Both Pratt (1991) and Jain (1988) in their analysis of the image transforms provide further insight on the transform domain statistics in an attempt to support the need for those transforms in order to decorrelate the input data before further processing for compression. In this analysis, the *first and second moments* of the transform coefficients have been analyzed. If a square image **I** of size $N \times N$ is considered to be a two-dimensional ergodic process with known mean and covariance function then its unitary transform

$$J(u, v) = \sum_{x=0}^{N-1}\sum_{y=0}^{N-1} I(x, y)T(x, y; u, v) \tag{2.126}$$

is also a stochastic process with a mean value

$$E\{J(u, v)\} = \sum_{x=0}^{N-1}\sum_{y=0}^{N-1} E\{I(x, y)\}\, T(x, y; u, v) \qquad (2.127)$$

with $E\{I(x, y)\}$ being the mean of $I(x, y)$. Thus, the covariance function can be written as

$$\Sigma_J(u_1, v_1; u_2, v_2) = \sum_{x_1}\sum_{x_2}\sum_{y_1}\sum_{y_2} \Sigma_I(x_1, y_1; x_2, y_2)T(x_1, y_1; u_1, v_1)T^*(x_2, y_2; u_2, v_2)$$

$$(2.128)$$

where $\Sigma_I(x_1, y_1; x_2, y_2)$ is the covariance function of $I(x, y)$. Also, the variance function of $J(u, v)$ is

$$\sigma_J^2(u, v) = \Sigma_J(u, v; u, v) \qquad (2.129)$$

These equations can be simplified in notation by adopting a matrix representation, thus

$$
\begin{aligned}
\mathbf{J} &= \mathbf{TI} &&\text{is the unitary transform} \\
\mathbf{m}_J &= \mathbf{Tm}_I &&\text{is the mean of the transform coefficients} \\
\mathbf{\Sigma}_J &= \mathbf{T\Sigma_I T}^{*T} &&\text{is the covariance matrix of the coefficients} \\
\mathbf{V}_J &= diag\,[\mathbf{\Sigma}_J] &&\text{is the vector of variances of the coefficients}
\end{aligned}
\qquad (2.130)
$$

Analysis of the performance of KLT reveals that this is the only unitary transform that performs a complete decorrelation for an arbitrary image achieving the best energy compaction of all unitary transforms (Pratt 1991). Apparently this is due to the fact that the variance function of the KLT coefficients equals the corresponding eigenvalue, that is

$$\sigma_J^2(u, v) = \lambda(u, v) \qquad (2.131)$$

Transformations other than the KLT result in at least some residual correlation between the transform coefficients, thus

$$\sum_{w=0}^{W} \lambda(w) \geq \sum_{w=0}^{W} \sigma^2(w) \quad W < N^2 \qquad (2.132)$$

In addition, in order to derive analytic representations for the first and second moments of the transform coefficients of arbitrary image transforms, it is usually assumed that images are WSS; thus the mean value of an image is a constant $E\{\mathbf{I}\}$ and the covariance function assumes the functional form $\Sigma_I(x_1 - x_2, y_1 - y_2)$. In addition, if the transform is orthogonal the summation $\sum_x \sum_y T(x, y; u, v)$ yields zero for all non-zero-th basis functions

$$E\{J(u, v)\} = E\{\mathbf{I}\}\sum_x \sum_y T(x, y; u, v) = 0 \quad \forall u, v \neq 0 \qquad (2.133)$$

Even though the input image is considered to be WSS this is not expected to be true for the transform coefficients, unless the transform kernel is space invariant. It can be shown that the Fourier transform is such a transform and an analytic representation for the first and second moments is possible, whereas for Hadamard, Haar and other transforms no closed form expressions have been developed for the covariance functions of those transforms.

Furthermore, based on the *central limit theorem* it is possible to approximate the probability densities of the transform coefficients by adopting a typical Gaussian distribution with the previously mentioned first and second moments. Thus the probability densities of the Fourier coefficients can be defined,

$$p\left(Re\left\{J_F(u,v)\right\}\right) = \sqrt{2\pi\sigma_J^2(u,v)}\, e^{\frac{-Re\left\{J_F^2(u,v)\right\}}{2\sigma_J^2(u,v)}}$$

$$p\left(Im\left\{J_F(u,v)\right\}\right) = \sqrt{2\pi\sigma_J^2(u,v)}\, e^{\frac{-Im\left\{J_F^2(u,v)\right\}}{2\sigma_J^2(u,v)}}$$

(2.134)

where $Re\{\}$, $Im\{\}$ represent the real and imaginary parts of the transform. The assumed Gaussian probability density of the real and imaginary parts imply that the magnitude of the transform is modeled by a Rayleigh distribution and the phase is modeled by a uniform distribution, that is

$$p\left(\|J_F(u,v)\|\right) = \frac{\|J_F(u,v)\|}{\sigma_J^2(u,v)}\, e^{\frac{-\|J_F^2(u,v)\|}{2\sigma_J^2(u,v)}}$$

$$p\left(\angle J_F(u,v)\right) = \frac{1}{2\pi} \qquad -\pi \le \angle J_F(u,v) \le \pi$$

(2.135)

The probability density of coefficients of unitary transforms other than the Fourier transform are usually modeled by Gaussian or Laplacian densities as follows,

$$p\left(J(u,v)\right) = \sqrt{2\pi\sigma_J^2(u,v)}\, e^{\frac{-J^2(u,v)}{2\sigma_J^2(u,v)}}$$

$$p\left(J(u,v)\right) = \sqrt{2\pi\sigma_J^2(u,v)}\, e^{\frac{-\sqrt{2}|J(u,v)|}{\sigma_J(u,v)}}$$

(2.136)

To illustrate the effect of the image transforms and to provide a representation of the covariance matrices of the transform coefficients Fig. 2.51 displays the covariance matrices of the transform coefficients of various image transforms. In order to make the illustration printable and readable the logarithm of the covariances of the coefficients are being displayed. In all these graphical representations of covariance matrices, a more distinguishable and highly contrasted diagonal denotes a better decorrelation by the corresponding transform. The figure includes the original image and the covariance matrix of the initial image pixels, which exhibit high correlation as indicated by the multiple bright regions throughout the whole surface of its covariance matrix. The DFT coefficients are represented only by their magnitude and not any information regarding the phase. In addition, the (1-step) Haar coefficients have

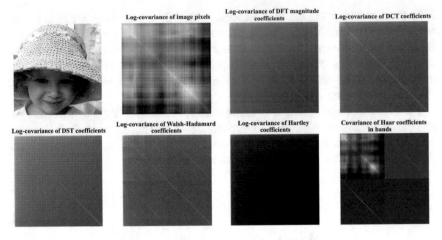

Fig. 2.51 Covariance matrices of various image transform coefficients

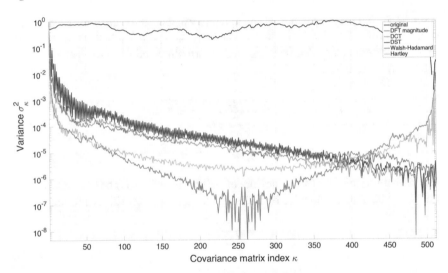

Fig. 2.52 Normalized log variance in each of the image transforms

been normalized individually in each of the four bands purely for illustration purposes; as expected the upper-left quarter of the matrix is equivalent to the covariance matrix of the initial image as it is simply a low-pass filtered replica of the original data, thus displays a large amount of correlation in bright regions. In addition, Fig. 2.52 shows a plot of the variances computed in each of the transforms, in normalized logarithmic scale; variances are shown as found in the covariance matrices without reordering or sorting (thus the 'mirroring' in the DFT magnitude is apparent).

2.3.2.2 Quantization

Quantization is usually defined as *the dividing of a quantity into a discrete number of small pieces, which are usually integer multiples of a common base quantity.* The oldest (and most common) example of quantization is that of *rounding*: every real number x may be rounded to the nearest integer $q(x)$ with a quantization error $e = |q(x) - x|$. Generally, it can be assumed that a quantizer consists of a set of intervals or 'cells' $S = \{S_i; i \in \mathfrak{I}\}$, where the set of indices $\mathfrak{I}$ is often a collection of consecutive integers (often starting with 0 or 1), along with a set of reconstruction values or points or levels $C = \{y_i; i \in \mathfrak{I}\}$, so that in overall the quantizer q is defined by the equation $q(x) = y_i$ for $x \in S_i$, which may eventually be represented by the formula (Gray and Neuhoff 1998)

$$q(x) = \sum_i y_i l_{s_i}(x) \tag{2.137}$$

where

$$l_{s_i} = \begin{cases} 1 & x \in S \\ 0 & \text{otherwise} \end{cases}$$

In order for this definition to have a substantial meaning, an assumption is made that S is a partition on the axis of the real numbers. This means that the intervals are independent and complete. The general definition also applies in the simplest case of rounding when $S_i = (i - 1/2, i + 1/2]$ and $y_i = i$ for all integers i. Even more generally, the intervals may take the form $S_i = (a_{i-1}, a_i]$ in which a_i (called *thresholds*) form an ascending sequence. The width of the interval S_i is the length of the $a_i - a_{i-1}$. The function $q(x)$ is usually called *quantization rule*. A simple quantizer with five levels is shown in Fig. 2.53. A quantizer is called *uniform* when the levels are equally spaced by a fixed distance δ among them, and the thresholds a_i are at the centers of the intervals. In all other cases the quantizer is called *non-uniform*.

The quality of a quantizer is evaluated by comparing the final outcome with the initial input. A common way of doing this is by defining a *distortion measure*—which quantifies the cost or the distortion that occurs in the reconstruction of the input to the quantizer output—and taking the average distortion as a measure of assessing the quality of the system, wherein lower average distortion means better quality. The most commonly used estimator is that of the squared error $d(x, \hat{x}) = |x - \hat{x}|^2$, ($\hat{x}$ representing the quantized value of x). In practice, the average value is a sample

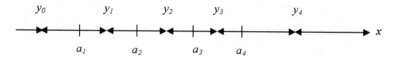

Fig. 2.53 A quantizer of five levels ($a_0 = -\infty$, $a_5 = \infty$)

average when the quantization is applied to sequences of real numbers, but in the general case data are taken as elements that share a common probability density function $f(x)$ corresponding to a random variable X, and the average distortion is converted to a statistical expectation (Gray and Neuhoff 1998)

$$D(q) = E[d(X, q(X))] = \sum_i \int_{S_i} d(x, y_i) f(x) dx \qquad (2.138)$$

When the distortion is estimated by means of a squared error, $D(q)$ is converted to MSE, a special case, which is often used.

It is desirable that the average distortion is maintained as low as possible by increasing the number of intervals (negligible average distortion may be achieved when there are many small quantization intervals). However, additional cost is imposed in the form of extra bits of information that must be created to describe the quantizer, which creates a digital representation problem in light of a limited capacity. A simple method for estimating this cost is as follows: the quantizer encodes an input x to a binary representation or codeword corresponding to a quantization index i, thus determining the quantization level to be used for the reconstruction. If there are N possible levels and all the binary representations or codewords are of the same length, the binary vectors will need ($\lfloor \log_2 N \rfloor + 1$) bits. This results to an estimate rate in bps

$$R(q) = \log_2 N \qquad (2.139)$$

This quantizer that leads to fixed length binary representations is called *fixed-rate quantizer.*

In overall, the target of quantization is the encoding of input samples, characterized by a probability density function, into a form with the least number of bits (low rate) in such a way as to ensure the reconstruction of the input signal at the greatest possible precision (with low distortion). What immediately becomes clear is that there is a trade-off in quantization: distortion against data rate (and vice versa). This trade-off can be quantified by a rate-distortion function $\delta(R)$, which is defined as the minimum distortion by a quantizer with a rate less or equal to R

$$\delta(R) = \inf_{q:R(q) \leq R} D(q) \qquad (2.140)$$

It is also possible to define, alternatively, the function $r(D)$, as the lowest rate of a fixed-rate quantizer with distortion less or equal to D.

Up to this point, the description was about the so-called *fixed-rate scalar quantizer*, in which each sample of the input is encoded independently into a binary string of fixed length. There, of course, are many other alternative forms of quantization, which achieve better results in terms of rate-distortion. Some indicative cases are (Gray and Neuhoff 1998):

• *scalar quantizer with memory*, where a prediction technique is applied to predict the current sample by storing one or more previous samples (like the Differential

Pulse Code Modulation (DPCM) method, which is considered by many to belong to the quantization rather than the coding methods)
- *variable-rate quantizer*, which has a partition at intervals and a dictionary of quantization levels just as the fixed-rate quantizer, but uses a variable length binary representations of the intervals. In this category belong the *entropy constrained scalar quantizers*, which are designed to introduce the smallest average distortion under a pre-conditioned entropy
- *vector (or multidimensional) quantizer*, which is the generalization of the scalar quantizer in n-dimensions, providing a general model of quantization, which can be applied to vectors without structural constraints.

To generate optimal scalar quantizers the Lloyd-Max method (Lloyd 1982) is typically applied, in which:

- to create an optimum quantizer with L quantization intervals, the quantization levels are initialized as y_i, $i = 1, .., L$ and the decision end-points are

$$x_0 = -\infty, \quad x_{L+1} = \infty, \quad x_i = \frac{y_i + y_{i+1}}{2}, \quad i = 1, ..., L \qquad (2.141)$$

- then the system performs iterations to achieve convergence or until the error is within desired limits, following the equations

$$y_i = \frac{\int_{x_{i-1}}^{x_i} x p(x) dx}{\int_{x_{i-1}}^{x_i} p(x) dx}, \quad x_i = \frac{y_i + y_{i+1}}{2}, \quad i = 1, ..., L \qquad (2.142)$$

This expression of y_i indicates that it should be selected iteratively as the centroid of probability densities in the quantization interval $[x_{i-1}, x_i]$. It is worth noting that when the probability density function satisfies the condition $d^2p(x)/dx^2 < 0$, which is satisfied for Gaussian and Laplacian distributions, the algorithm converges to a global minimum (in respect to distortion). The Lloyd-Max method generalizes for the case of optimum vector quantizers into the *generalized Lloyd-Max* method or the *Linde-Buzo-Gray (LBG)* method (Linde et al. 1980), in which the following occur:

- initialization of levels y_i, $i = 1, .., L$
- segmentation of the set of all learning vectors t_j into S_i sets, for which the vector y_i is the closest:

$$t_j \in S_i \iff d(y_i, t_j) \leq d(y_k, t_j), \forall k \qquad (2.143)$$

- redefinition of y_i as the center of mass of S_i:

$$y_i = \frac{1}{\#S_i} \sum_{x_i \in S_i} x_j \qquad (2.144)$$

- repetition of the last two steps until convergence

Although vector quantizers are generally better in performance, scalar quantizers are preferable in compression methods due to their significantly easy implementation and the much lower memory requirements. Scalar quantizers can be actually found at the heart of most wide-spread image compression standards, in many forms and alternatives, but with the same aim, in truncating the image data. It is the step within lossy image compression systems that generates the most (if not all) the degeneration of the original data, which is counter-balanced by the strong positive effect it produces for the entropy coding steps that usually follow. In the JPEG image compression standard, quantization is practically implemented in the form of an element-wise matrix operation on the 8×8 block it applies, using the following 8×8 reference quantization matrices for the luminance (q_Y) and chrominance (q_C) channels respectively (ISO-IEC-CCITT 1993b; Pennebaker and Mitchell 1993; Wallace 1991)

$$q_Y = \begin{pmatrix} 16 & 11 & 10 & 16 & 24 & 40 & 51 & 61 \\ 12 & 12 & 14 & 19 & 26 & 58 & 60 & 55 \\ 14 & 13 & 16 & 24 & 40 & 57 & 69 & 56 \\ 14 & 17 & 22 & 29 & 51 & 87 & 80 & 62 \\ 18 & 22 & 37 & 56 & 68 & 109 & 103 & 77 \\ 24 & 35 & 55 & 64 & 81 & 104 & 113 & 92 \\ 49 & 64 & 78 & 87 & 103 & 121 & 120 & 101 \\ 72 & 92 & 95 & 98 & 112 & 100 & 103 & 99 \end{pmatrix}$$

$$q_C = \begin{pmatrix} 17 & 18 & 24 & 47 & 99 & 99 & 99 & 99 \\ 18 & 21 & 26 & 66 & 99 & 99 & 99 & 99 \\ 24 & 26 & 56 & 99 & 99 & 99 & 99 & 99 \\ 47 & 66 & 99 & 99 & 99 & 99 & 99 & 99 \\ 99 & 99 & 99 & 99 & 99 & 99 & 99 & 99 \\ 99 & 99 & 99 & 99 & 99 & 99 & 99 & 99 \\ 99 & 99 & 99 & 99 & 99 & 99 & 99 & 99 \\ 99 & 99 & 99 & 99 & 99 & 99 & 99 & 99 \end{pmatrix}$$

$$(2.145)$$

The open source implementation of JPEG from the Independent JPEG Group (IJG) (Independent JPEG Group 2000) includes the following formula for the generation of the quantization matrices based on the reference matrices (2.145), depending on the coefficient of the desired quality defined by the user (f)

$$Q_\xi = q_\xi \times e^k, \qquad k = 6\ln2\,\frac{50 - f}{50} \qquad (2.146)$$

where Q_ξ the estimated quantization matrix, f the quality factor defined by the user and $ln \equiv \log_e$ the operator for the natural logarithm and $\xi = \{Y, C\}$. For f ranging in the percentage interval ([0, 100]), it holds that the intermediate variable $k \in [-4.159, 4.159]$, while its exponential representation $e^k \in [0.0156, 64.0000]$. As f increases, k and e^k decrease to allow for a consequent milder quantization. A graphical representation of e^k is shown in Fig. 2.54.

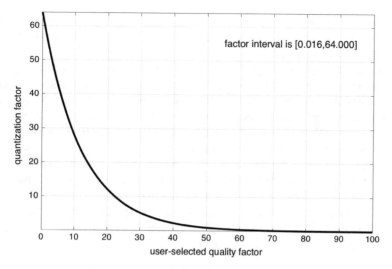

Fig. 2.54 Plot of the quantization factor e^k against the user-selected quality factor f

The quantization levels in the reference matrices (2.145) emerged from observing that after the DCT, which is applied to the original image data when compressing with JPEG, there is a particular spectral behavior, as expressed by the occurrence of large or small values to transform coefficients, which corresponds to specific frequencies, and therefore, other coefficients should be quantized with more and others with less tolerance.

The latest standard in image compression, JPEG2000, the quantizer is a typical *dead-zone uniform scalar quantizer* (Rabbani and Joshi 2002; Rabbani and Cruz 2001), as it is called due to that the central interval $(-1, 1]$ is twice as long as the other quantization intervals, and centered around zero (0). In this case, and because the transformation that is applied before quantization is the DWT, which leads to an analysis to multiple spectral bands, within the quantizer there is a special configuration on the spectral band in which it is acting (taking advantage of the features present in each zone). So for a given zone b a different quantization step Δ_b is defined. The choice of the quantization step in each zone is essentially defined by the user and may be based on modeling of the HVS, similarly to how the quantization matrices are defined in JPEG. The quantization rule in JPEG2000 is

$$q = sign(c) \left\lfloor \frac{|x|}{\Delta_b} \right\rfloor \tag{2.147}$$

where x is the sample at the input of the quantizer, $sign(x)$ the sign of the input sample and Δ_b the quantization step in the respective spectral band.

In conclusion, quantization is responsible for the distortion introduced in a compression process. So when *lossless compression is required quantization should not be applied*. In the process of quantization each sample or group of samples is mapped to a quantization level or a pointer to the dictionary of the quantizer and the input signal is trimmed substantially.

A simple example of image-domain-based quantization is shown in Figs. 2.55 and 2.56, which present typical examples of un-dithered uniform graylevel and color quantization into various number of levels each. In addition, Fig. 2.57 shows an example of the application of non-uniform, minimum variance color quantization. Each graylevel/color-quantized image is titled with the number of levels, the error estimation and the estimated quality in PSNR dBs.

A simple example of transform-domain-based quantization is shown in Figs. 2.56, 2.58, and 2.60, which present the effect of the quantization of the DCT transform coefficients using three different quantization matrices, based on the IJG JPEG quantization recommendation (2.145) and (2.146). The original image has been initially converter to the YC_bC_r color space and then transformed and quantized using an 8×8

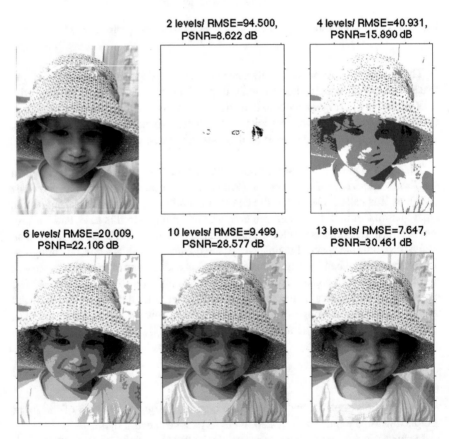

Fig. 2.55 Uniform quantization of a full-range graylevel image into various number of levels

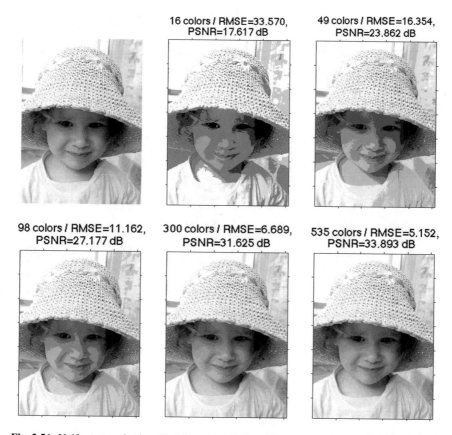

Fig. 2.56 Uniform quantization of a full-range color image into various number of colors

pixels block partitioning, then de-quantized and inverse transformed, to provide a reconstruction and an estimate of the distortion imposed by the quantization.

Figure 2.58 presents the result of quantization using quality factor 1, which produces a reconstructed image with an average per-channel quality of about 22 dB, with a maximum pixel difference of 190 and an average RMSE 20, as detailed in Table 2.4. (2.145) and (2.146) produce the quantization matrices in (2.148).

$$
q_Y = q_C = \begin{pmatrix}
255 & 255 & 255 & 255 & 255 & 255 & 255 & 255 \\
255 & 255 & 255 & 255 & 255 & 255 & 255 & 255 \\
255 & 255 & 255 & 255 & 255 & 255 & 255 & 255 \\
255 & 255 & 255 & 255 & 255 & 255 & 255 & 255 \\
255 & 255 & 255 & 255 & 255 & 255 & 255 & 255 \\
255 & 255 & 255 & 255 & 255 & 255 & 255 & 255 \\
255 & 255 & 255 & 255 & 255 & 255 & 255 & 255 \\
255 & 255 & 255 & 255 & 255 & 255 & 255 & 255
\end{pmatrix}
\tag{2.148}
$$

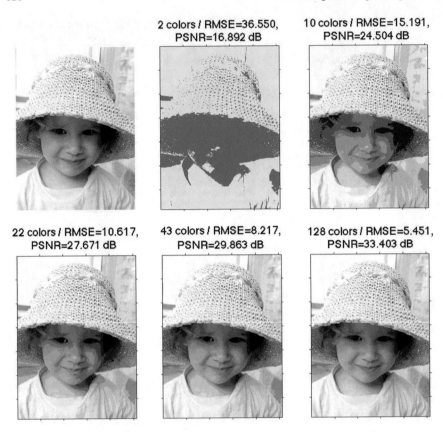

2 colors / RMSE=36.550, PSNR=16.892 dB

10 colors / RMSE=15.191, PSNR=24.504 dB

22 colors / RMSE=10.617, PSNR=27.671 dB

43 colors / RMSE=8.217, PSNR=29.863 dB

128 colors / RMSE=5.451, PSNR=33.403 dB

Fig. 2.57 Minimum variance quantization of a full-range color image into various number of colors

Fig. 2.58 Quantization in the DCT domain using quality factor 1

Table 2.4 Comparison between the original and the transform-domain-quantized image for quality factor 1

Channel	Max difference	RMSE	PSNR (dB)
Red	194	19.817	22.190
Green	185	16.983	23.531
Blue	198	23.359	20.762

Figure 2.59 shows the result of quantization using quality factor 25, which produces a reconstructed image with an average per-channel quality of about 26 dB, with a maximum pixel difference of 133 and an average RMSE 12, as detailed in Table 2.5. In this case the quantization matrices are differentiated and defined as shown in (2.149).

$$q_Y = \begin{pmatrix} 128 & 88 & 80 & 128 & 192 & 255 & 255 & 255 \\ 96 & 96 & 112 & 152 & 208 & 255 & 255 & 255 \\ 112 & 104 & 128 & 192 & 255 & 255 & 255 & 255 \\ 112 & 136 & 176 & 232 & 255 & 255 & 255 & 255 \\ 144 & 176 & 255 & 255 & 255 & 255 & 255 & 255 \\ 192 & 255 & 255 & 255 & 255 & 255 & 255 & 255 \\ 255 & 255 & 255 & 255 & 255 & 255 & 255 & 255 \\ 255 & 255 & 255 & 255 & 255 & 255 & 255 & 255 \end{pmatrix}$$

$$q_C = \begin{pmatrix} 136 & 144 & 192 & 255 & 255 & 255 & 255 & 255 \\ 144 & 168 & 208 & 255 & 255 & 255 & 255 & 255 \\ 192 & 208 & 255 & 255 & 255 & 255 & 255 & 255 \\ 255 & 255 & 255 & 255 & 255 & 255 & 255 & 255 \\ 255 & 255 & 255 & 255 & 255 & 255 & 255 & 255 \\ 255 & 255 & 255 & 255 & 255 & 255 & 255 & 255 \\ 255 & 255 & 255 & 255 & 255 & 255 & 255 & 255 \\ 255 & 255 & 255 & 255 & 255 & 255 & 255 & 255 \end{pmatrix}$$

(2.149)

Finally, Fig. 2.60 shows the result of quantization using quality factor 50, which produces a reconstructed image with an average per-channel quality of about 35 dB (with a very small RMSE of about 4.5 on the average), as detailed in Table 2.6. According to the IJG definition (2.145) and (2.146), in this case the quantization matrices are exactly equal to the reference matrices in (2.145).

It is really obvious that in all cases of the transform-domain-quantization a best reconstruction is guaranteed for the Green channel, then the Red and last the Blue, which is actually in line with the human vision modeling (the HVS model) and the predominant role of the Green channel.

Fig. 2.59 Quantization in the DCT domain using quality factor 25

Table 2.5 Comparison between the original and the transform-domain-quantized image for quality factor 25

Channel	Max difference	RMSE	PSNR (dB)
Red	118	12.358	26.292
Green	110	10.227	27.936
Blue	133	13.543	25.497

2.3.2.3   Encoding

The purpose of encoding (many times referred to simply as 'coding') is to exploit statistical redundancy among quantized samples (samples that have undergone quantization) to minimize the length of the final codestream (or bitstream). The stages preceding coding, namely transformation and quantization, limit the correlations on a strictly local level. In the ideal case all the input samples to the encoder are statistically independent. In this case these samples can be encoded independently and the only statistical redundancy that can be considered is the one related to the variance in the distribution of their probabilities. In the general case, however, it is almost impossible to ensure statistical independence of quantized samples. Nevertheless, so long as these inter-dependencies are limited to a narrow (spatial or temporal) interval, it is usually possible to design efficient coding systems with easily managed complexity. Known encoders produce codes that can either be of fixed or of variable length; among the most famous are the *Huffman* and *Arithmetic encoders*. Various widely used coding methods are listed in the following paragraphs, some of the accompanied by example implementations and examples; Huffman and Arithmetic coding are presented in more detail.

Fig. 2.60 Quantization in the DCT domain using quality factor 50

Table 2.6 Comparison between the original and the transform-domain-quantized image for quality factor 50

Channel	Max difference	RMSE	PSNR (dB)
Red	48	4.764	34.572
Green	43	3.932	36.238
Blue	67	5.230	33.760

Pulse Code Modulation

Pulse Code Modulation (PCM) was introduced during the development of a digital audio broadcasting standard (Sayood 1996). Today, the term is used for every encoding method related to encoding of an analog signal. The method is not connected to any specific method of compression: it simply implies the quantization and digitization of an analog signal. The signal range is divided into intervals, and each interval is assigned a unique index. For the encoding of an input sample, the interval to which it belongs is being tracked and the appropriate interval index is stored. The sampling of the signal occurs at regular intervals. Thus, the sequence of the signal in time or space can be stored as a sequence of digits (bits) of a pre-determent rate. An example of the application of PCM is shown in Figs. 2.61, 2.62, 2.63, and 2.64. The original signal is a 8 Kbps (8192 bps) one-dimensional signal.

Apparently in PCM the first process is sampling with the sampling rate being based on the Nyquist rate ($f_s \geq 2f_{max}$). The output of the sampling process is a discrete (in time) signal. This discrete signal is fed to a quantizer which, ultimately, makes the signal digital. At the final stage, an encoder converts the digital signal to a sequence of binary digits or bits. The quantization step size can be simply defined as $\Delta = \frac{\Re(x)}{q}$, with $\Re(x)$ the range of the input signal x and q the quantization levels; this

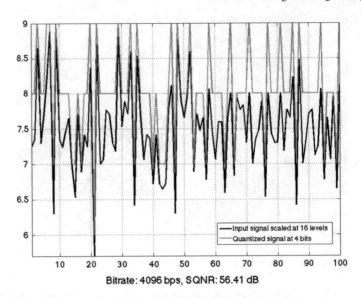

Fig. 2.61   PCM on an 8 Kbps signal using 4 bits at 4 Kbps rate and SQNR 56.41 dB

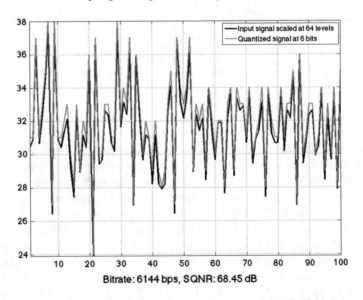

Fig. 2.62   PCM on an 8 Kbps signal using 6 bits at 6 Kbps rate and SQNR 68.45 dB

simplifies to $\Delta = \frac{2}{q}$ if $-1 \le x \le 1$. The error introduced by PCM is the quantization error (or quantization noise) $\epsilon = q(x) - x$ and it can be shown that the quality of the encoding can be estimated by the *signal to quantization noise ratio* as,

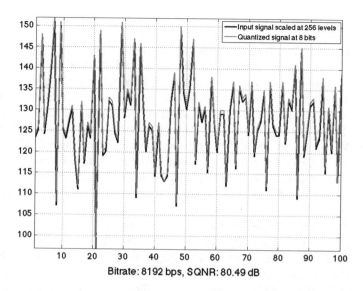

Bitrate: 8192 bps, SQNR: 80.49 dB

Fig. 2.63 PCM on an 8 Kbps signal using 8 bits at 8 Kbps rate and SQNR 80.49 dB

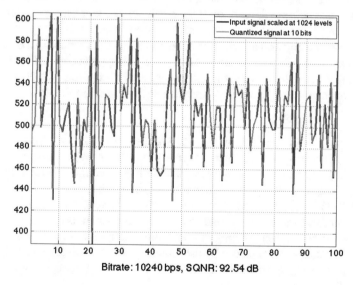

Bitrate: 10240 bps, SQNR: 92.54 dB

Fig. 2.64 PCM on an 8 Kbps signal using 10 bits at 10 Kbps rate and SQNR 92.54 dB

$$SQNR = 10 \log \left(\frac{\sum q_x^2}{\frac{\Delta^2}{12}} \right) \text{ dB} \qquad (2.150)$$

where $\sum q_x^2$ is the quantized signal power and $\frac{\Delta^2}{12}$ is the quantization noise power.

Differential Pulse Code Modulation

Differential Pulse Code Modulation (DPCM) is an encoding scheme based on PCM which adds a prediction functionality. Prediction in DPCM is considered in two approaches; in the first approach, prediction of the current value is made by knowing the previous value; in the second approach the difference relative to the output of a local model of the decoder process is considered (a decoder is incorporated in the encoder). In the simple (first) case the indices do not correspond to sample values (as in PCM), but to differences between successive samples (Sayood 1996). If, for example, a horizontal row of pixels of an image is encoded, an index may represent a difference in luminance between the current pixel to the previous one. It is known that there are many types of signals in which mostly small changes between successive samples appear.

When this method is applied to such signals, the indices that represent small differences are very frequent. By using entropy coding these indices can be greatly reduced in value and thus lead to better compression ratios, since, as will later be explained, a new distribution of samples emerges that approximates the *Laplacian distribution* or the so-called *two-sided geometric distribution*. This method is a simple example of predictive encoding, since, essentially, a predicting of the next value is being made according to the current. If the prediction is correct, the result is an index corresponding to a very small value, while if the prediction is wrong, the size of the index may be bigger than what the simple PCM would produce. An example of the application of DPCM on a full color (24-bpp) image is shown in Fig. 2.65. The DPCM image levels have been normalized for better illustration. The histograms of the luminance of the original image and the DPCM image are shown in Fig. 2.66.

Run-Length Encoding

Certain types of data may exhibit repetitive sequences of samples. This is strongly evident in binary images of digital documents, in which very large white regions (page background) may be found. A simple idea to efficiently compress such signals is to replace the sequence of repeated samples by a quantity that represents their number (Sayood 1996; Golomb 1966). According to the method of *Run-Length Encoding (RLE)*, an original sample represented by a particular symbol is considered the start of a sequence, which lasts as long as consecutive samples have the same value. In the end, this sequence is replaced by the sample value and the sequence length.

A typical example would be as follows: given a sequence of two symbols 'W' and 'B'

```
WWWWWWWWWBBBBBWWWWWWWWWWW
BBBWWWWWWWWWWBBBBBBBBBBB
WWWWWWWBBBBBBBBBBB
WWWWWWWWWWWWWWWWWWWWWW
BBBWWWWWWWWWWBB
```

Fig. 2.65 DPCM representation of a color image; the range of values is $[-255, 255]$

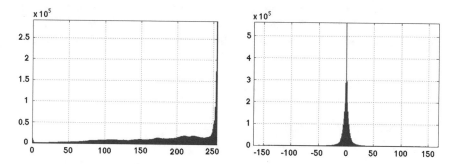

Fig. 2.66 Histograms of luminance channel in the original and the DPCM images

then according to RLE this sequence would be encoded as

```
W9B5W12B3W9B10W7B10W20B3W10B2
```

which represents the same message of the original 100 symbols using only 29 symbols (which is a 71 % compression). Clearly RLE results in very compact representations, especially in cases of large sequences of repeating symbols. It is effectively being applied in the standard JPEG encoding scheme for the efficient formation of the final bitstream of the compressed data, as will be shown in the corresponding section. In a way, RLE could be viewed as a simplified special case of a dictionary-based encoding scheme (described in the following paragraphs), in which only single symbols represent the dictionary and no patterns of consecutive different symbols (like words) are being exploited along with lookup tables for reference.

Table 2.7 Average results of RLE test on 100 natural and 100 text/graphics images

Image type	Gain (%)	Bitrate (bpp)
Natural images	17.3	6.61
Text/graphics images	97.7	0.18

A test on 100 graylevel natural images and on 100 graylevel and binary textual, line-art and synthetic graphics printed or handwritten images reveals the apparent advantage of using RLE in non-natural images as summarized in Table 2.7. In the reported results, 'gain' refers to the conservation of storage that can be achieved (compression), which attains the enormous amount of 97.7 %, (a data rate less than 0.2 bpp) for the case of non-natural images. The method is only of limited interest when applied to natural images, as only a 16 % gain has been achieved (6.61 bpp average data rate). Original images were treated as 8 bpp (luminance) images.

Figure 2.67 shows a typical example of two binary images where only the black pixels are considered to convey information (as in traditional fax applications). In these images, a simple RLE coding would result a significant amount of data compression up to a 87 %. The particular example used an alternative implementation

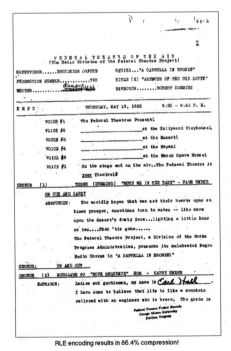

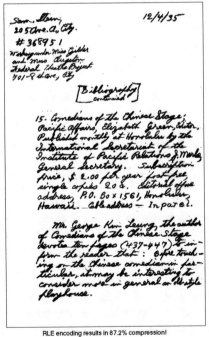

RLE encoding results in 86.4% compression!   RLE encoding results in 87.2% compression!

Fig. 2.67 Two typical binary images encoded using RLE

Fig. 2.68 RLE encoding rule and example

of RLE which is graphically depicted in Fig. 2.68, in which the encoding starts by the current row number (zero being the first row and column) complemented by the column positions than mark the beginning and ending of consecutive black pixels (the data).

Figure 2.69 presents a closeup in just 100 rows of the first image, along with a vertically stretched replica of the image rows for a better illustration of the corresponding row-wise and column-wise pixel counts (the histogram-like graphs at the bottom and right of the stretched image). Taking the first row in the image the RLE output of this single row is

```
500  486  490  495  497  502  506  547  549  557  559  563  567  609
612  623  629  664  666  673  676  680  685  722  725  738  745  790
792  797  799  847  851  930  934  941  944  949  955  991  995  997
1010 1052 1064 1119 1121 1170 1173 1181 1183 1231 1247
1249 1251 1292 1306 1376 1381 1397 1404 1443 1455 1459
1463 1537 1541 1548 1550 1589 1594 1604 1608 1649 1652
1658 1662 1745 1757 1808 1812 1869 1875
```

with the first entry denoting the image row index and each pair of entries denoting the location (column) of the beginning and ending of a series of black pixels. This single

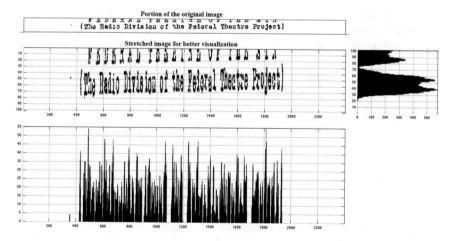

Fig. 2.69 Part of a binary image and the horizontal and vertical pixel counts

row is represented by 83 integers in the range [486, 1875], so apparently 16-bit integer representations are required. This means that this row requires $83 \times 2 \times 8 = 1,328$ bits to be perfectly reconstructed. The original image row is of 2,396 binary pixels (2,396 bits). Thus, RLE of this row results to a 44.6% compression, or a 1.8:1 compression ratio, which is relatively low, although expected, since this particular image row contains a significant amount of information (black pixels). In the overall image portion, 7,497 (16-bit) RLE values are needed to represent the 239,600 binary pixels, which amounts to 49.9% compression or an approximate 2:1 compression ratio. Apparently, other RLE rules and approaches are expected yield significantly different compression results.

Shannon-Fano Coding

For a given resource, the optimum compression rate that can be achieved is the one determined by the entropy of the source. Entropy, as already mentioned, is the measure of uncertainty of a source of information. Based on this principle, the basic idea behind the *Shannon-Fano Coding (SFC)* (Sayood 1996) is as follows: through the use of variable length codes, symbols are encoded according to their probabilities, using less bits to the symbols with the highest probability of occurrence, in compliance with the original Shannon theory. The method is considered to be suboptimal in the sense that it does not achieve the lowest possible code length but it guarantees that all code lengths are within one bit of their theoretical ideal set by the Shannon entropy. Although based on Shannon's theory, the method is co-attributed to Robert Fano, who published it as a technical report at MIT (Fano 1949).

SFC is a recursive process in which all symbols are assigned a codeword. Initially the symbols are arranged according to their probability in descending order, and then separated into two sets with total probabilities close to being equal. Then the first digit of the codeword of all symbols is being created, using "0" for the first set and "1" for the second. This process is repeated until no more sets remain with more than one members and all codewords have been assigned to all symbols. The process results in symbol with prefix codes (no other symbol has the same prefix).

The method is efficient in producing variable-length codes when the partitioning is as close as possible to being of equal probability. Although computationally simple, the method is almost never being used because it requires the a partitioning to equal probability sets is possible, in order to be efficient. Nevertheless, SFC is being used in the *IMPLODE compression method* in the ZIP compression.

The implementation of SFC is a simple tree-defining algorithm as follows:

1. Definition of the underlying probability mass function, by counting the occurrence of symbols
2. Ordering of the sequence of symbols in the order of descending probability
3. Division of the sequence into two parts having approximately equal total probability
4. Assignment of the binary digit "0" to the first set and of the digit "1" to the second set, so that all set symbols start with the same digit

5. Repetition of steps 3 and 4 for each of the two sets, to subdivide the sets and assign the appropriate additional digits to the corresponding codes until all symbols have become leaves on the tree that is being constructed

In a simple example of 10 symbols $S_I = \{A, B, C, D, E, F, G, H, I, J\}$ with corresponding occurrences $f_I = \{3, 8, 10, 5, 1, 2, 1, 7, 4, 9\}$, $I = 1, ..., N$, $N = 10$, the probabilities are shown in Table 2.8. The symbols are ordered in descending order of their probability and the sequence becomes $\{C, J, B, H, D, I, A, F, E, G\}$. Then the recursive splitting of the sequence into two sets takes place until all sets have only one symbol. Figure 2.70 shows graphically the overall process that results in the assignment of a single prefix code for each of the input symbols as shown in the table embedded in the figure. In this figure the red color denotes the "0" code paths,

Table 2.8 SFC example using 10 symbols with various probabilities

A	B	C	D	E	F	G	H	I	J
3	8	10	5	1	2	1	7	4	9
0.06	0.16	0.2	0.1	0.02	0.04	0.02	0.14	0.08	0.18

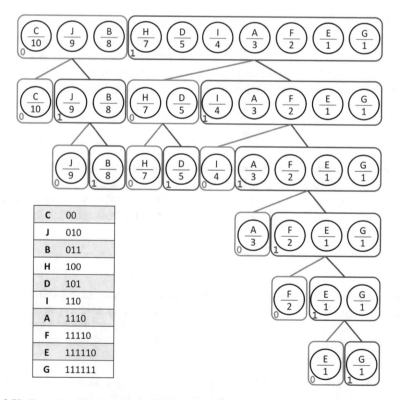

Fig. 2.70 Example of the formation of a Shannon-Fano tree of codes

whereas the blue color denotes the "1" code paths. Using (2.29) the entropy of this source is $H_i = 3.009$ bps. The data rate is estimated

$$\frac{\sum_{i \in S_I} bits(i) \times f_i}{\sum_{i \in S_I} f_i} =$$

$$= \frac{1}{50} [2\ 3\ 3\ 3\ 3\ 3\ 4\ 5\ 6\ 6][10\ 9\ 8\ 7\ 5\ 4\ 3\ 2\ 1\ 1]^T = 3.06 \text{ bps} \qquad (2.151)$$

Dictionary Methods

Dictionary Methods exploit the fact that many types of data exhibit repeated identical symbol sequences (patterns). Thus, it becomes possible to construct a 'dictionary' of 'model' sequences or structures from the original data and to replace the data with corresponding indices. Dictionary methods apply efficiently in text compression and generally ASCII files with a specific structure. They can also operate efficiently on graphics images or digitized text, but not in natural images (such as photographs). The most widespread form is expressed with the known Lempel-Ziv (LZ) algorithm (Ziv and Lempel 1978), and can be found in standard image storage formats such as TIFF for lossless compression (which yields a compression ration of around 2:1).

Huffman Coding

Huffman codes (Huffman 1952) are a subset of a large family of uniquely decodable variable-length codes. A Huffman code is generated by calculating the probabilities of the source symbols and the corresponding efficiency-indices so that the resultant code is minimized in length. It has been proven that Huffman codes have the minimum average length as compared to all other codes. Given a source with finite alphabet of symbols

$$S_I = s_0, s_1, ..., s_{k-1} \qquad (2.152)$$

following a corresponding probability distribution function f_I, it is reasonable to search for an optimal coding method, where the average length of the codeword be minimal through all uniquely decodable codes. This optimal encoder is not expected to be unique. David Huffman developed an algorithm for finding a set of lengths that satisfy the relation

$$\sum_{i \in S_I} 2^{-l+i} \leq 1 \qquad (2.153)$$

which minimizes the mean codeword rate. This equation expresses the characteristics which the lengths (l_i) must meet so that the codeword be uniquely decodable, as a necessary condition (Kraft–McMillan theorem (McMillan 1956)). Assuming the alphabet is sorted so that $f_I(s_0) \leq f_I(s_1) \leq ... \leq f_I(s_{k-1})$, Huffman coding is based on the observation that

> Among all the optimal codes, at least one has $l_{a_0} = l_{a_1} = ... = l_{max}$, the maximum codeword length, such that the codes c_{s_0} and c_{s_1} differ only in their last bit.

This observation indicates that the optimization problem can be reduced to finding only $K - 1$ codewords. This problem can now be expressed as follows: Determine the lengths that satisfy

$$\sum_{k=1}^{K-1} 2^{-l_{s_k}} \leq 1 \tag{2.154}$$

which minimize the quantity

$$R = (f_I(s_0) + f_I(s_1)) \cdot (l_{s_1'} + 1) + \sum_{k=2}^{K-1} f_I(s_k) l_{s_k} =$$
$$= f_I(s_0) + f_I(s_1) + \sum_{i' \in S_I} f_I'(i') l_i \tag{2.155}$$

where I' a new random variable with symbols alphabet

$$S_{I'} = s_1', s_2', ..., s_{K-1}' \tag{2.156}$$

and

$$f_{I'}(i) = \begin{cases} f_I(a_0) + f_I(a_1) & i = a_1' \\ f_I(i) & \text{otherwise} \end{cases} \tag{2.157}$$

Thus, the problem is identical to the original, but with a reduced alphabet. This reasoning leads naturally to the following algorithm, which iteratively reduces the problem of generating optimal codewords to the simplest problem, where there is only a binary alphabet (Cover and Thomas 2006; Taubman and Marcellin 2002b).

In order to create Huffman codewords (Taubman and Marcellin 2002b) classification of elements of the alphabet should take place, so that

$$f_I(s_0) \leq f_I(s_1) \leq \cdots \leq f_I(s_{K-1}) \tag{2.158}$$

If $K = 2$, $c_{s_0} =$ "0" and $c_{s_1} =$ "1" are instantly defined, otherwise,

- a new alphabet is being created $S_{I'} = s_1', s_2', ..., s_{K-1}'$, and a probability distribution function such that (2.157) holds.
- the creation algorithm is being applied recursively to find the optimal codewords $c_{s_1'}, c_{s_2'}, ..., c_{s_{K-1}'}$ for $S_{I'}', f_I'$
- the codeword $c_{s_1'}$ is being expanded by complementing either with "0" or "1" for the production of c_{s_0}, c_{s_1} respectively.

Re-using the simple 10-symbols example defined and presented for the case of Shannon-Fano Coding, and reviewing that the symbols are $S_i = \{A, B, C, D, E, F, G, H, I, J\}$ with corresponding occurrence frequencies $f_i = \{3, 8, 10, 5, 1, 2, 1, 7, 4, 9\}$, $i = 1, ..., N$, $N = 10$, as shown in Table 2.8, and that the symbols are ordered in a descending frequency of occurrence ordering $\{C, J, B, H, D, I, A, F, E, G\}$, then the

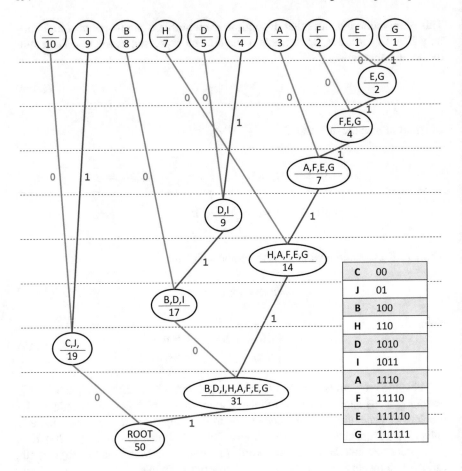

Fig. 2.71 Example of creating Huffman codes

method proposed by Huffman for generating the codes of those symbols is depicted in Fig. 2.71. In this figure the red color denotes the "0" code paths, whereas the blue color denotes the "1" code paths. The final resulting codes are shown as a table embedded in the same figure. Typically, using (2.29) the entropy of this source is $H_i = 3.009$ bps. The data rate of the Huffman-encoded symbols is estimated as

$$\frac{\sum_{i \in S_l} bits(i) \times f_i}{\sum_{i \in S_l} f_i} =$$

$$= \frac{1}{50} [2\ 2\ 3\ 3\ 4\ 4\ 4\ 5\ 6\ 6][10\ 9\ 8\ 7\ 5\ 4\ 3\ 2\ 1\ 1]^T = 3.06 \text{ bps} \quad (2.159)$$

Apparently, in this example the resulting compact representation is equivalent to the one provided by Shannon-Fano Coding, although the codes are different, and is very close to the entropy limit.

Golomb Coding

In 1966, Solomon Golomb presented a mental experiment in which a special roulette game is being played in a way that it produces binary outcomes (reporting whether a "0" is encountered or not—a "0" would have a probability of $1/37$ in this case). He realized that in cases of sequences of infinite outcomes, or more generally of largely un-balanced event probabilities, a Huffman coding would not be applicable (Golomb 1966). His insight was that a run length of n favorable events[10] of probability p between successive unfavorable events with probability $q = 1 - p$ would have a probability of $p^n q$, $n = 0, 1, 2, ...$, which is approximated by no other than the *geometric distribution*.[11] Golomb realized that a shift towards the study of distributions for the discrete case should initiate, at a time when Gaussian distributions dominated the representations of every process being studied. He tried to shape the form of Huffman coding when applied to the geometric distribution, which is a more accurate representation of processes in the discrete case, and especially in the case in which $p \gg q$. Apparently, Golomb was concerned about encoding run-lengths, so this method is primarily applicable to positive integer values, and may exhibit slightly different implementations whether the encoded inputs are natural numbers (including zero) or strictly integer numbers.

Golomb introduced a parameter m, which controls the coding process by dividing any input x into two parts, namely the *quotient q* and the *remainder r* of the division of the input by the parameter m as

$$q = \left\lfloor \frac{x}{m} \right\rfloor$$
$$r = x - qm$$

(2.160)

The quotient q undergoes a typical *unary encoding*, whereas the remainder r undergoes a *truncated binary encoding* and the two binary representations are concatenated to form the final Golomb code.

In *unary encoding* a natural number n is assigned a binary codeword of n 1's followed by a 0, thus $U(3) = 1110$, or $U(5) = 111110$. In the case of a strictly positive integer n the assignment corresponds to $(n-1)$ 1's, thus $U_{>0}(3) = 110$. The original Golomb coding accepts the first approach and considers natural numbers (including zero).

In *truncated binary encoding*, things get complicated, as a number is represented by its binary code that is truncated to the number of bits actually needed for a correct representation and one-to-one reconstruction, following a *prefix code* creation process. For example by encoding $n = 6$ coming from an alphabet of $k = 5$ bits,

---

[10]It should be noted that the definition of what is 'favorable' purely depends on the context.

[11]For simplicity, *geometric* is a distribution sharply concentrated around the mean value with only small values around it being probable (small variance).

truncated encoding outputs $T(6; 5) = 110$, whereas a typical *binary encoding* would output $B(6; 5) = 00110$. Practically, truncated binary encoding requires the size of the alphabet from which the input number is selected. Then it computes the bits needed to encode the length of the alphabet and an offset of numbers to advance while converting to the binary representation. Given an input number x and an alphabet of length m,

$$k = \lfloor \log_2 m \rfloor$$
$$o = 2^{k+1} - m$$
$$\text{CODE} = \begin{cases} \text{bin}(x, k), & x < u \\ \text{bin}(x + o, k + 1), & x \geq u \end{cases}$$

(2.161)

where $\text{bin}(\xi, b)$ is the binary representation or binary encoding of ξ using b bits. Apparently, the first o-bits of the code are assigned k bits whereas the last $(m - o)$-bits are assigned $(k + 1)$ bits, by skipping o numbers in the middle of the alphabet range. Any number less than the offset will be simply binary represented using k bits, whereas any number larger than the offset will be increased by this offset and then binary encoded using $(k + 1)$ bits. It should be noted that the offset purely depends upon the alphabet size.

Two MATLAB function included in this section implement the Golomb coding process; the first implements the truncated binary encoding algorithm and the second the complete Golomb coding procedure. These functions can either encode a single natural number or a vector of natural numbers. Obviously, the truncated encoding function is generic, and it can be used in other settings also. The results of the Golomb coding function have been tested against all examples provided by Golomb himself in the original 1966 paper (Golomb 1966).

```
function code = truncated_binary_code( x, varargin )
  if nargin==1,
    if length( x ) > 1
      m = length(x);
    else
      error('You need to specify the alphabet size!');
    end
  else
    m = varargin{1};
  end
  if find(x > m-1)
    error('Cannot compute code for X>M-1!');
  end
  % initialize
  code = cell(length(x),1);
  % compute the bits needed to encode the alphabet
  k = floor(log2(m));
  % compute the offset of numbers to skip while advancing
  o = 2^(k+1)-m;
  % since u is known then the assignmens is as follows:
  % the first o-bits are assigned typical binary codes of number X
  % whereas the last (m-o)-bits are assigned (k+1)-length binary codes of number
      (X+1)
  for i=1:length(x)
    if x(i) < o,
```

```
      code{i} = dec2bin(x(i),k);
   else
      code{i} = dec2bin(x(i)+o,k+1);
   end
end
```

```
function code = golomb_code( num, param )
   % get the two parts of the number to encode
   q = dix(num/param);
   r = rem(num,param);
   % produce unary of q
   ql = length(q);
   code1 = cell(ql,1);
   for j=1:ql
      code1{j} = '';
      for i=1:q(j)
         code1{j}(i) = '1';
      end
      code1{j}(end+1) = '0';
   end
   % produce the truncated binary code of r
   code2 = truncated_binary_code( r, param);
   % concatenate code1 and code2 to produce final code
   code = cell(ql,1);
   for j=1:ql
      code{j} = [ code1{j}, code2{j}];
   end
```

The Golomb coding method was independently rediscovered by Tanaka and Garcia in 1982 (Tanaka and Leon-Garcia 1982) and One et al. in 1989 as MELCODE (Ono et al. 1989).

Elias and Arithmetic Coding

A special case of codes, the *universal codes*, are those which map source messages to codewords with an average length that is bounded by a linear combination of the entropy, $L \leq \alpha(H + \beta)$; that is universal codes achieve an average length that is less than a multiple of the optimal code. Apparently, in universal codes, the amount of compression depends on the values of α and β, so these codes are usually defined as *asymptotically optimal codes*, with the obvious asymptotically optimal code corresponding to $\alpha = 1$, $\beta = 0$. In these codes it is sufficient to know the probability distribution to the extent that the source messages can be ranked in probability order, so that decreasing probability symbols get increasingly larger codeword lengths, thus achieving *universality*.

A universal code may be considered as providing an enumeration of a sequence of symbols, giving, actually, a representation of the integers that enumerate the sequence. Peter Elias, in his seminal paper in 1975 (Elias 1975), gave rise to a whole family of universal coding methods that map positive integers onto binary codewords.

The simplest of the Elias codes is the *Elias gamma code*, which simply maps a natural number $x \in N$ by a binary code that is constructed by concatenating $\lfloor \log_2 x \rfloor$ "0" with the binary representation of the number; $\lfloor \log_2 x + 1 \rfloor$ is the total number of bits required to represent the natural number. For example, the Elias gamma code for number $x = 15$ is produced by concatenating $\lfloor \log_2 15 = 3 \rfloor$ zeros with its binary representation 1111, thus $C_\gamma = 0001111$. Decoding starts by counting the number of zeros n at the beginning of each decoded sequence of bits. If there is no zero ($n = 0$) at the beginning then the number to be decoded is obviously "1"; in any other case ($n > 0$), the decoder reads the following $n + 1$ bits and decodes the corresponding binary number. A MATLAB implementation of the Elias gamma coding method is presented in the following listing.[12]

```
function y = elias_gamma_code( x )
  bts = floor( log2( x ) );
  dts = dec2bin( x );
  nbts = bts+1;
  nums = length( x );
  for i = 1:nums
    sbts = '';
    if bts(1,i)>0
      sbts = num2str( zeros(1,bts(i)));
      sbts = strrep( sbts, ' ', '');
    end
    sz = length(dts(i,:));
    bg = sz-nbts(i)+1; if bg<1, bg=1; end;
    snum = [ sbts dts(i,bg:sz)];
    y(i,1:length(snum)) = snum;
  end
```

A more complex and sophisticated Elias code is the *Elias delta code*, which builds upon the gamma code. Encoding of a natural number using delta coding requires to concatenate $C_\gamma \left(\lfloor \log_2 x \rfloor + 1 \right)$ with a modified binary representation of the number. The modification of the binary representation consists in discarding the first bit of the representation which is expected to be always a "1" for any number. For example, the Elias delta code for $x = 15$, the binary representation is $1111 \rightarrow 111$, whereas $C_\gamma \left(\lfloor \log_2 x \rfloor + 1 \right) = C_\gamma(4+1) = 00100$, thus $C_\delta(15) = 00100111$. To decode delta codes, the procedure begins with the decoding of the embedded gamma code, which is at the beginning of the delta code (using the procedure described for the gamma codes); this provides with knowledge regarding the number of bits b in the binary representation that follows. Then the decoder simply decodes the following $b - 1$ bits after complementing with a preceding "1". A MATLAB implementation of the Elias delta coding method is presented in the following listing.[13]

```
function y = elias_delta_code( x )
  bts = elias_gamma_code( floor( log2( x ) )+1 );
  dts = dec2bin( x );
  nums = length( x );
  for i = 1:nums
    sdts = deblank( dts(i, strfind(dts(i,:),'1'):end));
    sdts = sdts(2:end);
```

---

[12]This MATLAB function can be executed either for a scalar (single) or for a vector input and will respond with the appropriate results.

[13]This MATLAB function can be executed either for a scalar (single) or for a vector input and will respond with the appropriate results.

```
    b = deblank( bts(i,:) );
    d = sdts;
    snum = [ b d];
    y(i,1:length(snum)) = snum;
  end
```

Table 2.9 presents some representative examples of gamma and delta codes for selected natural numbers. Clearly the number of bits required by each of these codes (both gamma and delta) is

$$
\begin{aligned}
L_\gamma(x) &= 2\lfloor \log_2 x \rfloor + 1 \\
L_\delta(x) &= L_\gamma(\lfloor \log_2 x \rfloor + 1) + \lfloor \log_2 x \rfloor \\
&= 2\lfloor \log_2(\lfloor \log_2 x \rfloor + 1) \rfloor + 1 + \lfloor \log_2 x \rfloor
\end{aligned}
\tag{2.162}
$$

In the case of a random variable X with uniform distribution on $\{1, 2, ..., N\}$ and entropy $H(X) = \log_2 N$ bps, the expected lengths for the gamma and delta codes are respectively

$$
\begin{aligned}
E\left[L_\gamma(X)\right] &= \frac{1}{N} \sum_{x=1}^{N} \left(2\lfloor \log_2 x \rfloor + 1\right) \\
E\left[L_\delta(X)\right] &= \frac{1}{N} \sum_{x=1}^{N} \left(2\lfloor \log_2(\lfloor \log_2 x \rfloor + 1) \rfloor + 1 + \lfloor \log_2 x \rfloor\right)
\end{aligned}
\tag{2.163}
$$

An even more advanced method that is the third in the family of Elias coding methods, is the one that generates the *Elias omega codes*, or the *recursive Elias codes*, due to their recursive generation. These codes can also be applied to natural numbers and are more efficient in compact representations of small numbers. There are two alternative methods to generate the omega codes.

According to the first method, in order to encode a natural number x the code starts with a "0" and iteratively builds up as now bits are concatenated at the beginning of the code, as follows

1. Start by placing a "0" at the end of the code
2. Stop encoding if $x = 1$
3. Insert the binary representation of x at the beginning of the code
4. Change x, to be equal the number of bits just inserted minus one ($x = \lfloor \log_2 x \rfloor$)
5. Go back to step 2

The second method is an iterative application of Elias gamma coding and update of the input, which builds the code by changing bits at the beginning of it, as follows

1. Apply Elias gamma coding on x and generate code
2. Set $x = \lfloor \log_2 x \rfloor$ and go back to step 1 to replace the preceding "0"s until only one "0" is left
3. Move the left "0" to the end of the code

As an example, the Elias omega code for $x = 15$ is $C_\omega = 1111110$. Omega codes that can be generated by both these methods are listed in Table 2.9. A MATLAB implementation of the Elias omega coding method is presented in the following listing.[14]

```
function y = elias_omega_code( x, varargin )
   method = 1;
   if nargin>1,
      method = varargin{1};
   end
   if method == 1,
      % METHOD 1:
      % To code a number N:
      % Place a "0" at the end of the code.
      % If N=1, stop; encoding is complete.
      % Prepend the binary representation of N to the beginning of the code.
      %        This will be at least two bits, the first bit of which is a 1.
      % Let N equal the number of bits just prepended, minus one.
      % Return to step 2 to prepend the encoding of the new N.
      % As described in: https://en.wikipedia.org/wiki/Elias_omega_coding
      y = [ '0' ];
      while x ~= 1,
         dts = dec2bin( x );
         y = [ dts y ];
         x = length( dts )-1;
      end
   else
      % METHOD 2:
      % Call Elias gamma coding
      % Recursively apply Elias gamma coding on the num of zeros at front
      % Move the first zero to the end
      y = elias_gamma_code( x );
      N = strfind( y, '1' ); N = N(1)-1;
      while N>1
         yy = elias_gamma_code( N );
         y = [ yy y(N+1:end)];
         N = strfind( y, '1' ); N = N(1)-1;
      end
      y = [ y(2:end) y(1) ];
   end
```

Elias actually suggested the method now known as *arithmetic coding*; arithmetic coding was presented by Abramson in his *Information Theory and Coding* in 1963 (Abramson 1963), but remained almost a scientific curiosity until practical implementations were presented by researchers among which Rissanen, Pasco, Rubin and Witten (Rissanen 1976; Pasco 1976; Rubin 1979; Rissanen and Langdon 1979; Witten et al. 1987) can be identified as the most prominent. Arithmetic coding plays an important role in various image compression standards, such as the famous Joint Bilevel Image Group (JBIG) and JPEG (and respectively the newer Joint Bilevel Image Group 2 (JBIG2) and JPEG2000). There are two main factors in arithmetic coding: the probabilities of symbols and coding intervals. The probabilities of symbols define and yield compression efficiency and they determine the size of the intervals for the encoding process. Through an iterative process of updating prob-

---

[14]This MATLAB function can be executed only for scalar inputs.

Table 2.9 Elias gamma, delta and omega codes of natural numbers

x	$C_\gamma(x)$	$C_\delta(x)$	$C_\omega(x)$
1	1	1	0
2	010	0100	100
3	011	0101	110
4	00100	01100	101000
5	00101	01101	101010
6	00110	01110	101100
7	00111	01111	101110
8	0001000	00100000	1110000
9	0001001	00100001	1110010
10	0001010	00100010	1110100
20	000010100	001010100	10100101000
30	000011110	001011110	10100111100
40	00000101000	0011001000	101011010000
50	00000110010	0011010010	101011100100
100	0000001100100	00111100100	1011011001000

abilities of symbols and intervals, all symbols go through the encoder and the end result converges to a real number (theoretically of infinite accuracy between 0 and 1). An arithmetic encoder can be regarded as an encoding device that receives as input the symbols to be encoded and the corresponding estimate for the probability distribution, and produces a string of a length equal to the combination of the ideal code-lengths of the input symbols (Rissanen and Langdon 1979).

Simply described, in arithmetic coding *a message* is represented by the interval [0, 1) on the real axis, and is supposed to be governed by an underlying probability distribution. Each symbol of the message has a probability of occurrence (described by a source model) and is being processed sequentially. During this process each symbol narrows the overall interval according to its probability. Apparently, as the interval becomes smaller and smaller with the intercepted symbols, the amount of binary digits required to define it grows respectively. Since the narrowing of the interval depends upon the probability of the processed symbol, high probability symbols narrow the interval less than low probability, which implies that high probability symbols contribute fewer bits to the final code. Obviously, in arithmetic coding there exists no one-to-one correspondence between source symbols and output codes, as an entire sequence of symbols is encoded to a single arithmetic code. After the encoding of all message symbols the initial interval [0, 1) is diminished to an interval represented by the product of all probabilities of encoded symbols

$$P = p(s_1) \times p(s_2) \times p(s_3) \times ... \times p(s_N)$$

The interval precision expressed as the number of bits required to represent the interval is

$$-\log_2 P = -\sum_i \log_2(p(s_i))$$

The algorithm originally attributed to Elias (on a DMS) can be understood as a mapping of each symbol of a source to a distinct interval on the real axis $[c_n, c_n+a_n) \subseteq [0, 1)$, so that the length of this interval is equal to $f_I(s_n)$ the probability of the symbol s_n in the message I. The algorithm is recursive and is summarized in the following steps (Taubman and Marcellin 2002b)

- Initialize $c_0 = 0$ and $a_0 = 1$ so that the initial interval is $[0, 1)$
- For each symbol intercepted $n = 0, 1, ...$

  - Update $a_{n+1} \leftarrow a_n f_I(s_n)$
  - Update $c_{n+1} \leftarrow c_n + a_n F_I(s_n)$

where F_I is the cumulative distribution

$$F_I(s_j) = \sum_{k=0}^{j-1} f_I(s_k), \quad S_I = s_0, s_1, ...$$

In order for this to work, both the encoder and the decoder need to have access to the same distribution function f_I (and hence to the F_I), otherwise they should both be able to make the same assumptions about the underlying distribution. In practice the iterative process of the generation of arithmetic codes has been simplified as a recursive update of the interval start point and its length; thus for a message expressed by the sequence of symbols $S = \{s_1, s_2, ..., s_N\}$ governed by a source model that includes the symbol probabilities

$$p(l) = p(s_k = l), \quad l = 0, 1, ..., M - 1, \quad k = 1, 2, ..., N$$

M being the number of symbols the source can produce, and a cumulative distribution practically defined as

$$c(l) = \sum_{j=0}^{l-1} p(j), \quad j = 0, 1, ..., M, \quad c(0) \equiv 0, c(M) \equiv 1$$

the process defines the interval

$$\Phi_k(S) = [\alpha_k, \beta_k), \quad k = 0, 1, ..., N, \quad 0 \le \alpha_k \le \alpha_{k+1}, \quad \beta_{k+1} \le \beta_k \le 1$$

This interval is represented for simplicity as $|b, l\rangle$ with b its starting point (also referenced as the 'base') and l its length,

$$|b, l\rangle = [\alpha, \beta), \quad b = \alpha, l = \beta - \alpha$$

Then this interval is being iteratively narrowed according to the symbols intercepted using

$$\Phi_0(S) = |b_0, l_0\rangle = |0, 1\rangle$$
$$\Phi_k(S) = |b_k, k_k\rangle = |b_{k-1} + c(s_k)l_{k-1}, \, p(s_k)l_{k-1}\rangle, \quad k = 1, 2, ..., N \qquad (2.164)$$

A MATLAB implementation of the arithmetic coding method is presented in the following listings, first for the encoder and then the decoder.

```
%
% implementation of the Elias/Arithmetic encoding
%
% implementation is based on
%    Amir Said, "Introduction to Arithmetic Coding Theory and Practice",
%    Hewlett-Packard Laboratories Report, HPL-2004-76, Palo Alto, CA, April 2004
%    https://software.intel.com/sites/default/files/m/b/6/3/HPL-2004-76.pdf
%
function varargout = elias_arithmetic_encode( symbol_probabilities, symbol_indices )
   %% check input and initialize
   if (sum( symbol_probabilities )- 1) > eps,
      error( 'Probabilities of symbols do not add up to 1!');
   end
   c = [0 cumsum( symbol_probabilities )];
   l = 1;
   b = 0;
   %% run the main algorithm
   for i=1:length( symbol_indices )
      j = symbol_indices(i);
      b = b + c(j)*l;
      l = l*symbol_probabilities(j);
   end
   %% produce the output
   code = b+l/2;                          % compute the code
   bin_code = num2bin( code );            % convert the code to binary form
   %% handle the output
   switch nargout
      case 0
         fprintf( ['| The code is in [%.12f,%.12f] --> [%.12f]\n',...
            '| Represented by the binary code [%s]\n'],...
            b,b+1,code,bin_code);
      case 1
         varargout{1} = code;
      otherwise
         varargout{1} = code;
         varargout{2} = bin_code;
   end
```

```
%
% implementation of the Elias/Arithmetic decoding
%
% implementation is based on
%     Amir Said, "Introduction to Arithmetic Coding Theory and Practice",
%     Hewlett-Packard Laboratories Report, HPL-2004-76, Palo Alto, CA, April 2004
%     https://software.intel.com/sites/default/files/m/b/6/3/HPL-2004-76.pdf
%
function varargout = elias_arithmetic_decode( symbol_probabilities, elias_code,
    num_symbols )
    %% check input and initialize
    if (sum( symbol_probabilities )- 1) > eps,
        error( 'Probabilities of symbols do not add up to 1!');
    end
    c = [0 cumsum( symbol_probabilities )];
    s = [];
    %% run the main algorithm
    for i=1:num_symbols
        % find the upper bound of the interval that includes the current code
        s(end+1) = find(c<elias_code, 1, 'last' );
        % normalize the code to [0,1)
        elias_code = (elias_code-c(s(end))) / symbol_probabilities(s(end));
    end
    %% handle the output
    switch nargout
        case 0
            disp( s );
        otherwise
            varargout{1} = s;
    end
```

If for example, there is a source that produces four (4) symbols $S = \{1, 2, 3, 4\}$ with a corresponding probabilities $p(s_k) = \{0.2, 0.4, 0.3, 0.1\}$ and this source produces the message $M = \{2, 2, 1, 3, 3, 2, 1, 4, 3, 2\}$ then the encoding of this message would proceed as follows:

- The cumulative distribution is computed by the probabilities resulting $c = \{0, 0.2, 0.6, 0.9, 1\}$. The lower bound of the initial interval is set to zero (0) and the length of this interval to unity (1).
- The first symbol in the message is a "2" with a probability of 0.4, which points to the interval $|0.2, 0.4\rangle$.
- The second symbol is also a "2", which narrows the interval to $|0.28, 0.16\rangle$.
- The third symbol is a "1", which narrows the interval to $|0.28, 0.032\rangle$.
- The fourth symbol is a "3", which narrows the interval to $|0.2992, 0.0096\rangle$.
- The fifth symbol is also a "3", which narrows the interval to $|0.30496, 0.0028\rangle$.
- Then comes a "2", which zooms in the interval $|0.305536, 0.001152\rangle$.
- A "1", narrows the interval to $|0.305536, 2.304 \times 10^{-4}\rangle$.
- A "4", narrows the interval to $|0.30574336, 2.304 \times 10^{-5}\rangle$.
- A "3", narrows the interval to $|0.305757184, 6.912 \times 10^{-6}\rangle$.

- And the final symbol "2", narrows the interval to
  $[0.3057585664, 2.7648 \times 10^{-6})$. This corresponds to the interval
  $[0.305758566400, 0.305761331200)$.
- The easiest way to get a single number in the final interval is to get the middle point, which in this case is 0.3057599488, or, in binary form, is 0.0100111001000110 0100100010110100.

Typically this process is followed by another step to minimize the length of the binary code by estimating the number of bits to keep from the length of the final interval.

Decoding of this output to reconstruct the original message (the sequence of symbols $M = \{2, 2, 1, 3, 3, 2, 1, 4, 3, 2\}$) is straightforward. The output of the encoder (0.3057599488) passes though an iterative process that tracks the interval the code belongs to (in the cumulative distribution) and normalizes the interval to become $[0, 1)$. Apparently *the decoder has no clue on when to stop the process*, so additional data, like the number of encoded symbols, have to be stored and passed to the decoder for a proper decoding.

An in-depth analysis of the theory and practical application of arithmetic coding can be found in (Said 2004), where important implementation concepts are being presented.

Practical implementations of arithmetic encoders produce strings that exceed the ideal length (at approximately 6%). A typical practical arithmetic encoder is shown in Fig. 2.72. In general, the probability of a sample having a specific value is influenced by the values of adjacent samples. Thus, the probabilities of the symbols can be estimated, provided that the values of neighboring symbols are known. For a given neighborhood of symbols, every possible combination of symbols reflects a *contextual pattern*. The arithmetic coder is very efficient in coding sequences where the probabilities are changing for every contextual pattern. Figure 2.73 shows a causal neighborhood of pixels in an image.

Statistical Modeling

The entropy coder presented in previous paragraphs, along with others not referenced here, use a statistical model of the source that produces the symbols being encoded. Apparently the assumption of such an underlying model is crucial for any

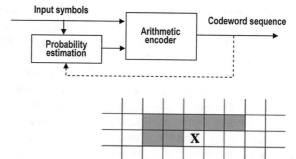

Fig. 2.72 Typical arithmetic encoder

Fig. 2.73 Causal region for the conditional probability estimation

coding process, as it is responsible for determining both the alphabet of the source symbols and the entropy bound for any possible coding rate achievable. Presupposing a *stationary model* where the probabilities of the symbols do not change then, according to the theory of Shannon, the entropy of the source is the ultimate limit for any possible data compression. Nevertheless, this model is far from being accurate and in practice many techniques are being employed to alter the description of the symbols in order to achieve a lower limit and be able to design far more efficient coding schemes. In the classic textbook on JPEG, Pennebaker and Mitchell (Pennebaker and Mitchell 1993) briefly and intuitively presented four possible ways to improve the entropy and achieve better compression rates by,

- *Changing the alphabet*
  According to this approach, the source alphabet may be altered using possibly very simple observations on the occurrences of the symbols in intercepted messages produced by the source. For example, in natural images it is typical that run-lengths of identical pixel values are very probably so instead of using an alphabet consisting of the values of those pixels, one might consider encode the state of intercepting the same value. This way, the 'same value state' would be assigned significantly less bits (being the most frequent) and the new alphabet would be expected to be of reduced entropy.

- *Removing redundancy*
  In addition to the identification of patterns in the occurrence of the symbols of the alphabet presented in the previous approach, it is highly likely that some of the state transitions (described by the newly adopted alphabet) are improbable (there is no way of them to happen), apparently like each state change to itself. Removing these cases from the alphabet, a new, reduced alphabet is created and the entropy is further reduced as redundant symbols are being discarded.

- *Grouping symbols*
  Pushing further the development of the alphabet created so far with the two previous approaches, another improvement is expected to stem from the grouping of consecutive occurrences of alphabet symbols. A complete alphabet for all possible combinations could be created this way, lowering even more the number of bits per symbol needed for highly likely combined symbol occurrences. Typically, the probability for the occurrence of two symbols in a row (joint occurrence) is represented by the product of the probabilities of the individual symbols, thus less probable symbol transitions are highlighted whereas high probable transitions are suppressed in terms of the bits needed to represent them, which has a significant positive impact in the entropy of the new complex alphabet.

- *Using conditional probabilities*
  If the assumed *statistical independence of the symbols of the alphabet* in the previous approaches is discarded, then the source is no longer memoryless, and becomes an *x-order Markov process*. In this modeling of the source, the occurrence of a symbol is expected to be influenced (up to an order) by previously intercepted symbols. By employing conditional probabilities of the symbols these influences are being captured and efficiently represented as newly defined (more complex)

alphabet symbols that can impose a further reduction in the entropy. Actually, it has long been recognized that this approach yield very efficient alphabets and significantly reduced redundancy that guarantees a more compact representation in a compression attempt. Widely used methods, like JPEG, heavily rely on this approach to define what is usually referenced *the context* of a decision during entropy coding.

The approaches are not simple theoretical curiosities, but rather important practical strategies to remove large portions of the redundancy inherent in any source of information that produces sequences of symbols to be encoded. By targeting to remove this redundancy, it is expected to achieve a reduction in the entropy of the source, and create a highly efficient alphabet for use in an encoding process. In addition, the approaches aim at providing a more accurate statistical model of the source (either by supposing a statistical independence or by using Markov modeling), which is of paramount importance in many data encoding schemes that rely on probabilities, such as the entropy coding schemes, which have gained high acceptance in numerous applications over many decades of data compression developments.

2.3.3   Lossy Coding and Rate-Distortion Theory

Transformation, quantization and encoding of arbitrary real numbers is at the core of any data coding method. Unfortunately, real numbers require an infinite amount of bits to be accurately represented. Thus, it is expected that a finite representation of a continuous random variable cannot be perfect. In such a context, one has to consider to trade-off between accuracy and size of any possible representation. Upon this basic notion, a *rate-distortion theory* has been created that relates the minimum expected distortion achievable at a particular data rate, given the source statistics and an arbitrary distortion measure. Assuming a source that produces a sequence of *independent identically distributed* random variables with probability $p(x)$, $x \in \mathcal{X}$, and assuming that the alphabet is finite, the encoder describes the source sequence X^n by and index $f_n(X^n) \in \{1, 2, ..., 2^{nR}\}$ and the decoder represents X^n by and estimate $\hat{X}^n \in \mathcal{X}$. In this respect, a *distortion function* or *distortion measure* is defined as a mapping from the set of source and reproduction alphabet pairs to the set of non-negative real numbers $d : \mathcal{X} \times \hat{\mathcal{X}} \to \mathcal{R}^+$. Accordingly, the distortion $d(x, \hat{x})$ is a measure of the cost of representing x by $\hat{x}$. Typical distortion measures usually being used are the *Hamming distortion*

$$d(x, \hat{x}) = \begin{cases} 0 & x = \hat{x} \\ 1 & x \neq \hat{x} \end{cases} \tag{2.165}$$

or the *squared-error distortion* $d(x, \hat{x}) = (x - \hat{x})^2$. In addition, the distortion between sequences x^n and $\hat{x}^n$ is defined as

$$d(x^n, \hat{x}^n) = \frac{1}{n} \sum_{i=1}^{n} d(x_i, \hat{x}_i) \tag{2.166}$$

thus the average per symbol distortion is equivalent to the distortion for a sequence.

The rate-distortion function $R(D)$ is the infimum of rates R such that (R, D) is in the rate-distortion region of the source for a given distortion D. Looking at it the other way around, the distortion-rate function $D(R)$ is the infimum of all distortions D such that (R, D) is in the rate-distortion region of the source for a given rate R, as already presented in (2.140). Accordingly, the *information rate-distortion function* $R^{(I)}(D)$ is defined as,

$$R^{(I)}(D) = \min_{p(\hat{x}|x):\sum_{(x,\hat{x})} p(x)p(\hat{x}|x)d(x,\hat{x}) \leq D} I(X; \hat{X}) \tag{2.167}$$

with the minimization over all conditionals $p(\hat{x}|x)$ for which $p(x, \hat{x}) = p(x)p(\hat{x}|x)$ satisfies the expected distortion constraint. Additionally, it can be proved that the definition of the rate-distortion function is equal to the latter information rate-distortion function definition, thus $R(D) = R^{(I)}(D)$ for an independent identically distributed source X with a distribution $p(x)$ and a bounded distortion function $d(x, \hat{x})$ (Cover and Thomas 2006).

In (Cover and Thomas 2006) is shown that for a binary (Bernoulli) source and considering a Hamming distortion, the rate-distortion function is defined as

$$R(D) = \begin{cases} H(p) - H(D) & 0 \leq D \leq \min\{p, 1 - p\} \\ 0 & D > \min\{p, 1 - p\} \end{cases} \tag{2.168}$$

whereas for a Gaussian source $\mathcal{N}(0, \sigma^2)$ with a squared-error distortion, the rate-distortion function is defined as

$$R(D) = \begin{cases} \frac{1}{2} \log \frac{\sigma^2}{D} & 0 \leq D \leq \sigma^2 \\ 0 & D > \sigma^2 \end{cases} \tag{2.169}$$

Apparently, there are cases in which 'good' approximations of an input are considered 'enough' as regards an expected outcome. In those cases, one trades off information 'quality' for a more compact (compressed) representation, as it is in those cases only that one is able to *surpass the entropy barriers*. Shannon already described these cases in his basic information theory, where he realized and put forward that in such conditions one should considered a balance between the compression rate achieved and the distortion that is imposed on the data.

As already mentioned on the general compression methods in the previous section, when applying lossy compression methods the degeneration of the original image is

Fig. 2.74 The basic
framework of lossy
compression

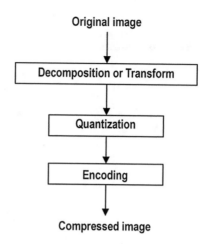

permitted in order to further reduce the length of the generated compression string. Distortions can either be visible or not. Generally, a higher compression is achieved as the quality of the final image is continuously reduced (or respectively as the requirements for image quality are being reduced the compression efficiency is increased). The basic framework of a lossy compression method is shown in Fig. 2.74.

It includes three basic processes: transformation, quantization and coding to an output code-string. The value of each process in the overall compression depends on the technique used. Of course, the more *intelligent*[15] a process, the better the quality that can be achieved for a given number of bits. The compression performance limit of each information source, as defined by its entropy, applies only to the case of lossless compression. In the case of lossy compression this is not true: the problematic in this case refers to *how much the compression ratio should at least be to keep the distortion within certain acceptable limits*. The level of distortion is usually a parameter defined by the user by controlling certain parameters, such as the level of quantization. This concern arises and is being studied through a class of theories known as *rate-distortion theory*. This theory puts theoretical performance limits for lossy compression according to some fidelity criteria. For a very large class of distortion measures and source models, the theory provides a rate-distortion function $R(D)$ (or respectively $D(R)$), as developed in previous sections on quantization, which has the following properties:

- For any given distortion level D, it is possible to find a coding method with efficiency close to R and average deformation close to D.
- It is impossible to find an encoder that achieves a compression ratio less than R, with distortion D (or less).

---

[15]The term is used metaphorically to denote the ability of a method to adapt to the input data and the exploitation of statistical, spectral or any other inherent redundancies.

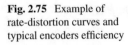

Fig. 2.75 Example of
rate-distortion curves and
typical encoders efficiency

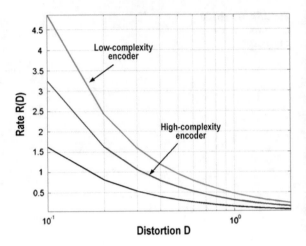

It turns out that the curve $R(D)$ is convex, continuous and decreasing function
of D. Figure 2.75 illustrates a typical rate-distortion function for a discrete source
with finite alphabet. The minimum degree of compression required for compression
without distortion is the value of R for $D = 0$, and is less than or equal to the entropy
of the source and depends on the distortion estimator D. In the image ideal curves
of encoders with high and low complexity are also shown. Generally, the better the
compression method the better the source statistics are being modeled and the better
the efficiency, which approaches the $R(D)$ threshold (Rabbani and Jones 1991a).

2.4 JPEG Compression

JPEG (Rabbani and Jones 1991a; ISO-IEC-CCITT 1993b; Pennebaker and Mitchell
1993; Wallace 1991) is a standardized image compression mechanism. Its name
comes from the *Joint Photographic Experts Group*, the name of the group that introduced this standardization. JPEG was designed for compressing of either color or
graylevel images of scenes that reflect the real world. It is a compression method
which guarantees good compression results, primarily when running on real-world
photos (or natural images). JPEG cannot provide good compression results on text
images, simple graphics and line art. Also, is is mainly used for compressing still
images (although there is a Motion-JPEG compression scheme for video sequences).
JPEG is a lossy compression method. It was designed to exploit known properties of
the HVS, with particular emphasis on its capacity to perceive changes in brightness
a lot better than changes in chromaticity. A useful feature of JPEG is that the degree
of distortion can be easily adjusted by changing the compression parameters. This
means that one may choose (at least approximately) between image quality or the size
of the final compressed file. Another important feature is that the JPEG decoders can

Table 2.10 Advantages and disadvantages of JPEG

Advantages	Disadvantages
Low memory requirements	A single image resolution (size)
Low complexity	A single image quality
Compression efficiency	Inability to predefine the compression rate
HVS modeling	No lossless compression
Robustness	No partial compression
	No region of interest
	Blocking artifacts in block boundaries
	Low noise resilience

use less accurate approaches in the calculations in order to achieve higher decoding speed (which, of course, affects the quality of the final image). The JPEG standard is applicable mainly in cases of compression of true color still images, achieving an average compression around 15 : 1. The advantages and disadvantages of JPEG compression standard are summarized in Table 2.10. The JPEG standard defines four modes of the basic algorithm:

- *Sequential coding*, in which each image block and each picture element is encoded following a scan path from top-left.
- *Progressive coding*, in which each image block and each picture element is encoded through multiple scans, to enable a progressive decoding of the image when there is a limitation in transmission times.
- *Lossless coding*,[16] in which the encoded image is coded so as to ensure accurate reproduction of the original image.
- *Hierarchical coding*, in which the image is encoded at multiple resolutions.

2.4.1 The Sequential JPEG

The sequential JPEG or baseline JPEG is the most basic of the modes that JPEG supports and it is the most usual mode of coding in JPEG. The block diagram of a typical sequential JPEG encoder is shown in Fig. 2.76. A decoder for such an encoder is follows the exact opposite path and the symmetric operations. Coupled, an encoder and a decoder are usually referenced as a *codec*. In the following paragraphs a description is given of the operation of each basic block of the baseline codec.

---

[16]There is a misunderstanding on this case, as JPEG does not support lossless compression. This mode does not have anything in common with the classical JPEG algorithm and can only support compression rates of about 2:1, using prediction in a causal neighborhood of pixels. It is essentially an entirely different method.

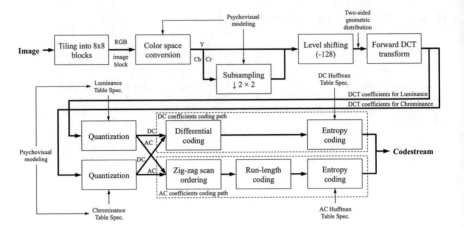

Fig. 2.76 Detailed block diagram of a sequential JPEG encoder

2.4.1.1 The Encoder

The encoding process in the sequential JPEG is performed in rectangular non-overlapping tiles of the image, usually called *blocks*, typically of 8×8 pixels in size. Initially, a color image is tiled and appropriately padded to the right and bottom, if needed. The following process is performed on each of the 8×8 image blocks separately. First, the color image undergoes a color space transformation, from the RGB color space, in which is usually represented, to the YC_bC_r color space, which, as previously analyzed, decorrelates the pixel values by providing a luminance-chrominance representation. According to the psychovisual modeling implied by the HVS modeling, the two chrominance channels are being subsampled by 2×2 (half samples in both directions), as it is expected that discarding data in these channels will not have a significant impact in the final reconstructed image quality, while it will greatly facilitate a better compression outcome. The luminance channel does not undergo any subsampling, due to its significance for the HVS. Then, the samples with values in the interval $[0, 2^b - 1]$ are shifted in the interval $[-2^{b-1}, 2^{b-1} - 1]$ by subtracting 128 from all samples. For each image channel in the case that the data rate is initially $b = 8$ bps which corresponds to all values being in the interval $[0, 255]$, these values are shifted to the interval $[-128, 127]$, thereby creating a two-sided geometric distribution with zero mean, apparently, increasing, in parallel, the number of bits needed for the new representation by 1 ($b' = 9$ bps). These shifted values of the luminance and subsampled chrominance pixels pass through DCT using (2.83) (which further increases the *bit-budget* of the data by outputing large and small floating point coefficients),

$$\left. \begin{aligned} J(u, v) &= \frac{2}{N} c(u) c(v) \sum_{x=0}^{N-1} \sum_{y=0}^{N-1} I(x, y) \cos \frac{(2x+1)u\pi}{2N} \cos \frac{(2y+1)v\pi}{2N} \\ c(\xi) &: \ c(\xi = 0) = \sqrt{\frac{1}{2}}, \quad c(\xi \neq 0) = 1 \\ N &= 8 \end{aligned} \right\} \Rightarrow$$

$$J(u, v) = \frac{1}{4} c(u) c(v) \sum_{x=0}^{7} \sum_{y=0}^{7} I(x, y) \cos \frac{(2x+1)u\pi}{16} \cos \frac{(2y+1)v\pi}{16}$$

$$(2.170)$$

The original discrete signal of 64 integer samples is a function of two dimensions x and y, whereas the real values of the DCT coefficients are a function of u and v. The first transform coefficient $J(0, 0)$ is called the *DC coefficient* and the remaining 63 coefficients are called the *AC coefficients*. In a classic case of a natural (continuous-tone) image tile of 8×8 pixels, many AC coefficients are expected to have low or zero magnitudes, thus, need not be encoded. This fact, complemented by the psychovisual modeling that guides to rely only on low-frequency coefficients, forms the foundation for achieving compression in JPEG.

In the next step, all coefficients are quantized using a quantization matrix, which is determined according to the compression ratio required and is different for the luminance and the chrominance channels. The quantization is nonlinear and has been described in the previously presented section on *Quantization*; it uses the reference quantization matrices in (2.145) and applies the rule described in (2.146) to convert the reference matrices to the matrices that correspond to the quality factor used in the JPEG compression.

The purpose of this quantization is to reduce the values of the transform coefficients and especially of those coefficients that are expected to have minimal contribution to the reconstruction of the image; an ultimate goal at this step is to increase the number of coefficients with zero value in a way that complies with the psychovisual studies. Quantization in JPEG is typically defined as,

$$J_q(u, v) = \left\lfloor \frac{J(u, v)}{Q(u, v)} \right\rceil \tag{2.171}$$

for each of the luminance-chrominance channels, where $\lfloor \rceil$ denotes the rounding operator, and $Q(u, v)$ are the elements of the quantization matrices, which are rounded and clipped within the interval $[1, 255]$.

After the quantization, the quantized DCT coefficients are being reordered according to a zig-zag ordering as shown in Fig. 2.77. This ordering is required for the next step, the entropy encoding, since the low-frequency coefficients that have high probability of being non-zero are being ordered before the coefficients of high frequencies, whose probability of being zero is high. A simple test on various natural (continuous-tone) images reveals that the zig-zag reordering is very close to being the

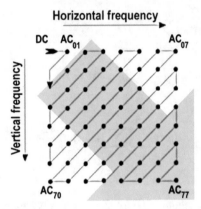

Fig. 2.77 The zig-zag ordering of the DCT coefficients

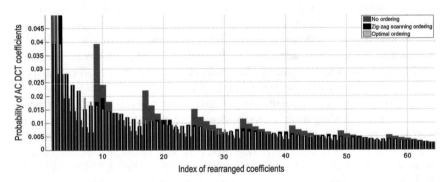

Fig. 2.78 Probability of non-zero AC coefficients of DCT and coefficients ordering

optimum possible reordering—actually the one being absolutely adaptive, adapted
to the dataset.

Figure 2.78 aggregates the results of such a test on the Wang database of 1,000 natural color images,[17] where three different orderings of the AC coefficients of the DCT
are being presented. The horizontal axis represents the index of the coefficients and
the vertical axis the probability of each coefficient over the total amount of 8×8 image
blocks in all 1,000 images; the dark grey bar-graph shows the probabilities without
any reordering; the black bar-graph shows the probabilities of the coefficients after
zig-zag ordering; the light grey bar-graph shows the 'optimum' reordering, where
all coefficient probabilities are in a monotonically decreasing order. Apparently the
zig-zag ordering is vary close to being 'optimum' and it does not need to be adaptive (it still applies for any other images). Purely for reference, Table 2.11 shows the
indices for the coefficients required by the zig-zag ordering and those imposed by the
optimum ordering (for this test set). The ordering is shown in a column-wise setting

---

[17]The Wang database is accessible at http://wang.ist.psu.edu/docs/related/.

Table 2.11 Comparison of the indices in zig-zag and optimum (adaptive) ordering

Zig-zag ordering								Optimum (adaptive) ordering							
1	9	2	3	10	17	25	18	1	3	2	4	5	6	10	9
11	4	5	12	19	26	33	41	8	7	11	12	13	14	15	19
34	27	20	13	6	7	14	21	21	20	18	17	24	25	23	16
28	35	42	49	57	50	43	36	22	26	33	34	32	27	35	28
29	22	15	8	16	23	30	37	31	40	39	36	38	41	30	47
44	51	58	59	52	45	38	31	37	29	42	46	48	45	49	43
24	32	39	46	53	60	61	54	52	51	53	44	50	54	57	56
47	40	48	55	62	63	56	64	58	55	59	60	61	63	62	64

just like the presented indices are in a column-wise ordering in the 8 × 8 grid, both beginning from the top-left corner and walking towards the right to the end of the columns before advancing one row.

The DC coefficients, which constitute the average of the 64 image samples, are being encoded using reversible encoding with prediction, DPCM, as shown in the Fig. 2.79. The reason for the use of such a coding scheme is that in adjacent image blocks the DC coefficients are expected to be highly correlated, and thus the value of their difference is expected to be very close to zero. Entropy coding of such small values are expected to produce small codewords, and, ultimately, the compression is significantly improved, as only the first DC coefficient has to be encoded to represent its full value and all others are encoded as very small differences.

Final stage in JPEG compression is the entropy coding, which achieves additional compression to create the final codestring via the coding of quantized DCT coefficients into a more 'compact' binary form. The JPEG standard specifies two entropy coding methods: *Huffman coding* and *arithmetic coding*. The basic sequential coding algorithm uses Huffman coding, mainly due to patents in the arithmetic coding proposed in JPEG. The Huffman encoder converts the quantized coefficients into a compact binary format by performing a two-step process, which creates a sequence of intermediate symbols before assigning the final Huffman codes to produce the codestream. The first step produces a set of *intermediate symbols*, which are a *category code* for the DPCM encoded DC coefficients, and a more complex *run-length and category code* for the AC coefficients.

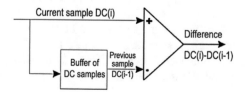

Fig. 2.79 Predictive coding (DPCM) for the DC coefficients

Table 2.12 Coding categories for DCT coefficients in JPEG standard for up to 12-bpp images

DC category	AC category	Range of values	
0	N/A	0	
1	1	(-1,1)	
2	2	(-3,-2)	(2,3)
3	3	(-7...-4)	(4...7)
4	4	(-15...-8)	(8...15)
5	5	(-31...-16)	(16...31)
6	6	(-63...-32)	(32...63)
7	7	(-127...-64)	(64...127)
8	8	(-255...-128)	(128...255)
9	9	(-511...-256)	(256...511)
A	A	(-1023...-512)	(512...1023)
B	B	(-2047...-1024)	(1024...2047)
C	C	(-4095...-2048)	(2048...4095)
D	D	(-8191...-4096)	(4096...8191)
E	E	(-16383...-8192)	(8192...16383)
F	N/A	(-32767...-16384)	(16384...32767)

Table 2.13 Reference table of Huffman codes for luminance DC coefficients

DC category	Code length	Huffman codeword
0	2	00
1	3	010
2	3	011
3	3	100
4	3	101
5	3	110
6	4	1110
7	5	11110
8	6	111110
9	7	1111110
A	8	11111110
B	9	111111110

Specifically, the DC coefficient differences are mapped to difference categories in accordance with the definition in the JPEG standard shown in Table 2.12. Then this category is checked against another reference table from the JPEG standard, which assigns Huffman codes to the category codes as shown in Table 2.13 for the luminance and in Table 2.14 for the chrominance channels respectively.[18] Thus, a DC difference of −13, would be mapped to the category 4 in the DC coefficients categories, according to Table 2.12, which outputs the Huffman codeword "101" for luminance (length of 3 bits). In addition, since a category 4 coefficient is being encoded 4 more bits are assigned, either as the 4 least significant bits of a positive

---

[18]These are Tables K.3 and K.4 in the JPEG standard.

Table 2.14 Reference table of Huffman codes for chrominance DC coefficients

DC category	Code length	Huffman codeword
0	2	00
1	2	01
2	2	10
3	3	110
4	4	1110
5	5	11110
6	6	111110
7	7	1111110
8	8	11111110
9	9	111111110
A	10	1111111110
B	11	11111111110

coefficient, or the 4 least significant bits of a negative coefficient minus one (1).[19] In this case the coefficient is negative, thus, the most appropriate complement of the codeword would be "(0011)-1" or "0010". So the final codeword for the DC difference -13 would be "1010010". During decoding, the codeword's first three bits "101" represent specifically and beyond any doubt (since the codes are prefix there is a one-to-one mapping with) the category in which the coefficient should be searched. Decoding "101" as category 4, guides to use the next four bits for the final decoding of the coefficient. Specifically, the next four bits "(0010) + 1" result in "0011", which is the twos-complement representation for the negative number -13.

Encoding of the AC coefficients is somehow more complex. As there is a large probability that null AC coefficients are to be encoded, their encoding considers run-lengths of zero coefficients in addition to the category coding applied to the DC coefficients. The encoding start by identifying the category of the non-zero AC coefficient using Table 2.12. Then the encoding assigns a codeword for the pair of the number of zero AC coefficients that precede the encoded non-zero coefficient and its category according to Table 2.15 (Tables K.5 and K.6 in the JPEG standard) for the luminance or Table 2.16 for the chrominance channels respectively. If a run-length of more than 16 zero AC coefficients are encountered, then the run-length reports a code for each 15 and continues normally for the next coefficients. Applying this coding method to the AC coefficient sequence,

$$0, 0, 0, 0, 0, 0, 0, 0, 123$$

---

[19]It should be stressed that the negative values are represented in the twos-complement format. Twos complement representation requires to flip all bits of the absolute value binary representation and then add 1.

Table 2.15 Reference table of Huffman codes for luminance AC coefficients

Zero runs	Category	Huffman codeword	Zero runs	Category	Huffman codeword
0	0	1010 (=EOB)			
0	1	00	8	1	111111000
0	2	01	8	2	111111111000000
0	3	100	8	3	1111111110110110
0	4	1011	8	4	1111111110110111
0	5	11010	8	5	1111111110111000
0	6	1111000	8	6	1111111110111001
0	7	11111000	8	7	1111111110111010
0	8	1111110110	8	8	1111111110111011
0	9	1111111110000010	8	9	1111111110111100
0	A	1111111110000011	8	A	1111111110111101
1	1	1100	9	1	111111001
1	2	11011	9	2	1111111110111110
1	3	11110001	9	3	1111111110111111
1	4	111110110	9	4	1111111111000000
1	5	11111110110	9	5	1111111111000001
1	6	1111111110000100	9	6	1111111111000010
1	7	1111111110000101	9	7	1111111111000011
1	8	1111111110000110	9	8	1111111111000100
1	9	1111111110000111	9	9	1111111111000101
1	A	1111111110001000	9	A	1111111111000110
2	1	11100	A	1	111111010
2	2	11111001	A	2	1111111111000111
2	3	1111110111	A	3	1111111111001000
2	4	111111110100	A	4	1111111111001001
2	5	1111111110001001	A	5	1111111111001010
2	6	1111111110001010	A	6	1111111111001011
2	7	1111111110001011	A	7	1111111111001100
2	8	1111111110001100	A	8	1111111111001101
2	9	1111111110001101	A	9	1111111111001110
2	A	1111111110001110	A	A	1111111111001111
3	1	111010	B	1	1111111001
3	2	111110111	B	2	1111111111010000
3	3	111111110101	B	3	1111111111010001
3	4	1111111110001111	B	4	1111111111010010
3	5	1111111110010000	B	5	1111111111010011
3	6	1111111110010001	B	6	1111111111010100

(continued)

Table 2.15 (continued)

Zero runs	Category	Huffman codeword	Zero runs	Category	Huffman codeword
3	7	1111111110010010	B	7	1111111111010101
3	8	1111111110010011	B	8	1111111111010110
3	9	1111111110010100	B	9	1111111111010111
3	A	1111111110010101	B	A	1111111111011000
4	1	111011	C	1	1111111010
4	2	1111111000	C	2	1111111111011001
4	3	1111111110010110	C	3	1111111111011010
4	4	1111111110010111	C	4	1111111111011011
4	5	1111111110011000	C	5	1111111111011100
4	6	1111111110011001	C	6	1111111111011101
4	7	1111111110011010	C	7	1111111111011110
4	8	1111111110011011	C	8	1111111111011111
4	9	1111111110011100	C	9	1111111111100000
4	A	1111111110011101	C	A	1111111111100001
5	1	1111010	D	1	11111111000
5	2	11111110111	D	2	1111111111100010
5	3	1111111110011110	D	3	1111111111100011
5	4	1111111110011111	D	4	1111111111100100
5	5	1111111110100000	D	5	1111111111100101
5	6	1111111110100001	D	6	1111111111100110
5	7	1111111110100010	D	7	1111111111100111
5	8	1111111110100011	D	8	1111111111101000
5	9	1111111110100100	D	9	1111111111101001
5	A	1111111110100101	D	A	1111111111101010
6	1	1111011	E	1	1111111111101011
6	2	111111110110	E	2	1111111111101100
6	3	1111111110100110	E	3	1111111111101101
6	4	1111111110100111	E	4	1111111111101110
6	5	1111111110101000	E	5	1111111111101111
6	6	1111111110101001	E	6	1111111111110000
6	7	1111111110101010	E	7	1111111111110001
6	8	1111111110101011	E	8	1111111111110010
6	9	1111111110101100	E	9	1111111111110011
6	A	1111111110101101	E	A	1111111111110100
7	1	11111010	F	1	1111111111110101
7	2	111111110111	F	2	1111111111110110
7	3	1111111110101110	F	3	1111111111110111

(continued)

Table 2.15 (continued)

Zero runs	Category	Huffman codeword	Zero runs	Category	Huffman codeword
7	4	1111111110101111	F	4	1111111111111000
7	5	1111111110110000	F	5	1111111111111001
7	6	1111111110110001	F	6	1111111111111010
7	7	1111111110110010	F	7	1111111111111011
7	8	1111111110110011	F	8	1111111111111100
7	9	1111111110110100	F	9	1111111111111101
7	A	1111111110110101	F	A	1111111111111110
			F	0	11111111001 (=ZRL)

the encoding process packs together the zeros before the 123 into the intermediate representation,

$$(8, 7) \ (123)$$

The zeros-runs/category code for (8, 7) in the Table of codes for luminance AC coefficients (Table 2.15) is

$$1111111110111010$$

whereas the binary code for 123 is 1111011, thus the codeword for the whole sequence would be

$$11111111101110101111011$$

which amounts to 23 bits for the encoding of 8 zeros and 1 integer.

If in the sequence of coefficients more than 16 consecutive zero coefficients are found, then the symbol (F, 0) is reported, which represents the maximum amount of consecutive zeros that can be grouped together into one codeword. Thus, the sequence

$$0, 0, 0, 0, 0, 0, 0, 0, 0, 0, 0, 0, 0, 0, 0, 0, 0, 0, 0, 0, 0, 0, 0, 0, 0, 0, 0, 0, 0, 0, 0, 0, 0, 23$$

would be considered as a grouping and coding of

$$(15, 0) \ (15, 0) \ (3, 5) \ (23)$$

and would be encoded as

$$11111111001 \ 11111111001 \ 1111111110010000 \ 10110$$

where the spaces are shown only to illustrate the four parts that are being encoded.

Table 2.16 Reference table of Huffman codes for chrominance AC coefficients

Zero runs	Category	Huffman codeword	Zero runs	Category	Huffman codeword
0	0	1010 (=EOB)			
0	1	01	8	1	11111001
0	2	100	8	2	1111111110110111
0	3	1010	8	3	1111111110111000
0	4	11000	8	4	1111111110111001
0	5	11001	8	5	1111111110111010
0	6	111000	8	6	1111111110111011
0	7	1111000	8	7	1111111110111100
0	8	111110100	8	8	1111111110111101
0	9	1111110110	8	9	1111111110111110
0	A	111111110100	8	A	1111111110111111
1	1	1011	9	1	111110111
1	2	111001	9	2	1111111111000000
1	3	11110110	9	3	1111111111000001
1	4	111110101	9	4	1111111111000010
1	5	11111110110	9	5	1111111111000011
1	6	111111110101	9	6	1111111111000100
1	7	1111111110001000	9	7	1111111111000101
1	8	1111111110001001	9	8	1111111111000110
1	9	1111111110001010	9	9	1111111111000111
1	A	1111111110001011	9	A	1111111111001000
2	1	11010	A	1	111111000
2	2	11110111	A	2	1111111111001001
2	3	1111110111	A	3	1111111111001010
2	4	111111110110	A	4	1111111111001011
2	5	111111111000010	A	5	1111111111001100
2	6	1111111110001100	A	6	1111111111001101
2	7	1111111110001101	A	7	1111111111001110
2	8	1111111110001110	A	8	1111111111001111
2	9	1111111110001111	A	9	1111111111010000
2	A	1111111110010000	A	A	1111111111010001
3	1	11011	B	1	111111001
3	2	11111000	B	2	1111111111010010
3	3	1111111000	B	3	1111111111010011
3	4	111111110111	B	4	1111111111010100
3	5	1111111110010001	B	5	1111111111010101
3	6	1111111110010010	B	6	1111111111010110
3	7	1111111110010011	B	7	1111111111010111

(continued)

Table 2.16 (continued)

Zero runs	Category	Huffman codeword	Zero runs	Category	Huffman codeword
3	8	1111111110010100	B	8	1111111111011000
3	9	1111111110010101	B	9	1111111111011001
3	A	1111111110010110	B	A	1111111111011010
4	1	111010	C	1	111111010
4	2	111110110	C	2	1111111111011011
4	3	1111111110010111	C	3	1111111111011100
4	4	1111111110011000	C	4	1111111111011101
4	5	1111111110011001	C	5	1111111111011110
4	6	1111111110011010	C	6	1111111111011111
4	7	1111111110011011	C	7	1111111111100000
4	8	1111111110011100	C	8	1111111111100001
4	9	1111111110011101	C	9	1111111111100010
4	A	1111111110011110	C	A	1111111111100011
5	1	111011	D	1	11111111001
5	2	1111111001	D	2	1111111111100100
5	3	1111111110011111	D	3	1111111111100101
5	4	1111111110100000	D	4	1111111111100110
5	5	1111111110100001	D	5	1111111111100111
5	6	1111111110100010	D	6	1111111111101000
5	7	1111111110100011	D	7	1111111111101001
5	8	1111111110100100	D	8	1111111111101010
5	9	1111111110100101	D	9	1111111111101011
5	A	1111111110100110	D	A	1111111111101100
6	1	1111001	E	1	11111111100000
6	2	11111110111	E	2	1111111111101101
6	3	1111111110100111	E	3	1111111111101110
6	4	1111111110101000	E	4	1111111111101111
6	5	1111111110101001	E	5	1111111111110000
6	6	1111111110101010	E	6	1111111111110001
6	7	1111111110101011	E	7	1111111111110010
6	8	1111111110101100	E	8	1111111111110011
6	9	1111111110101101	E	9	1111111111110100
6	A	1111111110101110	E	A	1111111111110101
7	1	1111010	F	1	1111111010
7	2	11111111000	F	2	111111111000011
7	3	1111111110101111	F	3	1111111111110110
7	4	1111111110110000	F	4	1111111111110111
7	5	1111111110110001	F	5	1111111111111000
7	6	1111111110110010	F	6	1111111111111001

(continued)

Table 2.16 (continued)

Zero runs	Category	Huffman codeword	Zero runs	Category	Huffman codeword
7	7	1111111110110011	F	7	1111111111111010
7	8	1111111110110100	F	8	1111111111111011
7	9	1111111110110101	F	9	1111111111111100
7	A	1111111110110110	F	A	1111111111111101
			F	0	1111111111111110 (=ZRL)

Finally, the symbol (0, 0) (i.e. no zeros and null coefficient) denotes the end of the sequence and is encoded as "1010" consuming 4 more bits in the final codestream (constant for each block in the image).

One might wonder why this coding stage is complicated, while it would be possible to simply apply Huffman coding of the quantized DCT coefficients. The answer is simple, if one considers the amount of data required to encode the coefficients; DC differentials are within the interval $[-2047, 2047]$ and AC coefficients in the interval $[-1023, 1023]$ for 8 bpp images, which corresponds to a requirement of code tables with 4,095 and 2,047 entries respectively. By adopting the encoding proposed in the JPEG standard, these tables are reduced to 12 and 160 entries respectively (with two more for the ZRL and EOB codes), which is a significant amount of complexity reduction.

Let us consider a full process example using a block selected from image 'lena' as shown in Fig. 2.80. This is a block from a color image so it is expected to have three channels, that is, three 8×8 matrices have to be encoded. First, the pixels of the image block undergo a color space conversion from RGB to YC_bC_r. Then the values are level shifted by subtracting 128. The level shifted values are transformed using DCT and quantized using the default quantization matrix in JPEG. The quantized transform coefficients are ordered according to the zig-zag scanning scheme and then each of them is entropy coded with the specific Huffman coding method in JPEG, using the

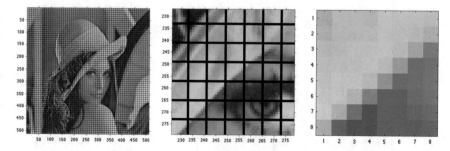

Fig. 2.80 Selected image block for JPEG compression experiments

default Huffman tables. The final produced codestreams $CODE_Y$, $CODE_{C_b}$, $CODE_{C_r}$ are shown with the individual codewords separated for illustration and verification.

$$R = \begin{bmatrix} 210 & 216 & 218 & 217 & 220 & 222 & 217 & 217 \\ 211 & 214 & 215 & 215 & 223 & 219 & 214 & 211 \\ 213 & 214 & 217 & 220 & 219 & 218 & 206 & 196 \\ 216 & 217 & 2188 & 219 & 217 & 206 & 183 & 192 \\ 221 & 223 & 222 & 218 & 201 & 184 & 187 & 198 \\ 222 & 223 & 213 & 193 & 176 & 182 & 187 & 196 \\ 221 & 209 & 179 & 154 & 168 & 184 & 181 & 192 \\ 211 & 165 & 153 & 154 & 175 & 175 & 173 & 180 \end{bmatrix}$$

$$Y = \begin{bmatrix} 164 & 177 & 178 & 176 & 186 & 184 & 183 & 181 \\ 170 & 177 & 178 & 181 & 185 & 183 & 179 & 168 \\ 177 & 179 & 184 & 185 & 187 & 182 & 163 & 123 \\ 180 & 182 & 185 & 187 & 181 & 152 & 112 & 111 \\ 185 & 187 & 187 & 178 & 144 & 109 & 111 & 113 \\ 187 & 189 & 173 & 128 & 102 & 100 & 109 & 113 \\ 191 & 169 & 116 & 89 & 102 & 98 & 105 & 109 \\ 166 & 103 & 88 & 89 & 98 & 100 & 98 & 102 \end{bmatrix}$$

$$G = \begin{bmatrix} 157 & 176 & 177 & 174 & 190 & 186 & 187 & 183 \\ 166 & 177 & 180 & 183 & 187 & 188 & 181 & 167 \\ 177 & 181 & 188 & 189 & 194 & 185 & 160 & 95 \\ 181 & 184 & 189 & 193 & 185 & 143 & 81 & 74 \\ 188 & 192 & 192 & 179 & 133 & 75 & 77 & 76 \\ 193 & 195 & 175 & 108 & 67 & 61 & 74 & 77 \\ 200 & 169 & 93 & 53 & 69 & 57 & 70 & 72 \\ 163 & 74 & 52 & 53 & 59 & 65 & 61 & 65 \end{bmatrix} \xRightarrow{\frac{RGB}{YC_bC_r}} C_b = \begin{bmatrix} 164 & 177 & 178 & 176 & 186 & 184 & 183 & 181 \\ 170 & 177 & 178 & 181 & 185 & 183 & 179 & 168 \\ 177 & 179 & 184 & 185 & 187 & 182 & 163 & 123 \\ 180 & 182 & 185 & 187 & 181 & 152 & 112 & 111 \\ 185 & 187 & 187 & 178 & 144 & 109 & 111 & 113 \\ 187 & 189 & 173 & 128 & 102 & 100 & 109 & 113 \\ 191 & 169 & 116 & 89 & 102 & 98 & 105 & 109 \\ 166 & 103 & 88 & 89 & 98 & 100 & 98 & 102 \end{bmatrix}$$

$$B = \begin{bmatrix} 153 & 172 & 173 & 169 & 181 & 173 & 176 & 169 \\ 164 & 170 & 169 & 181 & 180 & 168 & 167 & 136 \\ 176 & 175 & 180 & 176 & 171 & 173 & 135 & 86 \\ 176 & 176 & 180 & 174 & 163 & 109 & 80 & 84 \\ 175 & 169 & 173 & 161 & 93 & 79 & 88 & 81 \\ 174 & 176 & 147 & 81 & 74 & 67 & 80 & 82 \\ 179 & 147 & 74 & 69 & 82 & 62 & 78 & 77 \\ 135 & 70 & 64 & 70 & 73 & 68 & 74 & 72 \end{bmatrix} \qquad C_r = \begin{bmatrix} 152 & 146 & 146 & 147 & 142 & 145 & 142 & 144 \\ 148 & 145 & 144 & 142 & 144 & 143 & 143 & 150 \\ 144 & 143 & 141 & 143 & 141 & 143 & 150 & 173 \\ 144 & 143 & 141 & 141 & 144 & 158 & 173 & 179 \\ 143 & 143 & 143 & 146 & 161 & 176 & 176 & 181 \\ 142 & 142 & 147 & 167 & 175 & 181 & 177 & 180 \\ 139 & 147 & 167 & 171 & 171 & 183 & 176 & 180 \\ 151 & 168 & 172 & 171 & 178 & 176 & 176 & 178 \end{bmatrix}$$

$$Y = \begin{bmatrix} 36 & 49 & 50 & 48 & 58 & 56 & 55 & 53 \\ 42 & 49 & 50 & 53 & 57 & 55 & 51 & 40 \\ 49 & 51 & 56 & 57 & 59 & 54 & 35 & -5 \\ 52 & 54 & 57 & 59 & 53 & 24 & -16 & -17 \\ 57 & 59 & 59 & 50 & 16 & -19 & -17 & -15 \\ 59 & 61 & 45 & 0 & -26 & -28 & -19 & -15 \\ 63 & 41 & -12 & -39 & -26 & -30 & -23 & -19 \\ 38 & -25 & -40 & -39 & -30 & -28 & -30 & -26 \end{bmatrix} \qquad C_Y = \begin{bmatrix} 11 & 12 & 1 & 1 & 0 & 0 & 0 & 0 \\ 16 & -6 & -6 & -1 & -1 & 0 & 0 & 0 \\ -3 & -6 & 3 & 1 & 0 & 0 & 0 & 0 \\ 1 & 1 & 2 & -1 & 0 & 0 & 0 & 0 \\ -1 & 0 & -1 & 0 & 0 & 0 & 0 & 0 \\ 0 & 0 & 0 & 0 & 0 & 0 & 0 & 0 \\ 0 & 0 & 0 & 0 & 0 & 0 & 0 & 0 \\ 0 & 0 & 0 & 0 & 0 & 0 & 0 & 0 \end{bmatrix}$$

$$\xRightarrow[shift]{Level} C_b = \begin{bmatrix} -10 & -8 & -8 & -9 & -8 & -11 & -9 & -11 \\ -8 & -9 & -10 & -6 & -8 & -13 & -11 & -20 \\ -6 & -8 & -8 & -10 & -14 & -10 & -18 & -19 \\ -7 & -8 & -8 & -12 & -14 & -24 & -16 & -13 \\ -11 & -15 & -13 & -14 & -28 & -14 & -11 & -16 \\ -13 & -12 & -18 & -24 & -13 & -15 & -14 & -15 \\ -12 & -16 & -21 & -8 & -9 & -17 & -13 & -16 \\ -19 & -15 & -10 & -8 & -11 & -15 & -11 & -14 \end{bmatrix} \xRightarrow[Quantization]{DCT} C_{C_b} = \begin{bmatrix} -6 & 1 & 0 & 0 & 0 & 0 & 0 & 0 \\ 1 & 0 & 0 & 0 & 0 & 0 & 0 & 0 \\ 0 & 0 & 0 & 0 & 0 & 0 & 0 & 0 \\ 0 & 0 & 0 & 0 & 0 & 0 & 0 & 0 \\ 0 & 0 & 0 & 0 & 0 & 0 & 0 & 0 \\ 0 & 0 & 0 & 0 & 0 & 0 & 0 & 0 \\ 0 & 0 & 0 & 0 & 0 & 0 & 0 & 0 \\ 0 & 0 & 0 & 0 & 0 & 0 & 0 & 0 \end{bmatrix}$$

$$C_r = \begin{bmatrix} 24 & 18 & 18 & 19 & 14 & 17 & 14 & 16 \\ 20 & 17 & 16 & 14 & 16 & 15 & 15 & 22 \\ 16 & 15 & 13 & 15 & 13 & 15 & 22 & 45 \\ 16 & 15 & 13 & 13 & 16 & 30 & 45 & 51 \\ 15 & 15 & 15 & 18 & 33 & 48 & 48 & 53 \\ 14 & 14 & 19 & 39 & 47 & 53 & 49 & 52 \\ 11 & 19 & 39 & 43 & 43 & 55 & 48 & 52 \\ 23 & 40 & 44 & 43 & 50 & 48 & 48 & 50 \end{bmatrix} \qquad C_{C_r} = \begin{bmatrix} 13 & -4 & 0 & 0 & 0 & 0 & 0 & 0 \\ -4 & 2 & 1 & 0 & 0 & 0 & 0 & 0 \\ 0 & 1 & 0 & 0 & 0 & 0 & 0 & 0 \\ 0 & 0 & 0 & 0 & 0 & 0 & 0 & 0 \\ 0 & 0 & 0 & 0 & 0 & 0 & 0 & 0 \\ 0 & 0 & 0 & 0 & 0 & 0 & 0 & 0 \\ 0 & 0 & 0 & 0 & 0 & 0 & 0 & 0 \\ 0 & 0 & 0 & 0 & 0 & 0 & 0 & 0 \end{bmatrix}$$

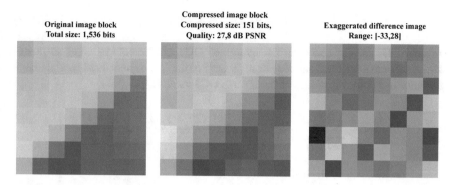

Fig. 2.81 Comparison of the original and reconstructed image block

$$C_Y^{zz} = \{11, 12, 16, -3, -6, 1, 1, -6, -6, 1, -1, 1, 3, -1, 0, 0, -1, 1, 2, 0, 0, 0, 0, -1, -1,$$

$$0, 0, 0, 0, 0, 0, 0, 0, 0, 0, 0, 0, 0, 0, 0, 0, 0, 0, 0, 0, 0, 0, 0, 0, 0, 0, 0, 0, 0, 0, 0, 0, 0, 0, 0, 0, 0, 0, 0, 0\}$$

$$\xrightarrow{\underset{ordering}{Zig-zag}} \quad C_{C_b}^{zz} = \{-6, 1, 1, 0, 0, 0, 0, 0, 0, 0, 0, 0, 0, 0, 0, 0, 0, 0, 0, 0, 0, 0, 0, 0, 0, 0,$$

$$0, 0, 0, 0, 0, 0, 0, 0, 0, 0, 0, 0, 0, 0, 0, 0, 0, 0, 0, 0, 0, 0, 0, 0, 0, 0, 0, 0, 0, 0, 0, 0, 0, 0, 0, 0, 0, 0\}$$

$$C_{C_r}^{zz} = \{13, -4, -4, 0, 2, 0, 0, 1, 1, 0, 0, 0, 0, 0, 0, 0, 0, 0, 0, 0, 0, 0, 0, 0, 0, 0,$$

$$0, 0, 0, 0, 0, 0, 0, 0, 0, 0, 0, 0, 0, 0, 0, 0, 0, 0, 0, 0, 0, 0, 0, 0, 0, 0, 0, 0, 0, 0, 0, 0, 0, 0, 0, 0, 0, 0\}$$

$$CODE_Y = 1011011\ 10111100\ 1101010000\ 0100\ 100001\ 001\ 001\ 100001\ 100001$$

$$\xrightarrow{\underset{(JPEG\ Tables)}{Huffman\ coding}} \quad 001\ 000\ 001\ 0111\ 000\ 111000\ 001\ 0110\ 1110110\ 000\ 1010$$

$$CODE_{C_b} = 100001\ 001\ 001\ 1010$$

$$CODE_{C_r} = 1011101\ 100011\ 100011\ 1101110\ 111001\ 001\ 1010$$

Apparently, the selected image block, which was a 3-channel 8×8 block of 8-bit pixels amounting 1,536 bits is now compactly represented by only 151 bits; this corresponds to a compression data rate of 0.786 bpp, or equivalently a compression ratio of about 10:1. Since the quantization was applied by using the reference quantization matrices (2.145) the reconstruction of the image by the 151-bits codestream is expected to produce only minimal or non-noticeable artifacts. Figure 2.81 shows side-by-side the original image block, the reconstructed block (after the JPEG compression) and the difference of the two.

2.4.1.2 The Decoder

In sequential decoding, all the steps of the encoding are being applied in the reverse order. Initially, the codestring representing the compressed data undergoes entropy decoding. The binary sequence is converted into a symbol sequence, and then the symbols are converted into DCT coefficients. Reversing the quantization is performed by a coefficient-wise multiplication with the quantization matrix,

$$J(u, v) = J_q(u, v) \times Q(u, v) \tag{2.172}$$

Inverse-DCT is performed on the quantized DCT coefficients in order to reconstruct the data from the frequency domain back to the two-dimensional spatial image domain. Using the definition of the inverse DCT in (2.85),

$$\left.\begin{aligned} I(x, y) &= \frac{2}{N} \sum_{u=0}^{N-1} \sum_{v=0}^{N-1} c(u)c(v)J(u, v) \cos \frac{(2x+1)u\pi}{2N} \cos \frac{(2y+1)v\pi}{2N} \\ c(\xi) &: \; c(\xi = 0) = \sqrt{\frac{1}{2}}, \quad c(\xi \neq 0) = 1 \\ N &= 8 \end{aligned}\right\} \Rightarrow$$

$$I(x, y) = \frac{1}{4} \sum_{u=0}^{7} \sum_{v=0}^{7} c(u)c(v)J(u, v) \cos \frac{(2x+1)u\pi}{16} \cos \frac{(2y+1)v\pi}{16}$$

$$\tag{2.173}$$

Finally, the decompressed data are level-shifted back to the interval $[0, 2^b - 1]$ (i.e. by adding 128 in the case of 8-bpp images) to reconstruct the original image data.

2.4.1.3 JPEG Compression Efficiency

The most basic estimator of the efficiency of a compression algorithm is the *Compression Ratio*, C_r, which is typically defined as the ratio of the original data to the compressed data and is expected to be greater that one,

$$C_r = \frac{\text{Original Data Size}}{\text{Compressed Data Size}} \tag{2.174}$$

Typically, the compression ratio is reported as a ratio $C_r : 1$, such as $2 : 1$.

As already pointed out, in image compression applications there is an interactive relationship presented as a trade-off between the compression ratio and the reconstructed image quality. A high compression ratio usually implies a low quality reconstructed image. Quality and compression may depend on the characteristics of the original image or the visual content of the scene. An image quality estimator, proposed by Wallace (1991), is the number of bits for a pixel (bpp) in the compressed image domain, the *Compression Rate*,

$$N_b = \frac{\text{Number of encoded bits}}{\text{Number of pixels}} \tag{2.175}$$

Table 2.17 presents four different image quality estimates derived by applying the estimator in (2.175). The results are accompanied by a subjective characterization

Table 2.17 Typical image quality related to various bit-rates

N_b [bits/pixel]	Image quality
0.25–0.5	Moderate to good
0.5–0.75	Good to very good
0.75–1.0	Excellent
1.5–2.0	Subtle difference from original

to 'moderate', 'good' or 'excellent', to denote what is roughly expected to be the judgment of an average observer.

Another estimator, which can be used for a variety of compression algorithms, is the classic RMSE, defined as a typical normalized Euclidean distance between the original and the compressed image,

$$RMSE = \sqrt{MSE} = \sqrt{\frac{1}{n}\sum_{i=1}^{n}(I_i - \hat{I}_i)^2} \qquad (2.176)$$

where I_i the original pixel values, $\hat{I}_i$ the compressed image pixel values and n the total number of pixels in the image.[20]

Figures 2.82 and 2.83 show examples of compression outcomes using the sequential JPEG encoding algorithm for a graylevel image (8 bpp) and a true color image (24 bpp). The results show the compressed image size, the compression ratio C_r, the compression rate N_b and the PSNR estimates, for five different compression quality factors. In the case of the color image, the quality estimate is based on the one proposed in (Kang and Leou 2003), in which the quality is assessed in the YUV color space, and the measurements of each channel ($PSNR_Y$, $PSNR_U$, and $PSNR_V$) are combined by using appropriate weights to create a single measurement value as,

$$PSNR = \frac{4 \cdot PSNR_Y + PSNR_U + PSNR_V}{6} \quad \text{dB} \qquad (2.177)$$

The examples clearly demonstrate the strength of JPEG in compressing continuous-tone images, either graylevel or true color. Even for compression ratios of more than 16:1 the quality of the reconstructed image peaks high above 30 dBs of PSNR (which is usually used as a 'psychological' limit for noticeable artifacts). The graylevel image reconstructed from the compressed stream using the reference quality factor (50) presents a ratio of 16:1 and a data rate of 0.49 bpp with a high 37.4 dBs quality. The corresponding color image reports an impressive 42:1 ratio at 0.58 bpp with a 38.1 dBs quality. In both cases the visual appearance of the images is acceptable for most observers. In the extreme case of using the lowest quality

---

[20]It is worth noting that there are cases in which it is possible high RMSE values to correspond to visually acceptable quality.

Initial uncompressed image:
3 145 728 bytes

JPEG quality factor 1:
44 667 bytes, 25.59 dB, 0.11 bpp, 70:1

JPEG quality factor 50:
196 528 bytes, 37.4 dB, 0.49 bpp, 16:1

JPEG quality factor 70:
279 917 bytes, 38.58 dB, 0.71 bpp, 11:1

JPEG quality factor 90:
615 304 bytes, 41.64 dB, 1.56 bpp, 5:1

JPEG quality factor 100:
1 568 654 bytes, 52.28 dB, 3.99 bpp, 2:1

Fig. 2.82 Baseline JPEG coding example for a graylevel image and five compression ratios

Fig. 2.83 Baseline JPEG coding example for a true color image and five compression ratios

factor (imposing the highest quantization), the images appear severely influenced by the compression but the basic image structure is still preserved and the quantization practically damages the tones of the images. The compression ratios are impressive at 70:1 and 163:1 for the graylevel and color image respectively. In addition, in the extreme case of using the highest quality factor, the outcome is a 2:1 and 4:1 compression for the graylevel and the color image respectively.

2.5   JPEG2000 Compression

JPEG2000 is the new standard (as of 2000) in digital image compression (Taubman and Marcellin 2002b; Rabbani and Joshi 2002; Rabbani and Cruz 2001; ISO-IEC 2000a; Christopoulos et al. 2002; Skodras et al. 2001; Christopoulos et al. 2000b; Santa-Cruz and Ebrahimi 2000a, b; Santa-Cruz et al. 2000). It appeared as a solution to issues in the existing standards, and as a response to the challenges posed by the new forms of information systems and the modern user requirements. The main axes on which the development of this new standard initiated were:

- Higher compression efficiency
- Multiple image resolutions within the same compressed file
- Presetting of the final data rate
- Quality scalability, and support for progressive decoding and scaling of the decoded image quality
- Lossy and lossless compression
- Partial or 'total' image compression
- Enhanced noise resilience
- Flexible codestream to support processing in the transform domain
- Robustness against recursive compression
- New type of file format to meet the requirements of modern digital photography, transmission among heterogeneous devices, better internal organization, code embedding, etc.

Even though it might sound paradoxical, the axis that played the least significant role in the development of JPEG2000 was that of the improved compression performance. Essentially, the motivation for this new compression standard was the addition of multiple functionalities and flexibility. The model on which this standard developed, was that of a typical compression system that is based on transform coding as typically depicted in Fig. 2.84.

The data transformation adopted and used in this standard was the wavelet transform (DWT). Quantization is this method is based on a conventional dead-zone scalar quantizer. The entropy coder is based on an implementation of a binary arithmetic encoder. In overall, the standard is quite open, and supports extensions and alternatives in almost all stages of the method, incorporating virtually all the solutions that were proposed at the stage of development by the researchers of the Working Groups, as well as other solutions that can meet specific future user needs. One of the

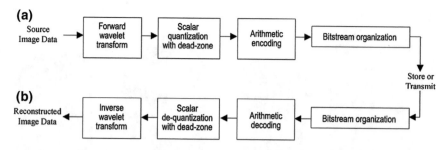

Fig. 2.84 The basic structure of the JPEG2000 codec, **a** the encoder and **b** the decoder

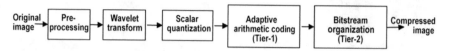

Fig. 2.85 The basic structure of the JPEG2000 encoder

most celebrated features is the flexible progressive coding, which can be performed either based on the position in the image, or on the size, or the quality, or the color component.

2.5.1 The JPEG2000 Encoder

A basic diagram of the encoder is shown in Fig. 2.85. A comprehensive graphical representation of the JPEG2000 encoding operation is shown in Fig. 2.86. Even thought the encoder usually constitutes an *informative* part of a compression standard, it is in this case highly sophisticated; it introduced several new concepts and ideas towards the optimization of the overall process, in addition to maintaining and supporting a number of options for alternatives. It is therefore worth studying the encoder proposed in JPEG2000 even for purely educational purposes, in order the grasp the insight of the researchers that made it possible.

2.5.1.1 Coding Preparation

At the beginning of the encoding process the image data undergo pre-processing. The pre-processing includes imperative (termed *normative* in JPEG2000) and optional (termed *informative* in JPEG2000) steps, including,

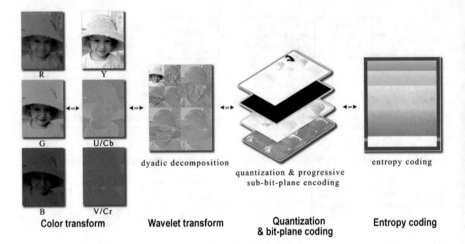

Fig. 2.86 Graphical representation of the JPEG2000 encoder operation

- *Tiling* (optional) of the image into non-overlapping regularly organized rectangular blocks for separate processing. This tiling is especially recommended for hardware implementations, where memory limitations and parallel processing requirements are of utmost importance. Figure 2.87 shows a random image tiling before the encoding process.
- *Level shifting of the values* of the samples (also termed *DC level shifting*) by subtracting a fixed quantity, the mean of the range of the values ($2^{b-1}, b = \text{bit} - \text{depth}$) from each sample, to generate a distribution symmetrically distributed around zero, in order to support the following functions without affecting the coding itself.
- *Color transformation* (optional) of the image for initial spectral decorrelation, provided that there are more than one color channels, and all the channels are of the same color depth (bpp) and dimensions. Figure 2.88 illustrates the color transformation from an initial RGB color space to a typical luminance-chrominance color space (either YUV or YC_bC_r). The transformations that are indicatively referenced in the JPEG2000 standard (informative) are the RGB to YUV (reversible transformation) and the RGB to YC_bC_r (irreversible transformation), defined as,

$$
\begin{pmatrix} Y \\ U \\ V \end{pmatrix} = \begin{pmatrix} 1/4 & 1/2 & 1/4 \\ 1 & -1 & 0 \\ 0 & -1 & 1 \end{pmatrix} \begin{pmatrix} R \\ G \\ B \end{pmatrix}
$$

$$
\begin{pmatrix} Y \\ C_b \\ C_r \end{pmatrix} = \begin{pmatrix} 0.29900 & 0.58700 & 0.11400 \\ -0.16875 & -0.33126 & 0.50000 \\ 0.50000 & -0.41869 & -0.08131 \end{pmatrix} \begin{pmatrix} R \\ G \\ B \end{pmatrix}
$$

$$(2.178)$$

Fig. 2.87 Tiling of an image before encoding

2.5.1.2 Image Transform

The wavelet transform (DWT) has been selected to replace the cosine transform (DCT) in JPEG2000, since it has been proven that it can be beneficial to coding for a number of reasons including,

- The DWT can be applied to the entire image and not only to distinct image tiles, and thus the appearance of discontinuities in the tile boundaries can be avoided
- The use of integer arithmetic DWT filters allows both lossy and lossless compression in the same codestream
- The DWT leads, by default, to multi-resolution representations, thus enabling a new mode of progressive coding
- The DWT results in representations of the image in various spectral bands and permits the application of different quantization in each band according to any selected HVS model

The application of the two-dimensional DWT is implemented as a separable transform (that is, two one-dimensional transforms), as shown in Fig. 2.50. The process is recursively applied to the LL band of the transform up to a particular desired level or up to the point in which it is not possible to further decompose the LL band (the

Fig. 2.88 Representation of
the color space
transformation

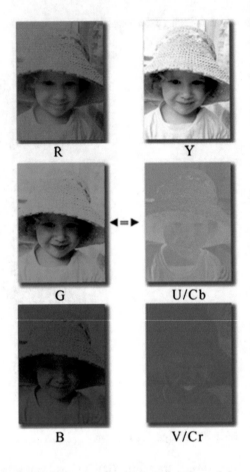

LL band consists only of one pixel). The result is a spectral representation of the original image into various zones as shown in Fig. 2.89.

After extensive study on various wavelet filtering options for use in theJPEG2000, researchers who participated in its development concluded in the adoption of two basic analysis-synthesis sets of filters, namely the *Le Gall or Integer (5, 3)* and the *Daubechies (9, 7) filter banks.*[21] The numbering in the names of the filter banks indicates the number of coefficients in the analysis and synthesis filters used in each pair. A representation of the application of the one-dimensional transform in terms of a filter bank pair is depicted in Fig. 2.90. By imposing the DWT to be biorthogonal, h_0 is orthogonal to to g_1 and h_1 is orthogonal to g_0. The absolute summations

---

[21] In the context of signal processing, a filter bank is a set of band-pass filters, each of which separates an input signal into one single frequency sub-band of the overall spectrum of the original signal. A graphic equalizer is usually referenced as one illustrative example of a filter bank, in which various components of the input signal are being attenuated separately and all components are recombined to form the modified signal at the output. Apparently this process includes two steps, a decomposition or an analysis and a reconstruction or a synthesis step.

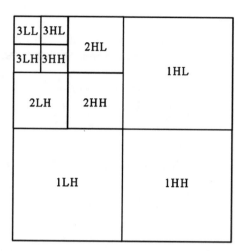

Fig. 2.89 Three-level 2-D DWT

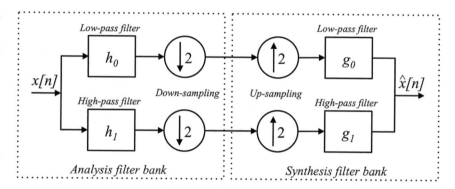

Fig. 2.90 DWT in terms of an analysis (*left*)—synthesis (*right*) filter bank

$$\left|\sum_n h_0[n]\right|, \qquad \left|\sum_n (-1)^n h_1[n]\right|$$

define the *DC gain* of the low-pass analysis filter and the *Nyquist gain* of the high-pass analysis filter respectively. The synthesis and analysis filters are related according to

$$g_0[n] = a(-1)^n h_1[-n]$$
$$g_1[n] = a(-1)^n h_0[-n] \tag{2.179}$$

where

$$a = \frac{2}{\left(\sum_n h_0[n]\right)\left(\sum_n (-1)^n h_1[n]\right) + \left(\sum_n h_1[n]\right)\left(\sum_n (-1)^n h_0[n]\right)}$$

a normalization factor.

According to this definition, the analysis Le Gall (5, 3) filter bank consists of a low-pass filter of five (5) coefficients and a high-pass filter of three (3) coefficients,

$$\text{low} - \text{pass filter} : h_1[n]$$

$$h_0 = \left\{ -\frac{1}{8}, \frac{2}{8}, \frac{6}{8}, \frac{2}{8}, -\frac{1}{8} \right\}$$

$$n = -2, -1, 0, 1, 2$$

$$\text{high} - \text{pass filter} : h_1[n] \qquad (2.180)$$

$$h_1 = \left\{ -\frac{1}{2}, 1, -\frac{1}{2} \right\}$$

$$n = -2, -1, 0$$

Using (2.179) it is possible to compute the synthesis filter bank, as

$$\text{low} - \text{pass filter} : g_0[n]$$

$$g_0 = \left\{ \frac{1}{2}, 1, \frac{1}{2} \right\}$$

$$n = -2, -1, 0$$

$$\text{high} - \text{pass filter} : g_1[n] \qquad (2.181)$$

$$g_1 = \left\{ \frac{1}{8}, \frac{2}{8}, -\frac{6}{8}, \frac{2}{8}, \frac{1}{8} \right\}$$

$$n = -2, -1, 0, 1, 2$$

This filter bank uses integer arithmetic, whereas the analysis Daubechies (9, 7) filter, with a low-pass filter of nine (9) coefficients and a high-pass filter of seven (7) coefficients,

$$\text{low} - \text{pass filter} : g_0[n]$$
$$g_0 = \{0.026748757410, -0.016864118442, -0.078223266528,$$
$$0.266864118442, 0.602949018236, 0.266864118442,$$
$$-0.078223266528, -0.016864118442, 0.026748757410\}$$
$$n = -4, -3, -2, -1, 0, 1, 2, 3, 4$$
$$\text{high} - \text{pass filter} : g_1[n]$$
$$g_1 = \{0.091271763114, -0.057543526228, -0.591271763114,$$
$$1.115087052456, -0.591271763114, -0.057543526228, 0.091271763114\}$$
$$n = -4, -3, -2, -1, 0, 1, 2$$

$$(2.182)$$

Using (2.179) it is possible to compute the synthesis filter bank, as low − pass filter : $g_0[n]$

$g_0 = \{-0.091271763114, -0.057543526228, 0.591271763114,$
$\qquad 1.115087052456, 0.591271763114, -0.057543526228, -0.091271763114\}$

$n = -4, -3, -2, -1, 0, 1, 2$

high − pass filter : $g_1[n]$

$g_1 = \{-0.026748757410, -0.016864118442, 0.078223266528,$
$\qquad 0.266864118442, -0.602949018236, 0.266864118442,$
$\qquad 0.078223266528, -0.016864118442, -0.026748757410\}$

$n = -4, -3, -2, -1, 0, 1, 2, 3, 4$

$$(2.183)$$

Figures 2.91 and 2.92 are graphical representations of the filter coefficients for the analysis and synthesis stage respectively, for both the filter banks in JPEG2000. It is noted that the analysis-synthesis filter bank scheme presented so far correspond to a one-dimensional one decomposition level wavelet transform. In case a multi-level decomposition is required then the transform is applied recursively on the low-pass filtered output at the analysis stage as shown in Fig. 2.93, where an N-level 1-D wavelet transform is presented. The output of the transform is a series of transform coefficients that includes $L_N, H_N, H_{N-1}, ..., H_2, H_1$. Since the two-dimensional DWT can be applied as a separable transform, it is usually applied as a sequence of two one-dimensional transforms, first applied on the columns of an image and then applied on the rows of the 'image' after the first step.

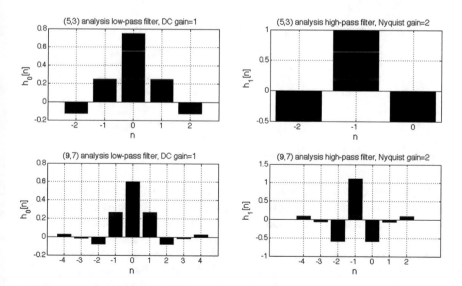

Fig. 2.91 Analysis stage filter banks (5, 3) and (9, 7) for JPEG2000

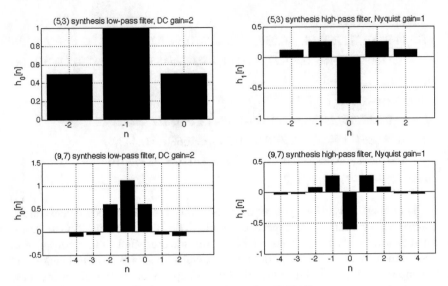

Fig. 2.92   Synthesis stage filter banks (5, 3) and (9, 7) for JPEG2000

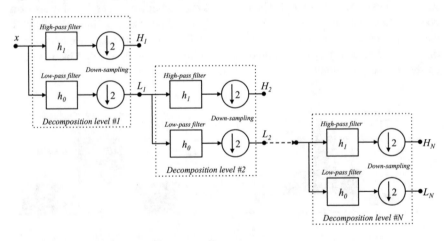

Fig. 2.93   Multi-level decomposition in 1-D DWT

Figure 2.94 shows a representation of the transform coefficients after the applications of one, two and three levels of two-dimensional DWT for a MRC image. It can de clearly seen in this representation that in each of the detail bands (all the bands except the LL at the upper-left part) the details captured correspond to a particular 'detail direction', vertical in the upper-right bands, horizontal in the lower-left bands and diagonal in the lower-right bands at each decomposition level. As depicted in Fig. 2.93 the decomposition at each new level is applied on the LL band (the low pass filtered part from the previous decomposition level). The double 'L' letters indicate the two-dimensional application of the low-pass analysis filter of the filter banks,

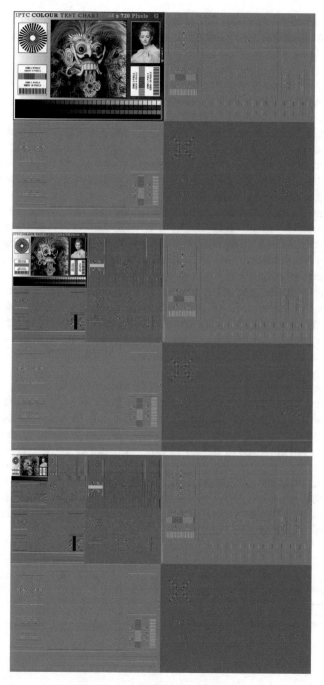

Fig. 2.94 Transform coefficients after a one/two/three-level 2-D DWT on a MRC image

column-wise and row-wise. In this example, the Le Gall (5, 3) filter bank has been used, and the resulting coefficient values have been scaled for a better illustration in Fig. 2.94.

2.5.1.3   Quantization

In sequential JPEG a uniform scalar quantizer is being used and a de-quantizer maps the quantized coefficients to the center of the quantization interval. For each of the DCT coefficients a different quantization step is being used in agreement with the HVS modeling, through the use of a quantization matrix of the same size as the part of the image gradually being encoded.

The same general principle applies in JPEG2000 with some variations introduced to meet the new requirements. One difference lies in the adoption of the central dead-zone quantizer. It has been shown (Sullivan 1996) that the optimal quantizer—from a rate-distortion point of view—for a continuous signal with Laplacian probability density (such as the the one of the transform coefficients) is a scalar quantizer with a central dead zone. The size of the dead zone as a fraction of the step increases with the reduction of the variance of the Laplacian distribution. Usually, its value does not exceed the value of two (2), being preferable to tend to one (1). In JPEG2000 the dead zone, typically, has two times the quantization step size, as shown in Fig. 2.95, although the standard supports any modification of the quantizer for each of the bands of the transform coefficients.

This specific quantizer is used in JPEG2000 because it exhibits efficient embedding features. This means that if a quantization index of M_b bits resulting from a quantizer with step size Δ_b is transmitted progressively starting with the most significant bit, the final index after decoding only N_b bits is identical to that which would have been produced by a similar quantizer with a step size of $\Delta_b 2^{M_b - N_b}$. This property ensures *progressiveness in quality*, which, from an optimization point of view, signifies that the decoder can stop decoding at any time having managed to reconstruct the same image that would have been produced by a quantization set to the same endpoint. It also allows the definition, in advance, of the compression data rate and the level of expected distortion.

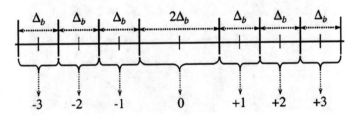

Fig. 2.95 Uniform scalar quantizer with central dead zone and step size Δ_b

Another significant feature of the JPEG2000 quantizer is that, in the inverse process, deviations from the middle point of the quantization interval are permitted for non-zero indices, in order to achieve a better adaptation to the skewness (asymmetry) in the probability distribution of the transform coefficients.

In the encoder, for each spectral band b of the DWT, a different quantization step size Δ_b is selected by the user, which is then used to quantize all the coefficients in that band. The selection of the step size may follow the perceptual significance of the content in that band (in terms of the HVS) (Albanesi and Bertoluzza 1995; Jones et al. 1995; O'Rourke and Stevenson 1995; Watson et al. 1997), or it may be a basis for achieving other goals, such as a better compression rate. The quantizer assigns a quantized value $q_b(u, v)$ to a transform coefficient $y_b(u, v)$ of band b, as shown in Fig. 2.95. Although, the quantization in an encoder is informative, it was initially proposed that a specific relation could be follows for quantization, that is

$$q_b(u, v) = sign[y_b(u, v)] \left\lfloor \frac{|y_b(u, v)|}{\Delta_b} \right\rfloor \tag{2.184}$$

b being the band and u, v, the spatial coordinates in the transform domain. The quantizing step size Δ_b is described by two bytes: a *mantissa* μ_b of 11-bits and an *exponent* ϵ_b of 5-bits, somehow complying with the ISO/IEC/IEEE 60559: 2011 or the IEEE Standard for Floating-Point Arithmetic (IEEE 754), as

$$\Delta_b = 2^{R_b - \epsilon_b} \left(1 + \frac{\mu_b}{2^{11}}\right) \tag{2.185}$$

where R_b is the number of bits representing the dynamic range of band b. It is noted that, when reversible encoding with a Le Gall (5, 3) filter-bank is being applied, then the quantization step size becomes unity (1).

The decoder becomes aware of the value of the quantization step size in two ways:

1. Transmission of the pair (ϵ_b, μ_b) for each band. This is called an *expounded quantization* and is similar to the approach being used in the older JPEG.
2. Transmission of the pair (ϵ_b, μ_b) only for the LL band as (ϵ_0, μ_0) and calculation of values for any other band. This is called a *derived quantization*, and involves the computation

$$(\epsilon_b, \mu_b) = (\epsilon_0 - N_L + n_b, \mu_0) \tag{2.186}$$

where N_L the total number of DWT decomposition levels, and n_b the decomposition level that corresponds to band b.

*The decoder needs not be signaled about the quantization step size in the case of reversible encoding (or lossless compression) as there is no quantization in this case ($\Delta = 1$).

In the decoder, when lossy encoding with a Daubechies (9, 7) filter-bank is applied, then a reconstructed coefficient $Rq_b(u, v)$ for a quantization step size Δ_b is estimated as

Fig. 2.96 Midpoint and
mass center reconstruction in
the quantization intervals,
assuming a two-sided
Laplacian distribution

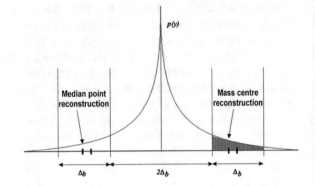

$$
Rq_b(u, v) = \begin{cases} \left[q_b(u, v) + \gamma \right] \Delta_b & q_b(u, v) > 0 \\ \left[q_b(u, v) - \gamma \right] \Delta_b & q_b(u, v) < 0 \\ 0 & \text{elsewhere} \end{cases} \tag{2.187}
$$

where $\gamma \in [0, 1)$ is a reconstruction parameter selected by the decoder that controls
the position in the quantization interval to report. Typically, when $\gamma = 0.5$ then
a midpoint de-quantization occurs. Values for $\gamma < 0.5$ introduce a bias towards
zero, which is expected to contribute to an improvement in the final decoded image
quality, in cases in which the probability distribution of the coefficients is Laplacian.
An empirical popular value is $\gamma = 0.375$, as this value corresponds roughly to
the position of the centroid (mass center) of the quantization interval as shown in
Fig. 2.96.

When a reversible Le Gall (5, 3) filter-bank is being used in the transform, the same
principle in (2.187) applies with $\Delta_b = 1$; when lossless compression is required then
the de-quantization involves virtually no operation, and the reconstructed coefficient
is assigned the same value as the quantized coefficient,

$$
Rq_b(u, v) = q_b(u, v) \tag{2.188}
$$

$Rq_b(u, v)$ being the reconstructed (de-quantized) coefficient.

2.5.1.4   Entropy Coding

The final stage in the coding process is the entropy coding, which includes a novel
bitsteam organization. In this stage the quantized transform coefficients pass through
an entropy encoder to produce the final compression bitstream. In order to meet the
requirements set for this standard, and especially to support codestream embedding
so that progressive transmission and decoding is possible, a specific type of entropy
encoder had to be selected. This is a bitplane encoder, which had already been tested
in several transform coding schemes based on DWT, like in the Embedded image

coding using Zerotrees of Wavelet coefficients (EZW) (Shapiro 1993) and Set Partitioning In Hierarchical Trees (SPIHT) (Said and Pearlman 1996). In these codecs the correlation among the bands is being exploited to improve the compression efficiency, which, unfortunately hampers the error resilience during data transmission and the embedding functionality for a flexible progressiveness. To tackle with this issue, in JPEG2000 each wavelet band is encoded independently. In addition, JPEG2000 employs block coding in the transform domain as in Embedded Block Coding with Optimized Truncation (EBCOT) (Taubman 2000b). In this coding scheme each band is divided into small non-overlapping blocks, named *codeblocks*, which are independently encoded. Their dimensions are determined by the encoder and are restricted to be powers of 2, to have a height greater or equal to 4 and the number of coefficients in a codeblock not to exceed 4096. This coding scheme introduces a number of advantages since it enables

- easy random access to an image region
- parallel implementations
- improved segmentation and rotation capabilities
- improved error resilience
- efficient control of compression rate
- flexibility in shaping progressive forms

As will be explained in the following paragraphs, by adopting an efficient data rate strategy that enables the optimization of the percentage of participation of each codeblock in the final bitstream, JPEG2000 achieves a better compression efficiency than other existing standards (Taubman et al. 2002c).

According to the entropy coding stage in JPEG2000, the quantized transform coefficients are being encoded bitplane-by-bitplane (bit-by-bit) as shown in Fig. 2.97 (Rabbani and Joshi 2002), starting from the Most Significant Bit (MSB). During the process, each coefficient is considered *insignificant* as long as its already processed

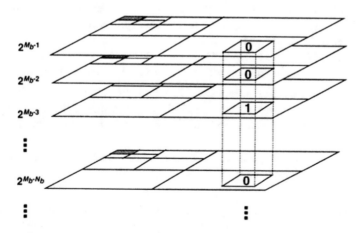

Fig. 2.97 Transform coefficients are encoded bit by bit starting from the MSB

bits are zero (for example, in Fig. 2.97 the coefficient is insignificant after the encoding of its first two significant bits). Once the process encounters a non-zero bit, then the coefficient becomes *significant* and its sign is encoded. Henceforth, the bits of this coefficient are called *precision bits*. Since the coefficients are expected to be of small values (due to the nature of the DWT to collect most of the energy of the image in low frequency coefficients), the quantized coefficients to be encoded are mostly *insignificant* in the early stages of coding, thereby producing very limited information about the specific bitplanes.

In JPEG2000 a very effective method is being used to exploit these redundancies, known as *adaptive binary arithmetic coding*.[22] One of the first implementations of an adaptive binary arithmetic coder was the *Q-coder* (Pennebaker et al. 1988) developed by IBM. A modified version of the Q-coder, the *QM-coder*, was selected for arithmetic coding in both JBIG and JPEG (Pennebaker and Mitchell 1993). Copyright issues, however, prevented its widespread in JPEG-based implementations. JPEG2000 adopted another modification of the Q-coder, the *MQ-coder*. This encoder is used in JBIG2 (ISO-IEC-ITU 2000) and its use was extended to JPEG2000.

Generally, the probability distribution of each bit symbol of a quantized coefficient is influenced by its previous coded bits, and the bits of the neighboring coefficients. The estimation of this probability is being done using contextual information generated by the current significance status of the coefficient, and the significance of the eight neighboring coefficients (in a typical 3×3 neighborhood), as defined by the current and previous bitplanes and based on the available, up to that point, encoded information. In arithmetic coding with contextual information, separate probabilities are being estimated and maintained for each context model, which are being updated based on a finite state machine, each time a symbol is encoded in the given context model. The MQ-coder selects among 46 modes for each context model: of these, modes 0 to 13 correspond to initialization and are used for fast convergence (fast attack) in robust probability estimation. Modes 14 to 45 correspond to probability estimates of the steady state. There is also a complementary non-adaptive mode (46), which is used for encoding symbols with equal probability distribution, and can neither enter or leave one of the other states. In practice, each band of coefficients is divided into codeblocks (typically, of size 64×64) and each bitplane of each codeblock is encoded in three passes (referenced also as *coding triplets*),

- *significance propagation pass*, in which the coefficients that are begin encoded are the ones that have not yet been marked as significant, and there are significant coefficients in their neighborhood; this is due to their high probability to become significant
- *magnitude refinement pass*, in which only significant coefficients are being encoded, so that their magnitude estimate is refined

---

[22]It is reminded that an *adaptive binary arithmetic coder* accepts the binary symbols of an input sequence, along with a corresponding probabilistic model, and outputs a codestream with a length of at most two bits greater than the combined ideal lengths of the code of the input symbols. By updating the probability estimate of symbols adaptivity is enabled (Pennebaker et al. 1988).

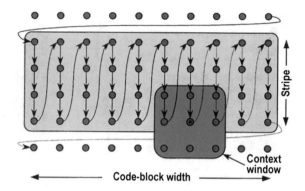

Fig. 2.98 Scanning flow in a codeblock

- *cleanup pass*, in which all non-processed coefficients of the bitplane are being encoded, keeping in mind that the first coding pass of the most significant bitplane is a cleanup phase; in addition, run-length encoding is also being applied at this phase

The order by which the coefficients are being encoded is shown in Fig. 2.98. The height of each vertical scan corresponds to four coefficients. At the end of this process that is called *Tier-1 coding*, a series of bitstreams for each encoded image codeblock are being produced, which should be properly arranged to form an *embedded bitstream*. This is the work of *Tier-2 coding* that follows.

The second phase of the final step of entropy coding, the *Tier-2 coding* consists virtually of a 'multiplexing' of the various bitstreams produced by *Tier-1 coding*, implemented through an efficient ordering. The aim of this step is to create a bitstream that allows easy access and flexible syntax control, guarantee progressiveness and enable region-of-interest coding. An important construct introduced in this phase is the *layer*, which is a collection of consecutive coding scans of all the codeblocks and coefficient bands. In this scheme, each codeblock contributes with a different number of coding passes. The organization of layers to achieve progressiveness in quality can be best shown through a graphic representations, as shown in Fig. 2.99 (Rabbani and Cruz 2001), which includes,

(a) a distribution and ordering of coding triplets by coefficient band and bitplane, where black squares correspond to the clean-up pass, the gray squares correspond to the significance propagation pass and white squares correspond to the refinement pass
(b) a selection of coding phases to fully reconstruct quality and resolution, indicated by a light gray cover over the total amount of data
(c) a selection of coding phases for lower resolution (small image) and maximum quality, indicated by a light gray cover over the the LL band and all coding passes and bitplanes
(d) a selection of coding phases for medium resolution and maximum quality, indicated by a light gray cover over the bands of the second decomposition level and all coding passes and bitplanes

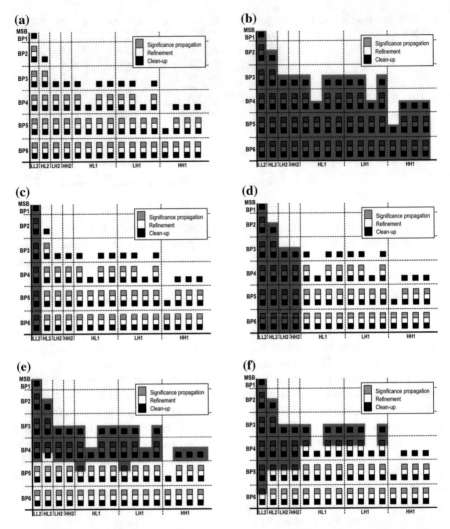

Fig. 2.99 Representation of various formations of layers by selecting results from Tier-1 coding

(e) a selection of coding phases for maximum resolution (full size) and a preset
objective quality (SNR), indicated by a light gray cover over a selected part of
the coding passes and bitplanes for all bands
(f) a selection of coding phases for maximum resolution and a preset subjective
quality (data related to the HVS), indicated by a light gray cover over a selection of the coding passes, bitplanes and bands, using more data from higher
decomposition levels

According to this organization, layers are being created using a collection of
consecutive coding passes throughout the coefficient bands. Multiple layers may be

Fig. 2.100 An example of bitstream organization by the formation of four layers

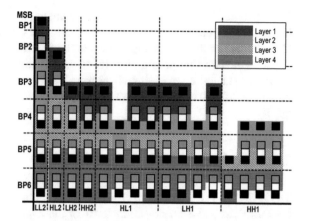

created, one after another improving the image quality. The total number of layers may range from $2^0 = 1$ to $2^{16} - 1 = 65535$, with a popular selection being 20. Figure 2.100 shows an example of an organization that uses four layers (Rabbani and Cruz 2001). In practice, the codeblocks belonging to a *precinct* are encoded together. Precincts are regions in all the coefficient bands that correspond to the same image-domain pixel locations, and can be of arbitrarily large size, as long as their dimensions are powers of 2. By default, precincts are of size 15×15, which corresponds to a division of a resolution level of a component into rectangles of $2^{15} \times 2^{15}$ samples. Further, as the encoding finalizes the bitstream, *packets* are being formed, which represent a specific tile, layer, component, resolution level and precinct. Packets include a *packet head* which encodes informative data regarding the localization of the encoded coefficients, the number of all-zero bitplanes to be skipped, the number of included coding passes and the corresponding length of the data for each codeblock.

Progressiveness in the final encoded bitstream can be defined in terms of four parameters, namely

- Image quality (SNR) corresponding to parameter *Layer—L*
- Image size corresponding to parameter *Resolution—R*
- Color component corresponding to parameter *Component—C*
- Position in the image corresponding to parameter *Position—P*

Definition of the progressiveness during the encoding, which eventually results in reorganization of the packets, is accomplished by selecting the value of a specific byte, so as to indicate the priority of these four parameters. The standard supports five types of progression, namely

- Layer - Resolution - Component - Position (LRCP)
- Resolution - Layer - Component - Position (RLCP)
- Resolution - Position - Component - Layer (RPCL)
- Position - Component - Resolution - Layer (PCRL)
- Component - Position - Resolution - Layer (CPRL)

These progression priorities are switchable among the different tiles that the image has potentially been split. This series of parameters is expressed in their execution from the end to the beginning. For example, in LRCP, first the algorithm runs for Position, then for Component, Resolution, and finally the Layer. Therefore a priority in quality (Layer) denotes that all packets with a certain level of quality (or compression rate) will be arranged before the packets corresponding to the next level of quality.

2.5.2 Enhanced Features in JPEG2000

The JPEG2000 standard, apart from the basic coding structure, supports a series of optimizations and extensions, like those of the *region-of-interest (ROI) coding* and the advanced *error resilience*, which might be of utmost importance, especially in image transmission applications, in cases with requirements for fast communication using a limited bandwidth, or in cases in which corruption by noise occurs in addition to a transmission typically suffering from congestion and outage. These two functionalities are being reviewed in the following sections.

2.5.2.1 Region-Of-Interest Coding

Region-Of-Interest (ROI) Coding allows an uneven distribution of image quality and an arbitrary reorganization of the bitstream according to regions of higher interest. A ROI is encoded in better quality than the rest of the image, which is consequently considered to be *the background* (sometimes denotes *BG* for simplicity). Two classes of ROI are defined in JPEG2000, namely *static ROI* and *dynamic ROI*.

- *Static ROI*, which is determined during the phases of encoding; this ROI is preferable in cases of storage, static transmission, remote sensing applications, etc.
- *Dynamic ROI*, which is defined interactively by a user on a client-server application scenario within a progressive transmission process. This ROI is preferable in telemedicine applications, mobile communications, mobile devices, etc., and can be implemented through the creation of dynamic coding layers to better approach the requirements of the user.

In practice, ROI coding is implemented by using a ROI mask, which is a binary map that determines which of the DWT coefficients contribute to reconstruct the ROI. This mask undergoes a transformation that resembles the dyadic decomposition in DWT, in order to be transformed to a representation that actually maps the ROI in the transform domain, in all bands. In the simple case of a rectangular mask, it is not even required that the mask be an image, since there is an easy way for its definition. Figure 2.101 represents the mask generation process (following the dyadic decomposition in DWT) in an appropriate form to become a map for the coefficients in the ROI. Coding of a ROI is accomplished by shifting (Fig. 2.102)

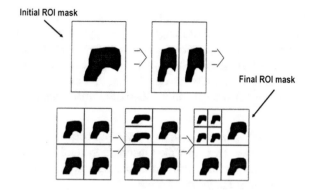

Fig. 2.101 ROI mask generation in JPEG2000

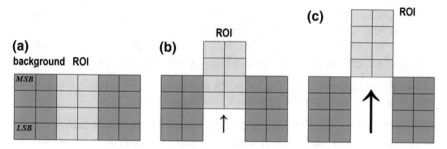

Fig. 2.102 Bit shifting for ROI coding: **a** definition of a ROI, **b** general scaling, **c** maxshift mode

only the transform coefficients that correspond to the ROI a number of bitplanes upwards (left shifting or multiplication by powers of two). As a consequence of this shifting, ROI coefficients are encoded first. The shifting of the bitplanes can vary for different ROIs, and the number of bitplanes that were shifted is stored in the header of the final bitstream to allow a proper decoding of the image. If the ROI coefficients rise above the MSB bitplane of the background, the method is called the *maxshift mode* (Rabbani and Cruz 2001; Christopoulos et al. 2000a), in which neither header data are required, nor the ROI mask itself, as the ROI is being decoded in its entirety before the decoding of the background, as the Least Significant Bit (LSB) of the ROI is higher than the MSB of the background.

2.5.2.2 Resilience to Transmission Noise

In general, typical image applications require the transmission of compressed image data through communication channels with various characteristics. For instance, wireless telecommunication networks are prone to random and sudden errors, whereas wired communications are vulnerable to data loss due to congestion; in addition, both are prone to outages. To address issues during image transmission, various mechanisms for error resilience and data correction have been defined in JPEG2000. Error resilience may be achieved through various approaches, such as data partitioning and re-synchronization, error detection and concealment, and priority-based Quality of Service (QoS) transmission. Both the syntax and the native tools for error resilience are implemented at the entropy coding and packet transmission levels (Liang and Talluri 1999; Moccagata et al. 2000). Summarizing, the errors that may occur are

- *corruption in the body of a packet*, in the form of numerically affected (corrupted) encoded data for a codeblock, in which incorrect symbols are decoded and thus the context regions are updated with erroneous data in the subsequent bitplanes, and as a result *significant distortions occur*
- *corruption in the packet head*, in the form of corruption of sensitive data, such as the packet size, which may lead to a decoding of data that, possibly, do not correspond to the packet, increasing, at the same time, the uncertainty about subsequent packets, and as a result *loss of synchronization occurs*
- *loss of data*, such as packet loss at the network level, which may lead to combined error effects affecting both the packet body and the packet head

JPEG2000 supports a number of protection mechanisms against those issues, protecting both the data of the codeblocks and their headers,

- *Segmentation symbols*: in this mechanism, a special symbol sequence is encoded at the end of each bitplane. If a wrong symbol sequence is decoded an error has been detected and at least the latest bitplane is corrupt.
- *Regular predictable termination*: the arithmetic encoder is terminated at the end of each coding pass using a special predictable termination. The decoder reproduces the termination and if it does not detect the same unused bits at the end, an error is detected in at least the latest coding pass.
- *Simultaneous use of both segmentation symbols and regular predictable termination*: guarantees a better response but has an impact in compression performance, as the number of excess bits needed becomes significant.
- *Re-synchronization marker*: a special synchronization marker, the Start Of Packet (SOP), precedes each packet head, with a sequence index. If a SOP marker with a correct index is not found in the decoder, an error is detected and the decoder waits for the next unaffected packet in order for the decoding to resume.
- *Relocation of packet head symbols*: usage of Packed Packet Headers, Main Header (PPM) or Packed Packet Headers, Tile-Part Header (PPT) markers for the relocation of all packet head symbols to the main image or the corresponding tile header and transmission through a channel with lower error rate.
- *Use of precincts*: the use of precincts in the encoding leads to a reduction of the spatial coverage of the packets and thereby limits the effect of errors in a particular location in the image.

Any of these mechanisms or their combinations may be employed in an error protection strategy designed by a user for the purposes of specific applications. Each of the mechanisms provides an additional level of protection introducing, though, a cost in compression efficiency of JPEG2000, as additional bits are required to enable each mechanisms. A user has to carefully study the scope of the application and the anticipated corruption, and balance the cost of introducing an appropriate mechanism with the data rate limits.

2.5.3 Brief Evaluation of JPEG2000

A typical example of a progressive decoding of a JPEG2000 compressed image up to full reconstruction (lossless) at 0.01, 0.025, 0.05, 0.1, 0.25, 2 bpp, is shown in Fig. 2.103. During this progression, the reconstructed image quality (measured as PSNR using (2.177)) increases from 27.55 dB up to 49.4 dB. It is apparent that

(a) 0.01 bpp, 27.55 dB   (b) 0.025 bpp, 30 dB   (c) 0.05 bpp, 31.9 dB

(d) 0.1 bpp, 33.1 dB   (e) 0.25 bpp, 37.2 dB   (f) 2 bpp, 49.4 dB

Fig. 2.103 Quality and data rate during decoding of a progressive-by-quality encoded image (0.01–2 bpp)

(a) level 1: 0.01 bpp   (b) level 8: 0.13 bpp   (c) level 12: 0.54 bpp

(d) level 14: 1.13 bpp   (e) level 15: 1.63 bpp   (f) level 16: 2.35 bpp

Fig. 2.104 Progressive image decoding of a maxshift mode ROI encoded image

the reconstruction that corresponds to 0.25 bpp marks the beginning of reconstructions with non-noticeable distortions, whereas at 2 bpp there is virtually no visible difference from the original.

In a typical example of ROI coding, Fig. 2.104 depicts the progressive decoding of an image, which has been encoded with layer (quality) priority and a ROI. The image was encoded using 20 quality layers and a circular ROI mask, while the coding strategy for the ROI was the *maxshift mode*. As shown in Fig. 2.104c, in which pixels of the background begin to be reconstructed, the reconstruction of the ROI has already completed. The uniform gray regions in the first two figures correspond to regions for which there were no available data (yet) for a reconstruction.

In a different example, priority was given to the progressiveness in *Position* and color *Component*, using a tiling into twelve image tiles, as shown in Fig. 2.105. In this example, the image with the same tiling was used to produce two compressed bitstreams, one with PCRL and the other with CPRL progressiveness. As shown

Fig. 2.105 Random image tiling into 12 tiles

in Fig. 2.106, the result of a gradual decoding is significantly different for the two progressive schemes, even when the samples that are considered correspond to the same compression rate (bpp). Again, the uniform gray regions in the images in Fig. 2.106 correspond to regions with no available data for reconstruction.

Numerous studies have been conducted to estimate the gain in compression efficiency attained by using the JPEG2000 standard (Skodras et al. 2001; Christopoulos et al. 2000b; Santa-Cruz and Ebrahimi 2000a, b; Santa-Cruz et al. 2000). Estimates tend towards a compression rate improvement of about 30 % compared to the conventional JPEG for the same objective image quality (Signal to Noise Ration (SNR)), while it is worth noting that the subjective quality is clearly better than that obtained by the conventional JPEG, since the distortion caused by the use of the DWT are less noticeable by human observers since they are of an entirely different nature compared to those caused by the DCT, which acts on small image blocks and introduces sharp and easily noticeable discontinuities at the block boundaries.

Figure 2.107 shows an example comparison between the two standards where the objective quality is compared for the same compression ratio. The supremacy in the image quality of JPEG2000 is apparent, although the PSNR estimate reports only a 4

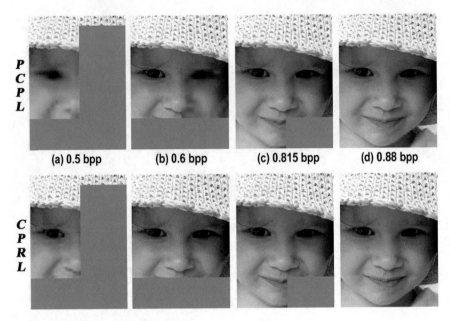

P
C
P
L

(a) 0.5 bpp   (b) 0.6 bpp   (c) 0.815 bpp   (d) 0.88 bpp

C
P
R
L

Fig. 2.106 Progressive decoding of two compressed bitstreams with different progressiveness priorities

(a)
JPEG (quality factor=10)
Size: 91666 bytes
Quality PSNR: 33.12 dB

(b)
JPEG2000 (data rate=0.2331 bpp)
Size: 91354 bytes
Quality PSNR: 37 dB

Fig. 2.107 Quality comparison between JPEG and JPEG2000 for the same compression rate

Table 2.18 Lossless compression ratios for various compression standards

Image	J2K	JPEG-LS	L-JPEG	PNG
Aerial2	1.47	1.51	1.43	1.48
Bike	1.77	1.84	1.61	1.66
Café	1.49	1.57	1.36	1.44
Chart	2.60	2.82	2.00	2.41
Cmpnd1	3.77	6.44	3.23	6.02
Target	3.76	3.66	2.59	8.70
Us	2.63	3.04	2.41	2.94
Average	*2.50*	*2.98*	*2.09*	*3.52*

dBs difference. The quantization is clearly visible in JPEG, whereas the JPEG2000 image is nearly indistinguishable from the original.

Table 2.18 summarizes the lossless compression performance results in an experiment using seven standard graylevel test images with distinct characteristics (shown in Fig. 2.108). In (Skodras et al. 2001), one of the many works on the presentation of the JPEG2000 standard, there is an extensive reference to the comparative results of the compression method in comparison with other known standards.

2.5.4 Progressive Transmission on the Web

This section presents, as an educative example on the functionalities offered by JPEG2000, the development of a Web application aiming at the progressive transmission of large images stored in multimedia databases.[23] The technology employed is based on the classic *client-server architecture* and *TCP/IP communication* through *sockets*, in which at the request of a client, the image database server can easily respond with partial image data deriving from single image files. In this method, a significant reduction in the size of the data transmitted through the network is accomplished, since the server transmits only the portion of the encoded image that corresponds to the image data that are of the client's interest. In addition, since the client employs caching mechanisms, there is a possibility of using existing data in the client (previously transmitted), so that additional network traffic reductions are possible. This was among the first works towards the development of systems supporting progressive transmission of images using the JPEG2000 standard and Web browser plug-ins technology (Politou et al. 2004). Although since that work other

---

[23]It is noted that this section is provided here purely to illustrate the potential in adopting the JPEG2000 coding strategy and to highlight an example of successful engineering. The system that is described was implemented in a *pre-HTML5 era*, so the need for all those complementing technologies for client-server communication was imperative. The value of this section is purely illustrative and educative.

Fig. 2.108 Graylevel image set used for a comparative lossless compression study

methods and approaches have already been proposed, they either did not rely on a common interface such as a Web browsers (Deshpande and Zeng 2001), or they were not based on a generally accepted communication protocol (Taubman 2002a).

One of the serious technical challenges on the Internet—that is constantly under study and development—is the limited bandwidth, the limitations on data transmission speed. The challenge of having a limited bandwidth, in conjunction with large storage requirements were, after all, the main reasons for the development of different compression methods. A typical example of a compression method with wide acceptance and dissemination on the Internet is JPEG (Pennebaker and Mitchell 1993). As known, using JPEG (for example, by running the classic cjpeg codec Independent JPEG Group 2000) with a quality factor of 15 on a typical digital photograph of say 24 MPixels (6,000 × 4,000 pixels), the transmission time may be significantly reduced from minutes to some seconds, maintaining a satisfactory digital image quality (around 30 dB PSNR). The emergence of JPEG2000 with a better compression efficiency and a set of enhanced functionalities widens the possibilities for higher quality applications.

One typical application of multimedia databases, are the databases in the cultural heritage domain, in which the content includes all kinds of digital media. Among those media, one with a significant importance and large volumes is the image. An image provides information about the shape, texture, color and other details, properties that often can not be described (at least easily) by simple text. Common problem in cultural databases is the difficulty in managing and publishing of information that is stored in large images. Significant work on this subject has been reported by many research groups, and has been a subject of research and development activities of the author of this treatise for many years.

Among the various works one particular will be the subject of this section, the "Ark of Refugee Heirloom" (Politou et al. 2002), which recorded the cultural identity of a specific part of a community. Content from that database has been used for a pilot implementation of a large volume image data transmission. The goal was to explore and propose a new, interactive and economic (in terms of data transmission), way of accessing multimedia data. The basic idea was to implement a progressive image transmission method for the Web, and to ensure an efficient way for data exploration. After extensive experimentation on various compression methods and the way an average user navigates in cultural content online, this investigation resulted in the following scenario for an efficient interactive environment,

- A visitor that enters the home page receives a list of small versions of all (or a part of) the images in the database, the 'thumbnails'
- The visitor selects one of the thumbnail images for further study and receives a medium sized image
- If the visitor wishes to examine the image in full resolution, the image is transmitted in two steps using quality refinement
- The transmission in all steps is differential, so as to transmit only the additional information that is required to achieve the requested quality and resolution at a specific step.
- The connection between the user (client) and the server is controlled by sockets, while the JPEG2000 decoding and presentation of the images is controlled by a web browser plug-in

In this scenario, each time the user requests additional information either to increase the resolution or the quality of an image, the required information is extracted and transmitted from the same single image file. This implies that the image encoding should have been done in such a way as to ensure the possibility of extracting various resolutions from the same file.

In order to proceed further with the description of the implementation, a description of the main pieces that complete the overall puzzle will be defined and put together. These *pieces* are the *syntax of the JPEG2000 file format*, the *client-server technology* and the *browser plug-ins*. The *puzzle* is a system to provide efficient progressive transmission under a Web-based image database navigation scenario.

First, the syntax in the JPEG2000 encoded file should be examined. This syntax defines all the basic properties of the encoded image, such as dimensions, the size of image tiles, the number and the rates of sampling of the spectral components, as

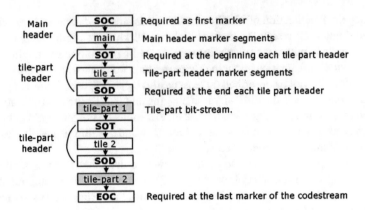

Fig. 2.109 Basic file structure of a JPEG2000 encoded image

well as the parameters of quantization and coding, the size of codeblocks and the transform that is applied. In addition, it includes information about the number of quality layers, resolutions and progressiveness priority, ROI encoding, as well as error correction settings. Most parameters can be selected at a tile level (ISO-IEC 2000a; Moccagata et al. 2000). In the simplest case, a file that represents an encoded JPEG2000 image has a structure that starts with a main header that is followed by a series of encoded information blocks as graphically shown in Fig. 2.109. The extension of this file, according to the standard (ISO-IEC 2000a) is jpc (**jp**eg **c**odestream). The main header contains global information necessary for the decoding of the whole file. Each image tile in the codestream consists of a header, which contains informations relating individually to the tile to which they belong. Each group of packets includes a sequence of encoded data (ISO-IEC 2000a; Moccagata et al. 2000). A progressive encoding would allow to be able to improve the quality, resolution, size and colors as more and more data are gradually being decoded from an encoded file that arrives at the receiver. The type or priority in progression is determined by the way in which information packets are arranged in the codestream. It is possible to determine progressiveness both at a global and at a tile-based level, which is expressed by the way tile-parts are being arranged. Each part has its own header (ISO-IEC 2000a; Moccagata et al. 2000). As mentioned in the previous paragraphs, the standard defined five different priorities for progressiveness.

In addition, the client-server technology is based on creating sockets to achieve communication between two remote computers. A socket operates as a phone. It is the terminal of a bidirectional communication channel. By connecting two sockets it is possible to transmit data to processes, even when operating on different computers, in a manner similar to the speech transmission in phones. In general, a software that is written to use the Transport Control Protocol (TCP) is developed using the client-server model, in which when two connected devices use TCP for data exchange, one of them assumes the role of the client and the other the role of the server. The application on the client is the one which starts what is called an *active open*. It creates

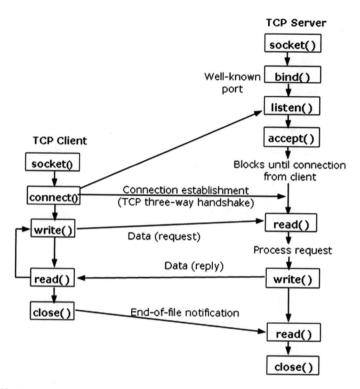

Fig. 2.110 Processes in a socket connection

a socket and tries to connect to a server. At the other end, the server application creates a socket that 'listens', waits, for incoming connections from clients, by performing what is called a *passive open*, as shown in Fig. 2.110. When the client initiates a connection, the server is informed that a process is trying to connect with it. By accepting the connection, the server completes the creation of a virtual circuit, which is nothing more than a logical communication path between two programs. It is important to note that the attempt to create a connection always creates a new socket; the original socket remains unchanged so that the server remains open to listen for new connections. When the server does not wish or is not required to await new connections then it may terminate the original passive socket (Comer and Stevens 2000).

Finally, a plug-in (a complementary software library) is a separate semi-autonomous part of software that behaves as part of a software, which it complements with new features and functionalities. In particular, a plug-in for a Web browser is a library that is activated and only works within a Web browser on the Web. The way the browser interacts with the plug-in is described by a set of programming tools, called the plug-in Application Programming Interface (API) (Netscape 1997). With the plug-in API it is possible to create dynamic plug-ins, which

- register one or more MIME types
- draw on a particular position on the canvas of the Web browser window
- receive and manage user input (eg. keyboard and mouse)
- receive and send data from/to the Web via URLs
- add hyperlinks or links, in general, to point to URLs

When a user opens a Web page that contains data types which activate a plug-in, then the Web browser responds by performing a number of predefined actions

- it searches for a registered plug-in that matches the specific requested MIME type
- it loads the dynamic plug-in into memory
- it activates the plug-in
- it creates an instance (copy) of the plug-in in memory
- it removes an instance of the plug-in from memory if not required
- it disables and removes the plug-in from memory when the removal of the last instance is requested

With the pieces of the puzzle defined, the presentation of the implementation begins with a *server application* and any additional *tools needed for image preparation* and concludes with a *client application*. The server application is essentially a service that runs on a Web server and has access to the images to be transmitted. The images were previously encoded using lossless JPEG2000, in an appropriate manner so as to ensure the progressive transmission in accordance with the predefined interactive navigation scenario. A special application was developed to properly prepare the images. On the other side of the connection, the client application is a browser plug-in, installed on the user's computer in order to enable processing of the user interactions and to decode and display the images requested by the user.

To be able to select multiple meaningful segments from a single encoded image file, a specific way of coding must be adopted. The data of the encoded bitstream must be arranged in code segments, so that the sequential 'addition' of arriving code parts to produce an increasing resolution (image size) as shown in Fig. 2.111. The shading in the various parts of this representation denotes the grouping of data that should be imposed for a progressive by resolution coding. Under the specific implementation scenario, and after a study on the requirements and practices in online multimedia databases, specific resolutions of the images were selected in this project to, somehow, optimize the experience of the users (based mainly on a better use of the network bandwidth). The final selection of the intermediate resolutions was based on empirical and experimental data from previous related activities (Cultural & Educational Technology Institute 2000a; Democritus University of Thrace 2000),

- 64×64 pixels for the lowest resolution (thumbnail size)
- 256×256 pixels for the 'medium' resolution (preview size)
- the full-sized original image

To achieve encoding of images so that they contain these specific resolutions, the number of levels of wavelet decompositions should be defined in advance. Specifically, to achieve the lowest resolution of 64×64 pixels, the number of wavelet decompositions n is estimated as

Fig. 2.111 Representation of code segments in a JPEG2000 compression bitstream: **a** the final bitstream and **b** the corresponding parts in the transform domain

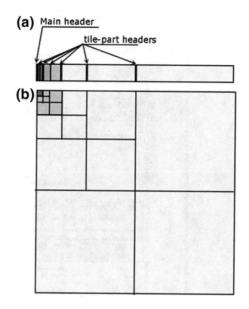

$$\left.\begin{array}{l} n_w \leq \log_2 \frac{w}{w'} \\ n_h \leq \log_2 \frac{h}{h'} \end{array}\right\} \Rightarrow n = \lceil max\{n_w, n_h\}\rceil \tag{2.189}$$

where w and h are the width and height of the original image, w' and h' is the required width and height respectively, n_w and n_h the number of divisions by two in width and height for the production of the desired size and $\lceil\ \rceil$ is the ceiling operator. In this particular case the value for both w' and h' was 64. This way the images are encoded with $n+1$ levels of resolutions (there will be $n+1$ Start Of Tile (SOT) markers in the code-stream), ensuring that the lower resolution is of 64×64 pixels at maximum. A representation of the structure is shown in Fig. 2.112. Encoding was performed via the *kakadu software* (Taubman 2000a), the code of which was incorporated into

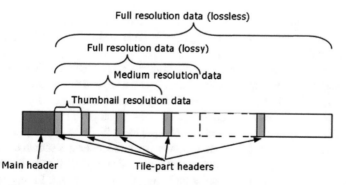

Fig. 2.112 Representation of the JPEG2000 bitstream for images with a predefined resolutions

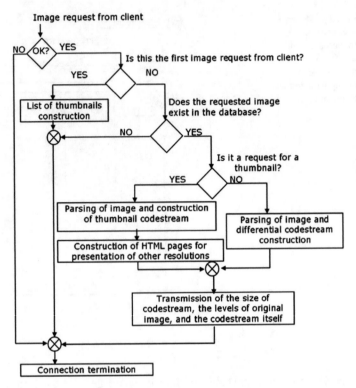

Fig. 2.113 Flowchart of a server response for progressive differential information transmission

the application created. The corresponding command line to compress the images
according to the scenario adopted is

```
kdu_compress -rate - Corder=RLCP ORGparts=R ORGgen_plt=yes Clayers=3
        Creversible=yes Clevels=n -i input_image -i output_code-stream
```

After imposing this encoding on the images of the database, the images are ready to
be used by the server to achieve progressive transmission on the Web according to
the application scenario.

As the server should be able to handle multiple and concurrent client connections
and requests, the server application was designed to operate in *multithreaded mode*.
It was equipped with the ability to send images of any resolution by analyzing the
syntax in the file of each encoded image. At the final stage of each processing cycle,
the application creates the required HTML pages to be delivered to the client. The
way in which the server application responds to a client request is shown in Fig. 2.113.
Each time a client requests the lowest resolution of an image, the application sends
the corresponding code-stream parts including the main header. In any other case
(the client has already received one of the resolutions an the image) in which the

client requests further image information, then the application sends the differential information by extracting the appropriate parts of the code-stream without resending previously sent information. It is in the client's application responsibility to manage the differential information for proper reconstruction of the requested image. To automate the process, the HTML pages in which the images are being presented, are created at the server side, always one step ahead of the corresponding request of the connected client. The process of creating low-resolution images on the server is based on marker operators and syntax control on the code-stream of an image. Whenever reading data from the main header of the file, the data are copied to a temporary storage in memory while replacing the values for the markers *XRsiz*, *YRsiz SIZ* (0xFF51) and the wavelet transform marker *COD* (0xFF52), with new values for proper decoding of the image corresponding to the requested resolution. Whenever reading data that correspond to the content of a tile, then they are copied in a new temporary file. At the end of the process, all temporary data are synthesized into a single stream and are sent to the client.

The client application is essentially a Web browser plug-in. The plug-in receives the encoded data from the server and decodes and presents the data to the user within the Web browser. The syntax of the HTML *EMBED* tag, with which objects are embedded to integrate a particular plug-in to an HTML page, is being employed

```
<EMBED
src="name.jpc"
onclick="(javascript routines)"
type="MIME type"
width="width"
height="height"
host="Server IP"
size="Requested resolution">
```

where *src*, *onclick*, *type*, *width* and *height* are parameters and events that determine typical properties of *EMBED* tags, while *host* and *size* are parameters of the plug-in,

- the parameter *src* contains the image file name being requested
- the *onclick* event is a general HTML event that determines how the Web browser responds to the use of the mouse buttons within a Web page
- the parameter *type* specifies the MIME type of the data, determined during the development of a plug-in and in the specific application it was defined as *application/x-Netscape-jpc*[24]
- the parameters *width* and *height* define the dimensions of the display area (within the Web browser), which the plug-in will span, and must be greater than or equal to the image to be displayed

---

[24]Remember this was an implementation in an early pre-HTML5 era.

- the *host* parameter specifies the IP address of the server with which a connection is being made for receiving data
- the *size* parameter specifies the desired resolution of the image for display

Once the Web browser detects the MIME type in the EMBED tag, the corresponding plug-in is called and control passes to the plug-in. Specifically, when a user opens a Web browser, all available plug-ins are loaded into memory. The appropriate plug-in is called whenever the Web browser faces content that requests its activation. If there is no other Web browser windows with the selected plug-in activated, then the plug-in is activated and binds global memory for the management of shared data. Then a new instance (copy) of the plug-in is created in the part of the memory controlled by the Web browser, to which the input parameters are eventually transferred. These parameters are no other than the ones specified in the *EMBED* tag that triggered the plug-in. By using these parameters, the Web browser binds an area on the canvas of the current window with the output of the plug-in. The dimensions of this area are defined by the parameters *width* and *height*. Parameters *src*, *host* and *size* are used subsequently according to the requests to the server. After the process of the received data, the client decodes and presents the data in the reserved area of the plug-in. When the user chooses to close the page (the window), the Web browser removes the active instance of the plug-in from memory, but does not completely releases the memory of the global data. This is only done when all Web browser windows with the active specific plug-in are closed. The entire process is represented graphically in Fig. 2.114.

According to the adopted case study scenario, the client-server connection starts with a user verification and continues with the client triggering the server to create the thumbnail images Web page. The page is sent to the client and control passes to the client, waiting user input. When the user requests information for an image then a series of processes is activated, which is shown in Fig. 2.115.

- if the image requested by the user already exists on the client (cache memory of the client Web browser) then no request is sent to the server and the client simply decodes and presents the stored data
- if the image requested by the user has lower resolutions in the cache memory of the client's Web browser, then the client sends a request to the server for differential transmission of information and the data obtained are added to those already existing for the creation of the next higher resolution; at the same time, the main header, which already exists on the client is updated to reflect the change in the resolution
- if the image is not on the client at any resolution (it is the first time the user requests data about this image), the lowest resolution is being received and presented

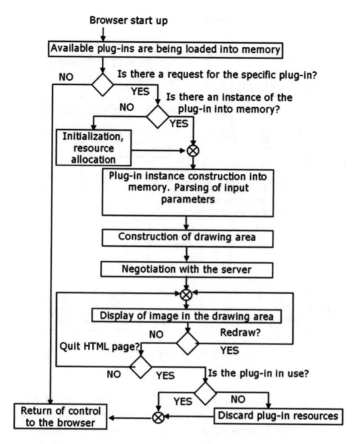

Fig. 2.114   Flowgram of the operation of a plug-in on a Web browser client

The described system has been put to the test by using a real multimedia data-base of cultural heritage, the one created for *The Ark of Refugee Heirloom* project (Cultural & Educational Technology Institute 2000a). A presentation of the results of this case study are in (Cultural & Educational Technology Institute 2000b) and are briefly presented here. A first step when a user is visiting *the Ark* for the first time, is the user authentication as shown in Fig. 2.116a. At the same time the server prepares the introductory page to present the list of images in low resolution (thumbnails). After the user authentication, the prepared page of thumbnails is sent from the server and displayed to the client as shown in Fig. 2.116b. If the user chooses to request further information regarding one of the images, the browser plug-in receives the additional information from the server and decodes and displays the medium reso-lution, accompanied by a link to the higher resolution as shown in Fig. 2.116c. If the

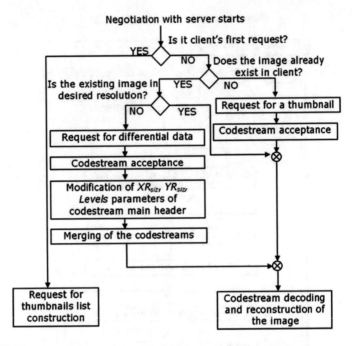

Fig. 2.115 Flowgram of the processes at the Web browser plug-in

user requests for even higher resolution, additional information that corresponds to the maximum resolution and a reduced quality is received and presented, as shown in Fig. 2.116d, providing, at the same time, a link to the full-resolution. At the final step, the user may request the full resolution and quality image (original image), as shown in Fig. 2.116e.

Table 2.19 provides comparative compression and transmission results for the image 'watch' (Fig. 2.116), which is a characteristic image from the database of the Ark in all resolutions.[25] Of interest in these results is the one that corresponds to the transmission of the maximum-resolution-low-quality (at 1 bpp), in which there is substantial differential data compression and transmission gain, while with the received image quality the user can have a *satisfactory* representation of the data. It is also worth noting that the JPEG2000 compression experiments at 1 bpp throughout the whole image set of the database resulted in an average quality of decoded image of about 35 dB PSNR, that backs the claim that it was a good selection for that progression step.

---

[25]Transmission times are reported for a noiseless transmission channel in a guaranteed constant 1 Mbps bandwidth.

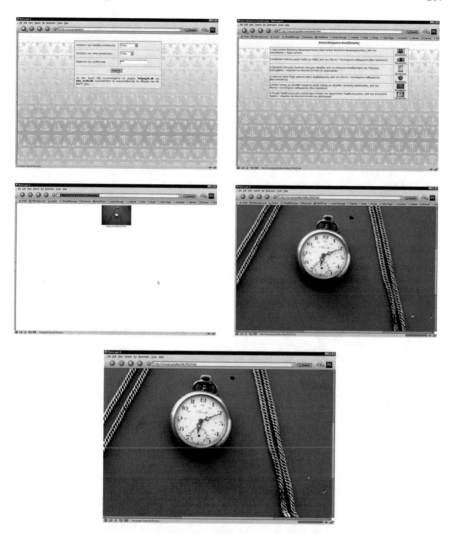

Fig. 2.116 Screenshots of the step-by-step interaction within the plug-in on the client

It is apparent that the JPEG2000 compression leads to image databases in which images can be compressed into multiple resolutions and quality layers within a single file each. These files consist of embedded data that can be used for the recovery and reconstruction of images in multiple ways. Some of the relevant features include progressive presentation in resolution and quality, zooming in a specific area, use of a dynamic Region Of Interest (ROI) and copyright control and management.

Table 2.19 Comparative compression and transmission results for four resolutions of the image 'Watch'

Image size (pixels)	Uncompressed data (Kbytes)	Compressed data (Kbytes)	Data rate (bpp)	Transmission data (Kbytes)	Data transmission	Compression ratio	Transmission times (s)
Thumbnail 43 × 28	3.5	2.27	16.21	2.27	15.57	1.54:1	0.02
Medium 169 × 110	54.5	26.59	12.20	24.32	10.71	2.05:1	0.19
Lossy-full 2696 × 1756	13,870	578	1.04	553.68	0.96	24.00:1	4.33
Lossless 2696 × 1756	13,870	4,152.29	7.48	3,598.61	6.23	3.34:1	28.11

References

Abramson, N. (1963). *Information theory and coding*. New York: McGraw-Hill.

Albanesi, M., & Bertoluzza, S. (1995). Human vision model and wavelets for high-quality image compression. In *Proceedings of the 5th International Conference in Image Processing and its Applications* (Vol. 410, pp. 311–315).

Bracewell, R. M. (1983). Discrete Hartley transform. *Journal of the Optical Society of America*, *73*(12), 1832–1835.

Christopoulos, C., Ebrahimi, T., & Lee, S. U. (2002). JPEG2000 Special Issue. *Elsevier Signal Processing: Image Communication17*.

Christopoulos, C., Askelof, J., & Larsson, M. (2000a). Efficient methods for encoding regions of interest in the upcoming JPEG2000 still image compression standard. *IEEE Signal Processing Letters*, *7*(9), 247–249.

Christopoulos, C., Skodras, A., & Ebrahimi, T. (2000b). The JPEG 2000 still image coding system: An overview. *IEEE Transactions on Consumer Electronics*, *46*(4), 1103–1127.

Comer, D. E., & Stevens, D. L. (2000). *Internetworking with TCP/IP: Client-Server programming and applications*. Prentice Hall. ASIN: B00LKKVXQ4.

Cover, T. M., & Thomas, J. A. (2006). *Elements of information theory* (2nd edn). John Wiley & Sons, ISBN: 978-0-471-24195-9.

Cultural & Educational Technology Institute, Greece. (2000a). *Ark of Refugee Heirloom—A cultural heritage database—Online*.

Cultural & Educational Technology Institute, Greece. (2000b). *Ark of Refugee Heirloom JPEG2000 prototype*.

Democritus University of Thrace, Greece. (2000). *Thracian Electronic Thesaurus*.

Deshpande, S., & Zeng, W. (2001). Scalable streaming of JPEG2000 images using hypertext transfer protocol. In *Proceedings of ACM* (pp. 72–281).

Elias, P. (1975). Universal codeword sets and representations of the integers. *IEEE Transactions on Information Theory*, *21*(2), 194–203.

Fano, R. M. (1949). *The transmission of information*. Technical report. 65. Research Laboratory of Electronics at MIT.

Fano, R. M. (1952). *Class notes for Transmission of Information, course 6.574*. Technical report. Cambridge, MA: MIT

Fisher, R. A. (1922). On the mathematical foundations of theoretical statistics. *Philosophical Transactions of the Royal Society, London A*, *222*, 309–368.

Golomb, S. (1966). Run-length encodings. *IEEE Transactions on Information Theory*, *12*, 399–401.

Gonzalez, R. C., & Woods, R. E. (1992). *Digital image processing* (3rd edn). Prentice Hall, ISBN: 978-0201508031.

Gray, R., & Neuhoff, D. (1998). Quantization. *IEEE Transactions on Information Theory, 44*(6).

Haar, A. (1910). Zur Theorie der orthogonalen Funktionensysteme. *Mathematische Annalen, 69*(3), 331–371.

Hartley, R. V. L. (1928). Transmission of information. *Bell System Technical Journal, 7*(3), 535–563.

Hartley, R. V. L. (1942). A more symmetrical Fourier analysis applied to transmission problems. *Proceeding IRE, 30*, 144–150.

Huffman, D. (1952). A method for the construction of minimum redundancy codes. *Proceedings IRE, 40*, 1098–1101.

Independent JPEG Group, IJG. (2000). *JPEG reference software.*

ISO-IEC. (2000a). *Information technology—JPEG 2000 image coding system—Part 1: Core coding system, ISO/IEC International Standard 15444–1.* ISO/IEC: Technical report.

ISO-IEC-CCITT. (1993b). *JPEG: Information Technology—Digital compression and coding of continuous-tone still images—requirements and guidelines, ISO/IEC International Standard, CCITT Recommendation T.81.* Technical report. ISO/IEC/CCITT.

ISO-IEC-ITU. (2000). *JBIG2, ISO/CEI International Standard 14492 and ITU-T Recommendation T.88.* Technical report. ISO/IEC/ITU.

Jain, K. A. (1988). *Fundamentals of digital image processing.* New Jersey: Prentice-Hall.

Jones, P., Daly, S., Gaborski, R., & Rabbani, M. (1995). Comparative study of wavelet and DCT decompositions with equivalent quantization and encoding strategies for medical images. In *Proceedings of SPIE* (Vol. 2431, pp. 571–582).

Kang, L. W., & Leou, J. J. (2003). A new error resilient coding scheme for JPEG image transmission based on data embedding and vector quantization. *Proceedings of IEEE International Symposium on Circuits and Systems—ISCAS2003* (Vol. 2, pp. 532–535).

Kullback, S., & Leibler, R. A. (1951). On information and sufficiency. *Annals of Mathematical Statistics, 22*, 79–86.

Lehmann, E. L., & Scheffe, H. (1950). Completeness, similar regions and unbiased estimation. *Sankhya, 10*, 305–340.

Liang, J., & Talluri, R. (1999). Tools for robust image and video coding in JPEG2000 and MPEG-4 standards. In *Proceedings of the SPIE Visual Communications and Image Processing Conference* (Vol. 3653, pp. 40–51).

Linde, Y., Buzo, A., & Gray, R. M. (1980). An algorithm for vector quantizer design. *IEEE Transactions on Communications, 28*(1), 84–95.

Lloyd, S. (1982). Least squares quantization in PCM. *IEEE Transactions on Information Theory, 28*(2), 129–137.

McMillan, B. (1956). Two inequalities implied by unique decipherability. *IEEE Transaction of Information Theory, IT-2,* 115–116.

Moccagata, I., Sodagar, S., Liang, J., & Chen, H. (2000). Error resilient coding in JPEG2000 and MPEG-4. *IEEE Journal of Selected Areas in Communications (JSAC), 18*(6), 899–914.

Netscape. (1997). *Plug-in guide.*

Ono, F., Kino, S., Yoshida, M., & Kimura, T. (1989). Bi-level image coding with MELCODE—comparison of block type code and arithmetic type code. In *Proceedings of IEEE Global Telecommunications Conference (GLOBECOM)* (pp. 255–260).

O'Rourke, T., & Stevenson, R. (1995). Human visual system based wavelet decomposition for image compression. *Journal of Visual Communication and Image Representation, 6*, 109–131.

Pasco, R. C. (1976). *Source coding algorithms for fast data compression.* Ph.D. thesis, Stanford University.

Pennebaker, W. B., & Mitchell, J. L. (1993). *JPEG still image compression standard.* New York: Springer.

Pennebaker, W. B., Mitchell, J. L., Langdon, G., & Arps, R. B. (1988). An overview of the basic principles of the Q-coder adaptive binary arithmetic coder. *IBM Journal of Research and Development, 32*(6), 717–726.

Politou, E., Tsevremes, I., Tsompanopoulos, A., Pavlidis, G., Kazakis, A., & Chamzas, C. (2002). Ark of refugee heirloom—A cultural heritage database. In *EVA 2002: Conference of Electronic Imaging and the Visual Arts* (pp. 25–29).

Politou, E. A., Pavlidis, G. P., & Chamzas, C. (2004). JPEG2000 and the dissemination of cultural heritage databases over the Internet. *IEEE Transactions on Image Processing, 13*(3), 293–301.

Pratt, W. (1991). *Digital image processing* (2nd edn). Wiley-Interscience Publication. ISBN: 0-471-85766-1.

Rabbani, M., & Cruz, D. Santa. (2001). The JPEG2000 still-image compression standard, tutorial session. In *IEEE International Conference on Image Processing—ICIP 2001.*

Rabbani, M., & Jones, P. W. (1991a). *Digital image compression techniques* (Vol. TT7). SPIE-Tutorial Texts in Optical Engineering. ISBN: 978-0819406484.

Rabbani, M., & Joshi, R. (2002). An overview of the JPEG2000 still image compression standard. *Signal Processing: Image Communication, 17*(1).

Rao, R. M., & Bopardikar, A. S. (1998). *Wavelet transforms: Introduction to theory and applications.* Prentice Hall. ASIN: B01A65JU7W.

Rissanen, J. (1976). Generalized kraft inequality and arithmetic coding of strings. *IBM Journal of Research and Development.*

Rissanen, J. J., & Langdon, G. G. (1979). Arithmetic coding. *IBM Journal of Resources and Development, 23*(2), 146–162.

Rubin, F. (1979). Arithmetic stream coding using fixed precision registers. *IEEE Transactions on Information Theory, 25*(6), 672–675.

Said, A. (2004). Introduction to arithmetic coding theory and practice. Technical report. Hewlett-Packard Laboratories Report, HPL-2004-76.

Said, A., & Pearlman, W. A. (1996). A new fast and efficient image codec based on set partitioning in hierarchical trees. *IEEE Transaction on Circuits Systems and Video Technology, 6*(3), 243–250.

Santa-Cruz, D., & Ebrahimi, T. (2000a). An analytical study of JPEG 2000 functionalities. In *Proceedings of IEEE International Conference on Image Processing—ICIP 2000.*

Santa-Cruz, D., & Ebrahimi, T. (2000b). A study of JPEG 2000 still image coding versus other standards. In *Proceedings of X European Signal Processing Conference* (Vol. 2, pp. 673–676).

Santa-Cruz, D., Ebrahimi, T., Askelof, J., Karsson, M., & Christopoulos, C. A. (2000). JPEG2000 still image coding versus other standards. In *Proceedings of SPIE, 45th annual meeting, Applications of Digital Image Processing XXIII* (Vol. 4115, pp. 446–454).

Sayood, K. (1996). *Introduction to data compression.* Morgan Kaufmann. ISBN: 978-1558603462.

Shannon, C. E. (1948). A mathematical theory of communication. *Bell Systems Technology Journal, 27*(379–423), 623–656.

Shapiro, J. M. (1993). Embedded image coding using zero trees of wavelet coefficients. *IEEE Transactions on Signal Processing, 41*(12), 3445–3462.

Skodras, A., Christopoulos, C., & Ebrahimi, T. (2001). The JPEG 2000 still image compression standard. *IEEE Signal Processing Magazine,* 36–58.

Sullivan, G. (1996). Efficient scalar quantization of exponential and Laplacian variables. *IEEE Transactions of Information Theory, 42*(5), 1365–1374.

Tanaka, H., & Leon-Garcia, A. (1982). Efficient run-length encodings. *IEEE Transactions on Information Theory, 28*(November), 880–890.

Taubman, D. S. (2000a). *Kakadu Software—A comprehensive framework for JPEG2000.*

Taubman, D. S. (2000b). High performance scalable image compression with EBCOT. *IEEE Transaction on Image Processing, 9*(7), 1158–1170.

Taubman, D. S. (2002a). Remote browsing of JPEG2000 images. In *IEEE International conference on Image Processing—ICIP2002* (pp. 22–25).

Taubman, D. S., & Marcellin, M. W. (2002b). *JPEG2000 image compression fundamentals, standards and practice.* Kluwer Academic Publishers. ASIN: B011DB6NGY.

Taubman, D. S., Ordentlich, E., Weinberger, M. J., & Seroussi, G. (2002c). Embedded block coding in JPEG2000. *Elsevier Signal Processing: Image Communication, 17*(1), 49–72.

Wallace, G. (1991). The JPEG still picture compression standard. *Communications of the ACM*, *34*(4), 30–44.

Watson, A. B., & Poirson, A. (1986). Separable two-dimensional discrete Hartley transform. *Journal of the Optical Society of America A*, *3*(12), 2001–2004.

Watson, A. B., Yang, G. Y., Solomon, J. A., & Villasenor, J. (1997). Visibility of wavelet quantization noise. *IEEE Transactions on Image Processing*, *6*(8), 1164–1175.

Witten, I. H., Neal, R. M., & Cleary, K. G. (1987). Arithmetic coding for data compression. *Communications of the ACM*, *30*(6), 520–540.

Ziv, J., & Lempel, A. (1978). Compression of individual sequences via variable-rate coding. *IEEE Transactions on Information Theory*, *24*, 530–536.

Chapter 3
Segmentation of Digital Images

—We bring the interpretation process into awareness through tricks. First, we degrade the image, making interpretation difficult. Second, we provide competing organizations, making possible several conflicting interpretations of the same image. Third, we provide organization without meaning to see how past experience affects the process.

Peter Lindsay and Donald Norman

3.1 Introduction

Most of the world's cultural and scientific heritage is still available in printed form. In the past decades a number of actions and projects targeted the digitization of the printed material and the disclosure of the wealth of this content through Web technologies. One of the obstacles in this distribution is the lack of an efficient method of managing the large volume of data involved in archives of scanned documents (Bottou et al. 1998), although Portable Document Format (PDF) seems to have conquered this realm. Many libraries and cultural institutions have already digitized and still digitize their collections, and major national and European research programs have already faced the issue of the digitization of documents of significant cultural value and the distribution through the Web. Meanwhile, companies, organizations and governments recognize the importance of the digital document towards a better organization and management, as well as a significant reduction in bureaucracy, while improving the services provided through the Web.

Color digitized documents or, more generally, mixed content images or MRC, differ from digital images of the natural world (referenced as *natural images* or *continuous tone images*) in that they contain—usually clearly defined—regions with distinct features, such as text, graphics and images of continuous tones. The compression of MRC was recognized as a special case in the field of digital image compression in the 1970s, when the growing use of fax machines demonstrated the capacity to digitize and transmit documents.

It is known that compression is directly related to the information it compresses and, therefore, there is now a plethora of compression techniques adapted to the

© Springer Nature Singapore Pte Ltd. 2017
G. Pavlidis, *Mixed Raster Content*, Signals and Communication Technology,
DOI 10.1007/978-981-10-2830-4_3

different types of images; there is compression for black and white (binary) images, for images of continuous tones and graphics. One of the key features that can be identified in MRC is that, in general, the text requires[1] a high spatial (dpi) but also a low color resolution (bpp), while, on the other hand, continuous tone images usually require higher color resolution (bpp) than spatial (dpi). Furthermore, an effective compression method for MRC must be adaptive to be able to exploit the inner nature of information redundancy for each case. Traditional compression methods such as the JPEG and the newer JPEG2000 (at least in their basic definition), presuppose that the image to be compressed is spatially homogeneous, and thus lead to low compression performance on MRC. The case of MRC requires, therefore, a different approach. An MRC compression scheme must be adjusted to the local content, to perceive the variations in the type of information handled. Two approaches may be identifying as possibilities for tackling with this challenge;

- one approach would be to *create an adaptive compression system*, which can identify the characteristics of the image and change its behavior during encoding;
- another approach would be to separate the image into layers, grouping similar characteristics, and using *a different compression method for each of the layers*, adapted to the characteristics of the respective layer. Thus, current compression methods such as JBIG (ISO-IEC-ITU 1993), JBIG2 (ISO-IEC-ITU 2000), JPEG (ISO-IEC-CCITT 1993a; Pennebaker and Mitchell 1993; Wallace 1991) and JPEG2000 (ISO-IEC 2000a; Taubman and Marcellin 2002; Christopoulos et al. 2002), can be applied without any adaptations or changes on the corresponding layers resulting in high compression efficiency.

From another perspective, the classification of ways to address efficient MRC compression reveals two more general approaches, that is,

- sequential or parallel processing of rectangular non-overlapping parts of an image, referenced as *block-based methods*;
- application of global approaches by creating layers of similar characteristics, referenced as *layer-based methods*.

Block-based methods, such as the ones described in Murata (1996), Harrington and Klassen (1997), Konstantinides and Tretter (1998), Ramos and DeQueiroz (1999), classify non-overlapping blocks in different classes and compress each class according to its characteristics. On the other hand, *layer-based methods* (Bottou et al. 1998; Buckley et al. 1997; Huang et al. 1998; DeQueiroz et al. 1999) divide MRC into different layers, such as the foreground and the background, and then encode each layer independently. Most methods of this class are in agreement with the *directive* of creating three layers (*foreground-mask-background*) of *ITU T.44 for MRC*. According to this partition into layers, the *foreground* contains the text and graphics while the background contains the continuous tone images and the rest of the document. The *mask* is usually a black and white binary image, which defines, in essence, which pixels of the MRC are the foreground and how they should be mixed for the final

---

[1]The term 'requires' is used to denote the requirements relating to the HVS modeling.

reconstruction of the decoded image. Despite the apparent differentiation of the two classes of methods, the approaches are, actually, closely related. Using additional information, a method of the first class can be transformed to a method of the second class. There can also be a combination of these methods to achieve better results in compression and reconstructed image quality.

Apparently, the performance of an MRC compression system is inextricably linked to an adopted *segmentation method*, which would lead, ultimately, to the desired partition into independent blocks, parts or layers. A *good* segmentation can improve both the compression ratio and the image quality (by leading to a reduced distortion because of the adapted compression method used). On the other hand, a *poor* segmentation may lead to increased reconstruction distortion, and in such a form that is easily identifiable by the HVS.

A large family of algorithms have been proposed for such a segmentation using features extracted from the DCT coefficients. For example Murata (1996) proposed a method based on the absolute values of the coefficients, while in Konstantinides and Tretter (1998) a measure of the coefficients' *activity* is calculated, on which the selection of a quantization table for JPEG compression is based. Other methods are based on the features extracted directly from the MRC. The document compression system Deja Vu Image Compression (DjVu) (Bottou et al. 1998) implements a scaled binary clustering algorithm to separate the foreground from the background. Huang et al. (1998) proposes an identification of text and graphics in scanned images of bank checks using morphological filters and global thresholding. Ramos and DeQueiroz (1999) proposes a measure of the *activity* within regions, for the classification of each region in (a) regions with edges, (b) smooth regions and (c) detail regions. In Cheng et al. (1997, 2001), Cheng and Bouman (1998) an MRC compression method is being described, belonging to the family of layer-based techniques, where the segmentation is optimized by estimating the optimal rate-distortion ratio provided that a particular codec has already been selected and to which the method tries to adapt. The authors of these papers are taking a 'relaxed' compatibility with the ITU T.44 directives, introducing additional and differentiated layers in the segmenting. The methods achieve very good results and even lead to improved quality compared to methods such as DjVu, JPEG and SPIHT (Said and Pearlman 1996).

Since, digital image segmentation is identified as of paramount importance in MRC compression, this chapter introduces to the field of segmentation and provides a brief description of known segmentation techniques and their applications.

3.2 Edges in Images and Edge Detection

One of the main features of the content of an image are the *edges*. The edges in an image collect substantially most of the information in the image—significance being defined in terms of the importance of edges in the perception of content in the context of the HVS. In the literature relating with the challenges of pattern recognition and segmentation, it is difficult to find an approach that disregards, either explicitly or

Fig. 3.1 Gradient and
direction of an edge

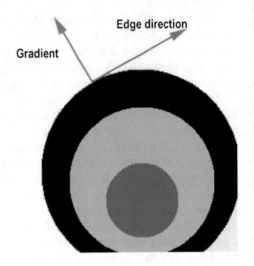

implicitly, the edges. Studies in modeling the HVS have highlighted the value of the edges in object segmentation and recognition by the visual perception, as pointed out by the literature in Chap. 1. *As edges in images can be defined any consistent pixel sequences, in which the luminosity or the chromaticity exhibit a strong relation on one direction and a strong differentiation in the normal direction.* An edge is a local property of a specific set of pixels and is estimated by the changes of the image in the neighborhood of the set of pixels at hand. An edge can be described by a vector of two components: the measure of a gradient and an orientation (angle). The direction of the gradient gives the direction of the maximum change (Fig. 3.1). The edges are often used in image analysis for identifying boundaries of regions and objects. The boundaries are normal (perpendicular) to the direction of the gradient of the image in the corresponding points. Figure 3.2 shows some typical edge profiles in images.

Fig. 3.2 Typical edge
profiles in images

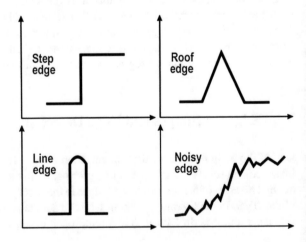

The various available edge detectors are typically optimized to the identification of a particular edge profile.

Since edges often occur in locations in images where boundaries of objects or distinct regions appear, edge detection is used in image segmentation to separate the various objects and regions. The representation of an image only by its edges has the additional advantage that the majority of data are significantly reduced without losing the main visual content. The edges are expressed by an appearance of high frequencies in the spatial frequency domain, and thus can be detected either by applying high pass filters in the Fourier domain or through the convolution of the image with specific kernels in the spatial image domain. In practice, edge detection is performed both in the image domain and/or in the frequency domain, depending on the application. In the image domain, the approach has a lower computational cost and leads, sometimes, to a better result, when the goal is, exclusively, a correct segmentation. Since the edges correspond to abrupt changes, it is readily understood that they can be detected by calculating *derivatives*. In Fig. 3.3 a representation of an edge and the first and second derivatives is shown. As shown in Fig. 3.3, the position of the edge can be detected by locating *inflection points* either by the position of the critical point in the first derivative (maximum in this case) or the location of the change in sign (zero-crossing) in the second derivative. For a discrete one-dimensional function $f[n]$, the first derivative is approximated by a difference

$$\frac{df[n]}{dn} \equiv f[n] - f[n-1] \tag{3.1}$$

Computation of this derivative is equivalent to the convolution of the function with the convolution kernel $[-1\ 1]$. Similarly, the second derivative can be estimated through the convolution of the function with the convolution kernel $[1\ -2\ 1]$. Different masks for edge detection based on the above equation allow us to calculate the derivatives of two-dimensional signals (such as images). There are two basic approaches (Davies 1990; Gonzalez and Woods 1992; Vernon 1991),

- *Prewitt's method*: convolution of the image with a set of masks (usually 8) each expressing a specific direction of the detected edge. For this purpose, the mask that gives the highest response to the central pixel determines the intensity and orientation of the edge. Two of the masks used to detect edges oriented at 0° and 45° are as defined in matrix form as

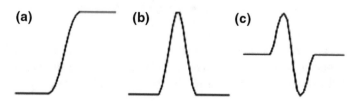

Fig. 3.3 Representation of **a** an edge, **b** its first and **c** its second derivative

$$\begin{pmatrix} -1 & 1 & 1 \\ -1 & -2 & 1 \\ -1 & 1 & 1 \end{pmatrix}, \begin{pmatrix} 1 & 1 & 1 \\ -1 & -2 & 1 \\ -1 & -1 & 1 \end{pmatrix} \tag{3.2}$$

Characteristic variations of Prewitt masks are the *Kirsch*, *Sobel* and *Robinson* masks.

The *Kirsch masks* for 0° and 45° are defined as

$$\begin{pmatrix} -3 & -3 & 5 \\ -3 & 0 & 5 \\ -3 & -3 & 5 \end{pmatrix}, \begin{pmatrix} -5 & 5 & 5 \\ -3 & 0 & 5 \\ -3 & -3 & -3 \end{pmatrix} \tag{3.3}$$

The *Sobel masks* for 0° and 45° are defined as

$$\begin{pmatrix} -1 & 0 & 1 \\ -2 & 0 & 2 \\ -1 & 0 & 1 \end{pmatrix}, \begin{pmatrix} 0 & 1 & 2 \\ -1 & 0 & 1 \\ -2 & -1 & 0 \end{pmatrix} \tag{3.4}$$

The *Robinson masks* for 0° and 45° are defined as

$$\begin{pmatrix} -1 & 0 & 1 \\ -1 & 0 & 1 \\ -1 & 0 & 1 \end{pmatrix}, \begin{pmatrix} 0 & 1 & 1 \\ -1 & 0 & 1 \\ -1 & -1 & 0 \end{pmatrix} \tag{3.5}$$

- *Gradient edge detection method*: it is the most widely used method for detecting edges. A convolution of the image with two masks is being done, one of which calculates the gradient in the horizontal and the other in the vertical direction. The magnitude of the gradient is estimated by

$$|G| = \sqrt{G_x^2 + G_y^2}$$

or alternatively

$$|G| = |G_x| + |G_y| \tag{3.6}$$

In many implementations, the magnitude of the gradient is what matters, ultimately. However, when the knowledge of the orientation of the edge is required, it can be revealed by

$$\theta = \tan^{-1}\left(\frac{G_y}{G_x}\right) - \frac{3\pi}{4} \tag{3.7}$$

The most common convolution masks that implement the method of gradients are those of *Prewitt*, *Sobel* (for horizontal and vertical direction) and *Roberts-Cross*, in which the result is calculated as $|G| = |P_1 - P_4| + |P_2 - P_3|$,

$$\begin{pmatrix} 1 & 0 \\ 0 & -1 \end{pmatrix}, \begin{pmatrix} 0 & 1 \\ -1 & 0 \end{pmatrix} \tag{3.8}$$

After calculating the magnitude of the first derivative the corresponding edges are identified by thresholding, according to which, some of the pixel values above a threshold are classified as edge pixels. An alternative to thresholding is to search for local maxima in the gradient image, leading to the creation of an edge range for each pixel.

A more efficient method for edge detection is that devised by *Canny* (Canny and Computational 1986). In this method, edge pixels are being identified in the gradient image through *non-maximal suppression* and *hysteresis tracking*. The Canny edge detection method was designed as an optimal edge detector in accordance with some criteria. It receives a graylevel image and generates a new image, which depicts the points at which intensity discontinuities can be identified. It is a multistep process. Initially, the image is smoothed by a convolution with a Gaussian kernel. Then a simple two-dimensional operator is applied on the image (similar to Roberts-Cross) for identifying regions in the image where the first derivative takes high values. The edges form 'ridges' in the gradient image. The algorithm checks all these 'ridges' and rejects the pixels that are not actually on them, eventually leading to the creation of an image with fine curves and lines. This process is called the *non-maximal suppression*. The detection process follows and is controlled by two thresholds, $T_1 > T_2$. The detection starts when on an edge a point is found with a value higher than T_1 and continues until a new point is detected with a value less than T_2. This lag (from the Greek 'υστέρηση' - hysteresis) helps deal with the image noise, so as not to drive the algorithm to divide the edges into multiple sections.

A method based on the second derivative of the image is the *Marr edge detector* (Marr 1982), also known as *second derivative zero-crossing detector*. Here, the second derivative is calculated using a Laplacian of Gaussian (LoG) (Haralick and Shapiro 1991) filtering approach. The Laplacian has the advantage of being an isotropic estimation of the second derivative of an image, i.e., the magnitude of the edge measure is taken regardless of the direction of the edge with the convolution of the image with a single mask. The edge locations are given by the points of sign changes in the resulting image. The range of the edges that can be detected, can be controlled by varying the variance of the Gaussian. The method searches for points in a Laplacian image, in which the value passes through zero (points where the value changes sign). Such points appear on edges or, in general, on points of the image where there are sudden changes in intensity. According to this definition, this method is usually considered to be a method of locating features rather than a method of finding edges. The sign changes are always displayed on closed curves (outlines), so, the output of the procedure is normally a binary image with curves of a width of a single pixel that reflect sign changes of the Laplacian. It is worth noting that the final result depends heavily on the application of the smoothing Gaussian filter. As this smoothing is intensified the less sign-change points are detected, and identified points correspond to decreasing image resolutions. After application of the LoG filter, detection of the edges is done in various ways. The simplest is by thresholding, wherein all sign-change points are recognized as boundaries that separate the background from the objects, in which a binary image emerges. A more efficient technique is to study all the points on both sides of the boundaries and to select those with the smallest absolute value of the Laplacian, closest to the point of sign change. When the sign change occurs between two pixels in the LoG image, an alternative output representation could be an image with a shifted grid by half a pixel down and to the right, relative to the original image. Such a representation is called a *dual-grid*. Of course, this method does not lead to a better spatial localization of sign-change points. Marr (1982) put forward the proposal that the HVS operates using an analogous approach, detecting sign-change points to a LoG version of the real image at different scales (changing the variance of the Gaussian). In normal operation, the method receives a graylevel image $I(x, y)$ and returns another graylevel image that corresponds to the Laplacian $L(x, y)$ as

$$L(x, y) \triangleq \Delta I(x, y) \triangleq \nabla^2 I(x, y) = \frac{\partial^2 I}{\partial x^2} + \frac{\partial^2 I}{\partial y^2} \tag{3.9}$$

In practice, (3.9) corresponds to the application of a convolution of the input image with a filter kernel, and, if the input image is represented by a set of discrete pixels (as it is the case), the approach simplifies to the definition of discrete convolution kernels that may approximate the second derivatives for the Laplacian. The most frequently used masks for a 4-neighborhood are in a 3×3 matrix form defined as,

$$h_{4,1} = \begin{pmatrix} 0 & 1 & 0 \\ 1 & -4 & 1 \\ 0 & 1 & 0 \end{pmatrix}, \quad h_{4,2} = \frac{1}{2} \begin{pmatrix} 1 & 0 & 1 \\ 0 & -4 & 0 \\ 1 & 0 & 1 \end{pmatrix} \tag{3.10}$$

The corresponding masks that take into account an 8-neighborhood are defined as

$$h_{8,1} = \begin{pmatrix} 1 & 1 & 1 \\ 1 & -8 & 1 \\ 1 & 1 & 1 \end{pmatrix}, \quad h_{8,2} = \frac{1}{3} \begin{pmatrix} 2 & -1 & 2 \\ -1 & -4 & -1 \\ 2 & -1 & 2 \end{pmatrix}, \quad h_{8,3} = \begin{pmatrix} -1 & 3 & -1 \\ 3 & -8 & 3 \\ -1 & 3 & -1 \end{pmatrix} \tag{3.11}$$

Using these kernels the method reduces to calculating convolutions of the image with the kernels. Since these kernels are used to approximate second derivatives, it is known that the outcome is significantly influenced by noise. This is the main reason for the application of smoothing with Gaussian filters before applying the Laplacian operator. The Gaussian smoothing operator is typically given for the two-dimensional case as

$$G(x, y) = -\frac{1}{2\pi\sigma^2} e^{-\frac{x^2+y^2}{2\sigma^2}} \tag{3.12}$$

with σ the standard deviation, whereas the overall process can be depicted in the form of convolutions as

$$\Delta[G(x, y) * I(x, y)] \tag{3.13}$$

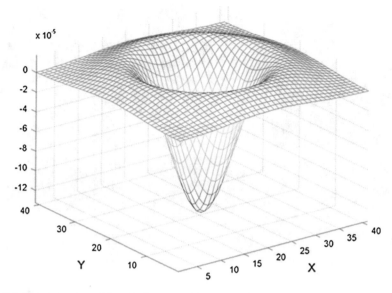

Fig. 3.4 Representation of the two-dimensional LoG function

Apparently, since convolution is associative, convolution of the Gaussian filter with the Laplacian could be performed at the beginning and then applied to the image,

$$[\Delta * G]I(x, y) \triangleq \mathrm{LoG}[I(x, y)] \tag{3.14}$$

The LoG function in two dimensions centered around zero and defined by a Gaussian standard deviation σ takes the form

$$\mathrm{LoG}(x, y) = -\frac{1}{\pi \sigma^4}\left[1 - \frac{x^2 + y^2}{2\sigma^2}\right] e^{-\frac{x^2+y^2}{2\sigma^2}} \tag{3.15}$$

Figure 3.4 shows a representation of this function for a kernel of size 40 where the standard deviation of the Gaussian is $\sigma = 7$. In discrete form, a 9×9 kernel that approximates this function for a Gaussian with $\sigma = 1.8$ is expressed as

$$\begin{pmatrix}
1 & 2 & 3 & 3 & 4 & 3 & 3 & 2 & 1 \\
2 & 3 & 4 & 3 & 3 & 3 & 4 & 3 & 2 \\
3 & 4 & 2 & -2 & -5 & -2 & 2 & 4 & 3 \\
3 & 3 & -2 & -1.2 & -1.7 & -1.2 & -2 & 3 & 3 \\
4 & 3 & -5 & -1.7 & -2.4 & -1.7 & -5 & 3 & 4 \\
3 & 3 & -2 & -1.2 & -1.7 & -1.2 & -2 & 3 & 3 \\
3 & 4 & 2 & -2 & -5 & -2 & 2 & 4 & 3 \\
2 & 3 & 4 & 3 & 3 & 3 & 4 & 3 & 2 \\
1 & 2 & 3 & 3 & 4 & 3 & 3 & 2 & 1
\end{pmatrix} \tag{3.16}$$

Fig. 3.5 Detecting edges in the **a** graylevel image: the result of **b** Prewitt, **c** Sobel, **d** Canny, **e** LoG, **f** zero-cross

Figure 3.5 shows an example of edge detection in a graylevel image. The methods presented include Prewitt, Sobel, Canny, LoG, zero-cross.

3.3 Segmentation of Images

As is pointed out in the previous paragraphs, whatever the approach to address the challenges of MRC compression, a necessity arises to apply a segmentation of the original image in order to identify the basic structural characteristics; this could aid in selecting the appropriate compression method in each case. In addition, in previous paragraphs was given an overview of the concepts regarding edges, which are among the main features that represent information within the images and must be taken seriously into account during image segmentation. The following paragraphs analyze segmentation of digital images in general, through a comprehensive reference to existing methods, advantages and disadvantages, and their application scope. What is important in the case of MRC compression is that a segmentation method should be able to successfully separate the text and graphics from the continuous tone sections, which may be complicated and contain regions with strong texture and color. Nevertheless, the general description of segmentation methods that follows along with their classification is given here regardless of their ability to assist in MRC compression.

Image segmentation is one of the primary steps in image analysis, especially for object identification and recognition. The main purpose is to identify homogeneous areas of the image as discrete parts or objects. A segmentation method in itself does not provide any insight on the actual content or the form of the objects. Usually it is based on finding maximum uniformity in luminance levels or structural relations within the regions being identified.

Many parameters relate to image segmentation. One of the most common problems is the choice of the most suitable method of separation of the objects from the background. The segmentation can not be successful if the differences between pixel values of the various objects are dim. For this reason, most of the time, segmentation is performed after applying a method to *improve* the image (image enhancement) in order to reduce noise and enhance the objects' differences. This way the effects of such an image enhancement are directly related to the result of segmentation. In this case, an *ideal image enhancement operator* is expected to maximize the differences in the objects' boundaries.

In general, key issues in segmentation are the selection of an appropriate segmentation algorithm, the measurement of its performance, as well as understanding of its implication on the final image resolution. A general categorization of segmentation methods differentiate pixel-based, region-based, contour-based and hybrid methods.

- *Pixel-based methods or global methods*: in these methods a function on the pixels is being used (e.g. brightness) and an operator is applied to threshold the image and provide the final segmentation.
- *Region-based methods*, in which an appropriate homogeneity criterion is being used to connect various pixels into regions. Typical region-based methods are,
  - *region growing* methods, in which starting from a seed region, adjacent pixels are tested based on a criterion of homogeneity and are absorbed (or not) by the region.
  - *split-and-merge* methods, in which sequential split and merge of regions takes place until the regions are homogeneous, and neighboring regions with similar characteristics merge.

- *Contour-based methods*, where information about discontinuities and edges is being exploited. Here belong methods that make use of the gradient, or Laplacian, or Laplacian of Gaussian (LoG), or even the Canny edge detection method.
- *Hybrid methods*, or variations of conventional methods, which rely on the use of appropriate functionals. The segmentation in this case arises from critical points.

In the following paragraphs an overview of segmentation methods is given. It should be noted that image segmentation has been approached under various perspectives (Pavlidis 1982); here, only a representative sample of these approaches is presented including

- Histogram thresholding
- Segmentation based on edge detection
- Tree representation techniques

- Region growing methods
- Clustering-based segmentation
- Probabilistic and Bayesian methods
- Segmentation using neural networks
- Other-hybrid methods

3.3.1   Segmentation Based on Histogram Thresholding

One very basic segmentation method is segmentation based on histogram threshold-ing. Many objects (and image regions accordingly) are characterized by a nearly con-stant reflectivity or light absorption surface properties. Thus, a brightness value can be determined as a threshold for the segmentation of objects from the background. This thresholding method is a quick method with relatively low computational require-ments. Nevertheless it is still widespread, especially for simple applications. It may even take place in real-time using appropriate hardware. A full segmentation of image I into a finite set of S regions $R_1, ..., R_S$, such that

$$I = \bigcup_{i=1}^{S} R_i \quad R_i \cap R_j = \varnothing, \quad i \neq j \tag{3.17}$$

Generally, a two-level thresholding is the transformation of an input image I into a binary output image J defined as

$$J(i,j) = \begin{cases} 1, & I(i,j) \geq T \\ 0, & I(i,j) < T \end{cases} \tag{3.18}$$

where T is a threshold, $J(i,j) = 1$ for the objects, and $J(i,j) = 0$ for the background. Important facts about thresholding include,

- when the objects in the image are not in contact with each other and their brightness (or color) is clearly different from the one in the background then it is expected that the thresholding can be applied successfully;
- a proper choice of the threshold may lead to a successful segmentation;
- the selection of the threshold may either be manual or automatic, as the result of a threshold detection method;
- only in 'unusual' cases the segmentation with a global threshold is expected to be successful, since significant variations in brightness of both the background and the objects is usually observed;
- a segmentation using various thresholds (or adaptive thresholding) may lead to better segmentation results, by introducing a spatial parameter to determine the appropriate threshold according to characteristics of the image regions being studied

- thresholding can be applied not only to images with brightness information but also to gradients, texture property values, or other possible image analysis features.

  The general thresholding relation in (3.18) can be expressed in three variations.

- *Zonal thresholding*, wherein an image is segmented according to a number of brightness values in a set of values D as

$$J(i,j) = \begin{cases} 1, & I(i,j) \in D \\ 0, & \text{elsewhere} \end{cases} \tag{3.19}$$

  This segmentation can be also used as a boundary detection method.
- *Multi-thresholding*, when the result of the segmentation is not a binary image but an image with a limited number of gray values, which is expressed as

$$J(i,j) = \begin{cases} 1, & I(i,j) \in D_1 \\ 2, & I(i,j) \in D_2 \\ \vdots \\ n, & I(i,j) \in D_n \\ 0, & \text{elsewhere} \end{cases} \tag{3.20}$$

- *Semi-thresholding*, which is sometimes used to aid in the analysis of images, by aiming to remove the background and maintain parts of the image that describe objects,

$$J(i,j) = \begin{cases} I(i,j), & I(i,j) \in D \\ 0, & \text{elsewhere} \end{cases} \tag{3.21}$$

In Ohlander (1975) a histogram thresholding technique was proposed that is very useful for the segmentation of continuous tone color images. It is based on color histograms. The image is thresholded to the most clearly identifiable peak in each of the color histograms in a recursive manner that processes all clearly recognizable peaks. The peak separation criterion is based on the highest-to-lowest-peak ratio, which must be greater or equal to two. Regions with texture are separated from uniform areas through the use of the Sobel operator and the identification of regions where there is a high activity of edges (using the simple rule of identifying more than 25 edge pixels in a 9×9 window). These areas are designated as 'active' and are not subject to segmentation by thresholding, except in specific appropriate cases.

In Cheriet et al. (1998) a general iterative method of image segmentation that extends the method of Otsu (1979) was proposed. This method was developed in the framework of the segmentation of document images, and particularly the segmentation of digitized bank checks. In this method the object with the highest brightness is segmented at each iteration. Obviously, the method is independent of the number of objects in the image. The thresholding is understood here as the separation of pixels into two groups: the *objects* and the *background*. In each iteration of the algorithm the histogram of the image is computed and the regions corresponding to the highest peak are separated from the rest of the image. The process is repeated until there are no peaks in the histogram. Similarly to the Otsu technique, the method produces better results in images of two classes (bi-modal histogram images). The method can also help extract and analyze information from document images, as it preserves the topological characteristics of the extracted information for further analysis. After extensive testing of the method, the authors concluded that the method works efficiently (93–100 %) when the objects in the images are of a lower brightness in relation to the brightness of the background.

In a number of applications, a histogram thresholding is not possible because the histogram exhibits just one peak. In other cases the image may be of such quality that no pre-processing can improve the objects-background contrast sufficiently. Histograms of one peak are usually found in images where the background dominates, as in medical images and simple documents. Similarly, in aerial and satellite images, which can include numerous objects, the histogram may also have a single peak, due to the wide brightness range on each object and possible overlaps. In Bhanu and Faugeras (1982) a gradient based algorithm was proposed to alleviate this problem and was compared with the non-linear algorithm proposed in Rosenfeld et al. (1976). The process is based on matching a likelihood value at each pixel and then using the gradient, taking into account the correlation among the pixels in the 8-neighborhood. The process is iterative, and the pixel values are changed so that the resulting image histogram is no longer of a single peak. Compared with the method proposed in Rosenfeld et al. (1976), the authors of Bhanu and Faugeras (1982) asserted that their technique provides better control over the use of parameters specified by the user. The user controls the degree of smoothing at each iteration, and the initial probabilities. An extension of this approach was proposed in Bhanu and Parvin (1987), where the segmentation was based on a recursive split and merge approach. This method was intended to prevent the use of any inductive or other measures to partition the image, so no robust merging step is required for the final elimination of boundary problems. Further, the process does not require a detailed detection and peak selection process.

Li et al. (1997) proposed that the use of two dimensional histograms are more useful in identifying segmentation thresholds. In these histograms both the values of the pixels and the average value of their neighborhood is being considered. The authors demonstrated that the application of the Fisher linear discriminant analysis[2] on the histogram leads to an optimal projection, wherein the data blocks are sharply defined and thus allow for an easier and more accurate segmentation. The experimental results demonstrated that while the computational requirements are similar to those of the one dimension histograms, the segmentation results are better.

---

[2]Fisher linear discriminant (Brown 1999), is a clustering method, in which data of high dimensionality are projected on a line and a clustering is applied on the one dimensional space. The projection maximizes the distance between the average values of the classes, while, at the same time, it minimizes the variance within each class. The Fisher criterion that is being maximized for all line projections is defined for two classes as $J(w) = \frac{|m_1 - m_2|^2}{\sigma_1^2 + \sigma_2^2}$, where m is the mean and σ^2 the variance of the classes.

3.3.2   Segmentation Based on Edge Detection

Segmentation of images using edge information is one of the primary methods of segmentation and is still very popular. The whole idea is based on the detection of edges in images using various edge detection operators. Whatever the detected edges, they signify changes or discontinuities in brightness, color or texture. The image that results from an edge detection method can be used as the segmentation result. Nevertheless, usually, further sequential processing should be applied to join any disconnected edge pixels so that those pixels correspond more precisely to object boundaries in the image. The ultimate goal is to achieve at least a partial segmentation that reflects the edges that correspond to boundaries of objects in the images. The main problems encountered in the application of this process arises from the presence of noise or low contrast between the objects and the background. There is a number of categories of segmentation based on edges, including:

- *Thresholding based on the edges*: the approach relies on thresholding imposed upon an edges image. An edges image comprises non-zero values, with small values corresponding either to insignificant changes in brightness or to noise due to quantization and small illumination irregularities. Thresholding using a global threshold is usually inefficient in this case.
- *Edge mitigation*: the approach relies on contextual information in a predefined neighborhood to assess the certainty of edges. Any detected edges are significantly influenced by the noise in the image. The quality of the edges image can be significantly improved if edges are post-processed with information in their neighborhood. Thus, the final decision on the actual existence of an edge is extracted by assessing relevance information in a predefined neighborhood. The process is usually recursive, whereby the confidence interval for each edge in the region is iteratively improved.
- *Detection and following of edges*: the approach relies on following the edges (boundaries) of a region using 4-connectivity or 8-connectivity (usually in blocks of 3×3 pixels). Is can either be *inner, outer*, or *extended* detection of boundaries that leads respectively to boundaries that are inside, outside, or on the boundaries of the region. The extended edge following leads to the definition of a common boundary between adjacent regions (Fig. 3.6).
- *Edge detection as a graph search*: it is used when additional information is required for the detection of edges; for example, when the starting and ending point of a border curve is known and the precise position of the curve is not known. Quantities such as the smoothness or the curvature may be used as auxiliary information. A *graph* has a general structure of n_i nodes and links with corresponding weights (n_i, n_j). The edge detection process using graphs is transformed to a search of the best path in a graph of paths with weights that run from a start node to an end node. Figure 3.7 shows an image of edge directions and the graph that corresponds to the edges of the image, while Fig. 3.8 shows an application example of the graph search for identifying boundaries of a coronary vessel in graylevel medical image. Note that the graph methods ensure that the identified contours are globally

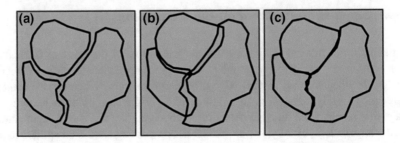

Fig. 3.6 Region boundaries after **a** inner, **b** outer and **c** extended edge following

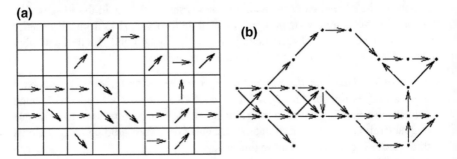

Fig. 3.7 Graph representation of an edges image: **a** quantized edge directions and **b** corresponding edges graph

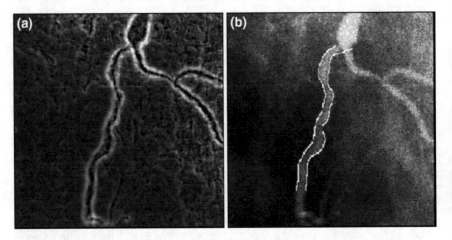

Fig. 3.8 Application of the graph search in medical imaging for determining boundaries of a coronary vessel: **a** edges image, **b** identified boundaries

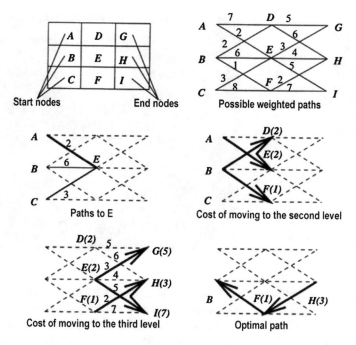

Fig. 3.9 Simple example of boundaries detection in a 3 × 3 image block

optimal. Without a priori knowledge of the starting and ending points, though, graph methods turn out to be rather complex to be considered.

- *Edge detection as an application of dynamic programming*: dynamic programming[3] is an optimization method that searches for optimal points in functions, in which all variables are not interdependent. As an example one may consider the boundaries detection procedure described in Fig. 3.9. The problem is reduced to finding the path with the least cost, by following an optimization principle, whereby *whatever the choice of the route to a point E, there is an optimal path between E and the end node*. In other words, *if an optimal path passes through a point E, then the paths start-to-E and E-to-end are also optimal*. Figure 3.10 shows an example of applying the dynamic programming method in the recognition of boundaries in X-ray medical images. Some key points to be highlighted include,

  - the complete graph (paths-costs) must be available to implement dynamic programming;
  - the objective function must be monotonic and separable (at least for the implementation of the known algorithm).

---

[3]Dynamic Programming is a method for problem solving, in which the solution path is designed in reverse (from the required outcome backwards to the beginning).

Fig. 3.10 Detection of
pulmonary slots using
dynamic programming on a
medical image showing a
section of a human lung in
x-rays. Recognized slots are
depicted in *white* (*bottom
picture*)

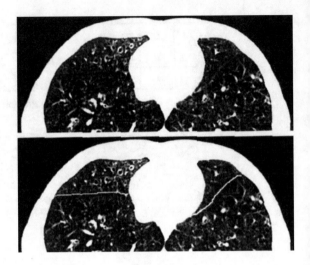

- *Hough transform* (Hough 1959): the approach relies in the identification of canonical objects, like lines and curves that can be mathematically expressed. If an image consists of objects of known shapes and sizes, then the segmentation can be considered analogous to the problem of identifying these objects in the image. One way of solving this problem is through the use of a mask of specific size and shape and search for corresponding regions in the image. This approach, however, presents several problems due to differences in shape, size and orientation, necessitating a method independent of those limitations. A robust such method is the application of Hough transform, which can lead to satisfying results even when there is overlap or occlusion of the objects in the image. This transform (at least in its line-detection version) is a method of mapping the two-dimensional Cartesian space to a corresponding space of polar-coordinates, in which each point reflects a line segment in the original image (Fig. 3.11). An example of its use in Magnetic resonance imaging (MRI) is shown in Fig. 3.12, where the problem of segmenting of the right and the left hemisphere of the brain is being tackled. Figure 3.13 shows an example of the application of the circular Hough transform in the detection of circles in an edges image with overlapping objects and occluded surfaces.
- *Definition of regions based on boundaries*: the approach relies on efficient algorithms to describe regions defined by incomplete boundaries. When the segmentation leads to incomplete description of the boundaries of objects, the problem of proper identification of regions becomes quite complicated. To address such problems, region detection methods using incomplete boundary descriptions were developed, which do not always lead to good segmentation but are practically very useful. These methods typically apply multiple thresholding to achieve the final segmentation and recognition of regions.

A description of how neighboring pixels may be used in segmentation is given in Ahuja et al. (1980). For each pixel a neighborhood is initially defined, within a certain

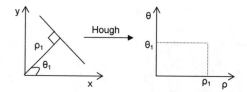

Fig. 3.11 Hough transform as a mapping of the Cartesian $x - y$ space into the polar $\rho - \theta$

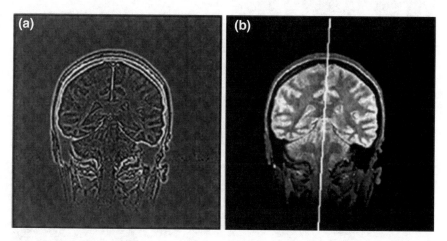

Fig. 3.12 Use of the Hough transform on **a** an MRI edges image for the segmentation **b** into the *right* and *left* brain hemisphere

predetermined window size. Thus a vector is defined for neighboring pixels having as elements either the pixel values themselves or the average pixel values in windows of sizes 3×3 and 5×5. This paper examined both representations. The aim was to identify a table of weights, which when multiplied with these vectors would result in a cut-off value (threshold), which in turn could be used to separate the pixels into two classes. In the experiments described, the Fisher linear discriminant was used. The authors reported that in the presence of noise, the use of brightness as a single characteristic actually provided better results. The authors concluded the the results led to that the brightness values of the pixels and their neighbors are a good set of features for their efficient classification.

A set of algorithms for segmentation of continuous tone images using boundaries analysis was proposed in Prager (1980). The goal of the method was the precise localization of the boundaries of the objects present in an image. Initially, smoothing and noise removal takes place. Then, the edge intensity is estimated at each point in the image. An edges elimination process follows, to suppress the overestimated results from the previous step. Subsequently, the edges are combined and features are calculated. These features include the length, contrast, frequency, mean value, variance, and the location of each line segment. Further processing follows to remove unwanted segments and to generate a quantity to estimate the certainty

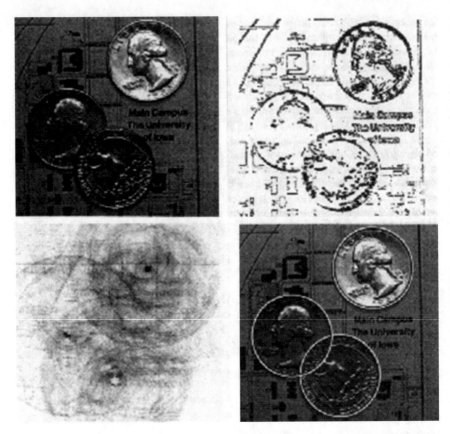

Fig. 3.13 Use of the circular Hough transform to detect *circle* in an edges image; (*top-left*) original image; (*top-right*) binarized image; (*bottom-left*) the image in the Hough transform space; (*bottom-right*) image with boundaries superimposed on the identified objects

for each remaining line segments. The system output is a set of line segments that is accompanied by additional features (such as length and certainty). According to the authors the advantages of the method are two: first, the results of each step are clearly defined (modular architecture) and, secondly, each step may be omitted or even replaced.

In Perkins (1980) another segmentation technique based on edges is proposed. The authors presupposed that a segmentation based on edges is not successful due to the introduction of small gaps, the treatment of which may lead to a merging of disparate regions. To avoid this problem, a *growing-shrinking technique* was proposed, whereby edges are extended to eliminate small gaps and then are contracted after the labeling of the recognized regions. The size of the expansion is controlled so that small regions are excluded from the process. The process involves the use of the Sobel edge detection filter for the extraction of the intensity and the direction of the edges. The edges are thinned and the result is thresholded automatically to exclude any gaps. These gaps are separating different intensity regions but small gaps may still appear. Segmentation is performed by extending the active edge regions, labeling of the uniform regions, and shrinking of the edge regions. The authors report successful results for continuous tone images and electronic schematic diagrams.

An adaptive thresholding method based on calculus is described in Chan et al. (1998). An inductive algorithm is described that runs in seven steps. Initially, the image is normalized by a mean value filter. What emerges is a gradient magnitude image. Thresholding is then applied along with thinning to find boundary points. Then the original pixels of the image are selected on those identified boundary points as local thresholds. The surface of thresholds is completed and is used to segment the original image. The noise is removed from the segmented image using a method that involves derivatives. This method is recursive and ends when a criterion of iterations or time is satisfied. The experimental results showed that the segmentation is more successful when there is a large number of objects in the image (since it created better threshold surfaces). The method proved particularly effective in image segmentation for Optical Character Recognition (OCR).

3.3.3 Segmentation Based on Tree Representations

In Cho and Meer (1997) a segmentation method is proposed, in which the result is a collection of results of multiple segmentations on an image. Instead of using statistics to characterize the spatial structure of a pixel neighborhood, for each pair of adjacent pixels their statistics are used to determine local homogeneity. Several initial segmentations are performed on the same original image by changing statistical characteristics of the hierarchical Region Adjacency Graph (RAG) technique based on a pyramidal image representation. From the collection of initial segmentations for each pair of pixels a probability is extracted, which contains comprehensive local information. The final segmentation results from the processing of that probability field with the same pyramidal RAG technique. The pairs with high probability of occurrence are grouped under the concept of local homogeneity. The technique can be used for the extraction of high probability regions from the probabilities field. Then, Bayesian networks can be used to extract features from the image. Such features are the variance in the magnitude of the regions, the ratio between width and length of regions and the average intensity value of pixels. Finally, post-processing is performed to undo the over-segmentation. The RAG of the final segmentation provides the spatial relationships between regions and can be further used for interactive image analysis. The process is unsupervised.

In Yeung et al. (1998) is proposed a technique for video segmentation through clustering and graph analysis. The method extends to a Scene Transition Graph (STG) representation for analyzing time-structures in a video. Initially, similar scenes are being recognized in order to reduce the amount of information to be processed. Then the dissimilarity between scenes is recognized based on the heterogeneity on all pairs of images of the two scenes. So automatic segmentation of scenes and parts of the history in a video can be achieved. The images are segmented based on the featured video. The scenes are then grouped based visual similarity and the temporal dynamics of the scenes. Then a STG is designed and analyzed, and then used for the compact representation of the structure of scenes and the time-based flow. This method makes it possible for a decomposition to hierarchical story units, each consisting of groups of similar scenes.

3.3.4 Region-Based Segmentation

The segmentation techniques presented so far aim at finding boundaries between objects by detecting edges. Even though it is relatively simple to create regions based on the edges and boundaries, as well as to detect boundaries of existing regions, the segmentation results based on boundaries are not the same as those based on regions. Regions-based techniques are more robust to noise, in cases in which the edges and boundaries are difficult to identify. An important property of these regions is their homogeneity, and it is this property that is being used in region growing techniques (the idea of which is to separate images into areas of maximum homogeneity). Homogeneity criteria may be the luminance, color, texture, shape, or any model (semantic approach). These properties affect the type, complexity and size of the a-priori information required for proper segmentation. The mathematical representation of regions in an image was given in (3.17). In the case of region-based segmentation, additional conditions apply, in which

$$
\begin{aligned}
&I = \bigcup_{i=1,\dots,N} R_i \\
&R_i \cap R_j = \varnothing && i \neq j \\
&H(R_i) = TRUE && i = 1,\dots N \\
&H(R_i \cup R_j) = FALSE && i \neq j, \quad R_i, R_j - \text{neighboring}
\end{aligned}
\tag{3.22}
$$

where I the image, i the regions index, N the number of regions, and $H(R_i)$ the segmentation rule as a logical (true or false) estimate of the homogeneity of the region R_i.

Regions that result after the segmentation must be homogeneous and maximal in area, in the sense that the homogeneity criterion should be waived in any merger of neighboring regions. It should be noted that the algorithms of this family can easily be generalized to cases of three-dimensional images (volumetric data). In this family belong techniques like,

- *Region growing*. The process starts with single pixels to represent entire regions and evolves iteratively into producing regions by grouping neighboring pixels based on merging criteria. Figure 3.14 shows an example of segmentation based on region growing methods on a simple graylevel image; the process of recursive region growing and boundary merging is depicted using pseudocolor representations to identify the different regions.

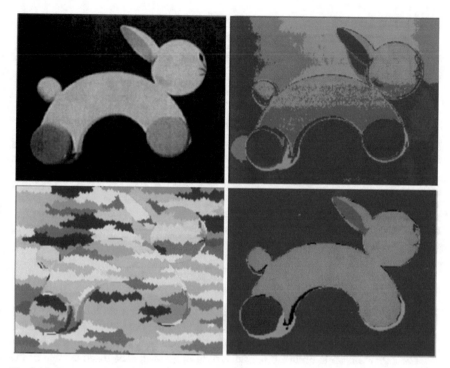

Fig. 3.14 Example of segmentation by a region growing method (image courtesy of Radek Marik)

- *Region splitting.*[4] It is the reverse process of region growing and begins with the whole image as a single region. Recursively it splits the single region into smaller regions obeying certain homogeneity criteria similar to those applied in region growing. Although the method seems to be complementary to the previous, its result are not the same, even when the same homogeneity criteria apply. In the extreme example of the segmentation of a checkerboard image, when the homogeneity criterion is the variation in average luminance in a quadrant, region growing leads to a correct result, whereas region splitting concludes that the image cannot be segmented as it is represented by a single homogeneous area, as shown in Fig. 3.15.
- *Region split-and-merge.* A combination of the methods of region growing and region splitting may lead to a better outcome by exploiting the advantages of both methods. This approach acts on pyramidal representations of images. The regions are rectangular and represent data of the respective level of the pyramidal representation. If a region at some level is not homogeneous, then it is divided into four regions in a higher resolution to the next pyramid level. Similarly, if

---

[4]Region growing methods are bottom-up approaches, since they start by a single pixel and scale up to the overall image, whereas region splitting methods are top-down approaches, since they start by examining the whole image and scale down to the single pixel.

(a)   **(b)**   **(c)**

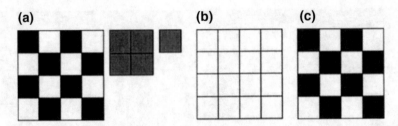

Fig. 3.15 Different segmentation results by growing and splitting methods; **a** the original image and the resolutions of the pyramid based on the homogeneity criterion, **b** the result of region splitting, **c** the result of region growing

Fig. 3.16 Region split-and-merge in a hierarchical data structure

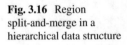

Split

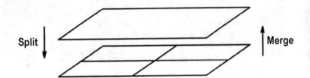

Merge

four regions derived from the same parent node exhibit a similar homogeneity measure, then they are merged into one region in the next level (Fig. 3.16). The process seems similar to the construction of a quad tree, where each leaf reflects a homogeneous area, or an element in a pyramidal level. The disadvantage of the method lies in its very nature and is none other than the hypothesis of the square regions. An example of this tree structure is given in Fig. 3.17.

- *Watershed segmentation.* The ideas of both watersheds and basins are widespread in the area of topography. Watershed lines separate water collection basins. To transfer these ideas in image segmentation, the image data are regarded as topographical surfaces, where the intensity of the pixels correspond to elevation data. Thus, the edges of regions correspond to watersheds whereas the inner surfaces of the areas with low gradient value are the water basins. In Fig. 3.18 an example of watershed segmentation is shown for the one-dimensional case, where local minima correspond to the water collection basins and the local maxima correspond to the watersheds.

While the applicability of this topographical approach to images is apparent, the development of algorithms to implement the watershed segmentation is highly complex. Figure 3.19 shows the result of watershed segmentation on an edges image obtained by applying the Sobel edge detection method to an MRC graylevel image. The edges image is shown here after a histogram equalization for a better illustration and the final segmentation is presented with pseudo-coloring for a better separation of the identified regions. Apparently, this method may lead to over-segmentation, and to be useful, it must be complemented by segmentation restriction criteria.

In the following, brief descriptions of segmentation methods based on region handling (and especially region growing) are presented. These algorithms start with

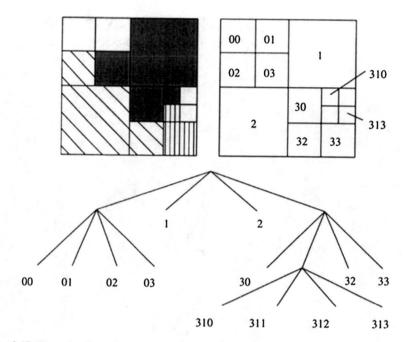

Fig. 3.17 Example of creating a segmentation tree

Fig. 3.18 Example of
watershed segmentation for
one-dimensional data

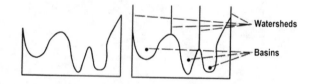

regions of one or more pixels, called seed pixels, and implement recursive region
growing based on homogeneity criteria. If the neighboring pixels are similar to the
seeds then they are grouped into one region. The process iterates until all image
pixels become part of at least one region.

In Chang and Li (1994) a framework of region growing for image segmentation
is proposed. The process is guided by analysis of spatial characteristics and there is
no requirement for a-priori knowledge regarding the image or any pre-configuration.
The algorithm is also known as the *Fast Adaptive Segmentation* algorithm. The
original image is divided into small primary regions, which are considered to be
homogeneous. These regions are then merged to create larger ones until no more
merging is possible. Two regions are merged when satisfying homogeneity criteria and if the value of the edge that joins them is weak. Main focus in this work
was in the assessment of how different merging criteria may influence the quality
of segmentation and the time required. The best segmentation efficiency that the
method was able to achieve was about 86 %. The algorithm automatically calculates

Fig. 3.19 Watershed segmentation after applying Sobel edge detection on a graylevel image

segmentation thresholds based on an analysis of local characteristics. The algorithm is robust and produces high quality segmentation in graylevel textured images. By appropriate substitution of the measured characteristics, the method can be applied to other cases. The major drawback of this method lies in the limited use of the homogeneity criterion in very small regions.

In region growing, the seed regions can be selected either automatically or manually. Automatic selection may be based on finding the pixels of interest, based on e.g. pixel intensity criteria. It can also be defined by the peaks of the histogram of the image. On the other hand, these regions can be manually selected for each object in images. In Adams and Bischof (1994) the growing efficiency of the seed regions in graylevel image segmentation was studied, wherein these regions are selected by the user. Each seed region is a connected component that comprises one or more points and is represented by a set S. The set of direct neighboring pixels is estimated for these points. Then the adjacent pixels are analyzed and if they intersect a region of the S set, then a measure d is computed (a distance between a pixel and the intersected region). If the neighboring pixels intersect several regions, then the sum is taken for which the measure d is maximum. The new situation of regions for a set becomes the input to the next iteration of the process. The process iterates until all pixels have been clustered in regions; at each iteration, a pixel is merged with an adjacent region if it is *similar* to that region. The algorithm is dependent on the manner and order of processing the pixel. The authors indicate that the method is sufficiently robust, fast and free from parameters, but is dependent on the processing order of the pixels.

In Mehnert and Jackway (1997) an improvement of the above algorithm was proposed, in which the manner and sequence of processing of the pixels was made indifferent. The study presented a new technique for Improved Seeded Region Growing (ISRG). This algorithm still provides the advantages of the previous method and has the additional characteristic of being independent of the pixel processing sequence. If two or more pixels have the same minimum value of similarity, then they are all examined in parallel. No pixel can be 'labeled' and no region to be updated until all pixels with the same priority are processed. If a pixel can not be labeled (i.i. belongs to two or more neighboring regions with the same probability) then it is called 'tied' and is not considered in the region growing process. After all pixels of the image are labeled, then the tied pixels are reviewed independently, to determine if it is possible to include them into some region. Therefore, an additional criterion is used (e.g. inclusion to the neighboring region with the largest mean value).

In Basu (1987) a general semantic framework for identifying regions was developed, in order to describe a set of image models that use it. This set is generated on an empirical basis using simple and intuitive properties of regions. It can be extended to include new semantic content without changing the overall conceptual and computational framework of the method. When a region is recognized, each pixel is selected as elementary. An ideal region is taken from a set of pixels by approximating the intensity values using a linear or a quadratic model. A set of attributes for each pixel of the set is being calculated along with each pixel in the ideal region. Specifically, four properties (contrast, variance, total average intensity value and representative average value) are being used. The distance is computed between pixels of the set and the ideal region, as a measure of dissimilarity between them. Placing a pixel in a region is defined by that measure. The authors conclude that a significant improvement in results arises when using semantic information. There is no optimal set of properties that can be used for all images. For each image a number of different properties or combinations should be tested to achieve the best possible segmentation.

In Beaulieu and Goldberg (1989) a hierarchical optimization algorithm is proposed for region merging, based on stepwise optimization, and produces a hierarchical decomposition of images. The algorithm begins with an initial separation of the image into regions. At each iteration, two regions are merged provided that they minimize some criterion. The algorithm checks the entire image, before each merging, to find the best pairs of regions. Thus, most similar regions are merged. Gradually, the algorithm merges regions and produces a partition sequence. This sequence corresponds to the hierarchical structure of the image. This process is also called *agglomerative clustering*. The authors conclude that the results obtained from the final clustering are far better than conventional non-hierarchical methods, such as the known k-means algorithm. The advantage of the algorithm to non-hierarchical is that no seed points are required.

In some studies, information about the edges or gradients has been used in conjunction with region growing for image segmentation. Such a method is proposed in Gambotto (1993), where an algorithm combines region growing with edge detection. The process is iterative and uses both methods in parallel. It starts with an initial set of seed points, which are inside the actual region boundaries. The pixels that are adjacent get merged throughout the iterations when certain similarity criteria are satisfied. A second test uses the mean value of the gradient in order to limit a non-ending expansion. At the end, a cleanup step is performed. The analysis is based on the cooperation between the region growing and edge detection algorithms. Since the growing appears as a continuously merging of new sections in a region, some pixels in the previous and the next regions may be misclassified. A rule of a minimum neighborhood is applied for local re-classification of these pixels. The algorithm was tested on a set of images obtained using X-rays. The authors concluded that their method behaves similarly to the *snake* methods, with the difference that it can be used in the segmentation of convex regions with unknown and complex shapes.

In Hojjatoleslami and Kittler (1998) a region growing method is proposed that uses gradient information for the determination of boundaries. The method has the ability to find the boundaries of the dark-to-light regions on textured background. It is based on a region contrast measure corresponding to the variance of the intensity levels in the region as a function of the developing boundaries during segmentation. This helps identify the best external boundaries of the regions. The application of a reverse testing using a gradient measure results in the maximum gradient boundaries for the region being grown. The uniqueness of the approach is that in each step at most a single pixel will meet the requirements for merging in the region. The growth process is directional so that the pixels incorporated following a sequence, and discontinuity measurements are taken at a pixel basis. The authors conducted numerous experiments on natural (continuous tone) and artificial images, and concluded that the results are more satisfactory than conventional thresholding methods.

The algorithm is unaffected by noise even when noise is significantly high. The main advantage of the method is that no a priori knowledge of the regions is needed.

Lu and Xu (1995) propose a region growing method applicable to texture segmentation, in which a two-dimensional self-recursive model is applied. A neural network is used to implement the parameter estimation process for the model and to assess the local texture features during the process of segmentation. The segmentation process involves two steps, the initialization and the clean-up step. In the initialization, the image is divided into individual sections in the form of a predetermined window size. Each section is assumed to be part of the internal surface of a region if it belongs to the same category with the four adjacent pixels (horizontal and vertical). Otherwise it is considered indefinite. During the clear-up, all original parts extend in parallel. Each indefinite section is divided into smaller sections of equal size. Each smaller section is compared to the adjacent sections using local information extracted from the neural network. The process ends when each pixel is categorized into one of the original parts. The algorithm was tested on a total of 38 different texture images from the Brodatz set (Brodatz 1966) and shown to function efficiently, using low computational resources.

In He and Chen (2000) a *resonance* algorithm is proposed for segmentation, which does not differ significantly from the seeded region growing. It is emphasized that the method is much more robust to intensity variations than conventional algorithms, acting directly on graylevel images. The basic idea is based on the distribution of power from a source in other parts of a system. The method considers each pixel as one mass connected to a baseplate via a spring. In the image there are many such mass-spring pairs, which can be regarded as being immersed into water. When an external force acts on a pair then a shift is observed. If the external force is matched to the natural frequency of the mass, the mass echoes and vibrations are transmitted to other mass-spring pairs, which start to oscillate. This phenomenon is evident in a small radius and it becomes weaker with distance. It should be stressed that a seeded segmentation algorithm based on this theory should only be used in *feature images*[5] and not directly to the images themselves. The experiments were performed in the Brodatz texture image set. The average intensity to every 8×8 window is computed and a corresponding feature image is produced. Since the average value depends on the intensity, it is expected that the effective capacity of the algorithm can be evaluated experimentally by the use of such a feature. The results show a performance of the algorithm better than that obtained by the fuzzy c-means algorithm or histogram analysis methods.

Singh and Al-Mansoori (2000) compares region growing with gradient methods for the detection of regions of interest in digital mammographies. These regions of interest are the basis for the application of techniques based on shape and texture to detect tumor masses. This study also proposed a two-step process, where gradient techniques are applied prior to the application of region growing, so that fewer regions for further analysis will occur. Initially the image is enhanced by histogram

---

[5]The term *feature images* denotes images that consist only of detected features and not typical pixel values.

equalization and fuzzy enhancement methods. After image enhancement, images are subject to region growing or gradient based segmentation. The segmented image is analyzed to assess regions of interest and the results are compared with results from previous diagnosis by expert radiologists. The authors conclude that the region growing methods give fewer regions of interest than methods based on gradient, without lowering the segmentation efficiency.

3.3.5   Clustering-Based Segmentation

Image segmentation may be performed efficiently through pixel clustering. The cluster analysis allows dividing the data into semantic subgroups and can be applied to image segmentation for classification purposes. This analysis requires either the definition of seed regions by the user or the usage of non-parametric methods of finding the obvious regions without using seed points. Clustering is mainly used in applications such as segmentation and unsupervised learning (Jain and Dubes 1988). A number of issues regarding clustering worth to be studied, such as how many clusters are best and how to assess the quality of the clustering. In many segmentation techniques such as the fuzzy c-means clustering, the number of clusters should be known or selected a priori. To alleviate this, several techniques that do not require initialization have already been proposed (see Pauwels and Frederix (1999) for a study of the limitations of the classical clustering methods and a description of non-parametric clustering without initialization).

In Pauwels and Frederix (1999) the non-parametric clustering in image segmentation is investigated. A robust and versatile method is proposed that can handle very diverse clusters using intermediate levels processing. Initially, a density image is generated by a convolution of the original image with a Gaussian kernel. Then a neighborhood is defined for each pixel. The neighborhood consists of 1 % of the total image pixels. After the convolution, the candidate clusters are recognized with the aid of the gradient, and each point is associated with the maximum density point in the neighborhood. The process leads to over-clustering assembling the data into a set of small clusters around local maxima. A hierarchy of clusters is created using the density of the data for the systematic merging of neighboring clusters. The sequence of merging is determined after comparing the local maxima in the neighborhood with the density at boundary points. The assessment of the validity of clusters is done by using two mathematical quantities, *isolation* and *consistency*. Using this algorithm it is possible to export regions of high differentiation and consistent semantic content. The authors conclude that this algorithm produces a smooth version of the result of the inverse normalization algorithm without corrupting the information in the image. It can be applied to graylevel images such as continuous tone images, photographs of persons, manuscripts, etc. In the experiments presented the created clusters were homogeneous and objects were separated clearly from the background. One of the advantages of this method is that it automatically determines the number of regions or objects and therefore does not require prior knowledge of the number of clusters.

In Kurita (1991) an efficient clustering algorithm was developed for region growing applications. The algorithm is a typical example of hierarchical clustering. It starts with an initial partition of the image into N sections and gradually reduces this number by merging pairs based on some criterion. The merging process is repeated until the required number of regions have been reached. The algorithm uses sorted linked lists for maintaining correlation of neighboring regions, and a stack structure for storing the unevenness of all possible pairs of regions. A set of experiments in remote sensing images yielded good segmentation results in flat areas, while segmentation of convex regions resulted in multiple regions (over-segmentation). The authors concluded that the algorithm implements the stepwise optimization method for region growing.

In Frigui and Krishnapuram (1999) a study of three basic issues on conventional clustering is presented. These issues relate to the sensitivity during the initialization, the difficulty in defining the number of clusters, and the susceptibility to noise. The proposed method, Robust Competitive Agglomeration (RCA) begins with a large number of clusters to reduce the sensitivity to initialization and determines the final number of clusters through a competitive combination. The method combines the advantages of hierarchical clustering and partitioning clustering. To prevent sensitivity to noise, ideas from the statistical robustness are incorporated. Fuzzy logic is being used to resolve overlapping clusters. To address regions of ambiguity and reduce sensitivity to initialization, finite rejection is used. The algorithm was tested on noisy synthetic data with ellipsoidal and linear clusters, and a multitude of natural (continuous tone) images. The results demonstrate that the method can provide a robust estimate of the original parameters even when the clusters differ significantly in size and shape and the data are noisy.

In Ohm and Ma (1997) a segmentation based on cluster characteristics is proposed for image sequences. The algorithm analyzes specific features from an image sequence and controls their reliability locally to create parts that might belong to an object. The segmentation process is based on clustering that takes into account different characteristics such as color and motion. The approach is similar to vector quantization. Different weights are being applied to different features according to their reliability. The pixel-feature vector is then compared with a set of feature vectors of a cluster and thus falls into a class of features. The entire set of feature vectors is updated for each image in the sequence, which is being used for the segmentation of the next image. After a scene change, the whole process begins again with a new set of feature vectors calculated from the first image of the scene. The authors conclude that the algorithm is a hybrid combination of iterative methods and produces a dense vector field. It can automatically provide a flexible definition of the weights for different features.

In Ng (2000) an extension to the conventional k-means algorithm is described, which proposes a conversion of the separation rule in order to control the number of cluster members. With the introduction of appropriate restrictions in the mathematical expression of the problem, the authors have developed an approach that allows the use of the model of k-means for the efficient formation of clusters with a pre-selected number of members in each cluster. The aim is to minimize an objective function, usually a function of distances, calculated for all feature vectors from the respective centers of the classes.

A comparison between different clustering techniques and the study of their behavior is very important. It is important that an optimal clustering technique is able to select the correct number of clusters. In Dubes and Jain (1976) there is a detailed comparison of three categories of clustering methods. A total of eight methods are compared in terms of clustering efficiency, squared error minimization, hierarchical clustering and graph-theoretical clustering. Important issues on the use of clustering methods, such as user options, computational cost, inputs and outputs and comparisons of methods with different outcomes are considered. The authors suggest that the different techniques can be compared in terms of four factors, including how the next clustering is selected or updated, what are the criteria for the creation of new clusters, or the criteria for deletion or merging of clusters, and what the initialization process should be. Methods of clustering, in which performance is based on a criterion of squared error minimizing are negatively judged. In addition a set of nine criteria are being defined for comparing methods in the sense of how clusters are created, and based on the data structure and sensitivity of the method to changes in data that do not significantly alter their nature. This paper compares the different methods based on those criteria and eventually provide a scoreboard.

The validity of the clusters is also important. In Yarman-Vural and Ataman (1987) several issues in clustering methodology are being studied, including the creation of clusters, the number of members in each cluster, the inductive cluster partitioning and the noise influence, in proper clustering. Validity criteria are analyzed, including maximum likelihood and the sum of squared error. Maximum likelihood was found to provide better results when the number of clusters varies. The paper also proposes a method of improving conventional clustering for efficient clustering in noisy data. The algorithm ISODATA is modified and the resulting method improves clustering efficiency up to 20 % in a dataset with Gaussian data. Similar efficient performance is also reported for image analysis applications.

In Dubes (1987) the utility of the above method is investigated for the validity of clustering, and it is being compared with the proposed improved Hubert's statistic. Results in a Monte Carlo study[6] are then being produced by varying parameters such as size, number of standards, a sampling window and number and distribution groups.

In Zahid et al. (1999) an inductive method is proposed to impose fuzzy clustering validity. Its principle relies on fuzzy segmentation. The main focus of the study is the analysis of the physical structure inherent in the data. The measure of assessing the quality of fuzzy clustering, corresponds to a real number at the output of the fuzzy clustering algorithm being used. Two types of outputs are considered. For each output a combination of criteria of fuzzy segmentation and fuzzy density is

---

[6]In essence, Monte-Carlo statistics support the creation of a probability density function for the study of the effect of noise in the data. Monte-Carlo processes may prove useful in the study of the characteristics of a distribution, which are affected by noise, along with the study of characteristics that are crucial for the interpretation of the data.

assessed. Maximizing the criteria leads to good clustering. This maximum value allows the definition of the correct number of clusters. The first function calculates a ratio of fuzzy segmentation and fuzzy density by estimating geometrical properties and properties of members of data. The second function computes the same ratio by using only the properties of members. The goal of the clustering phase is to create a clearly defined fuzzy partition, which can be as close as possible to the physical structure of the data. Experimental results show that the method gives a better estimate than other clustering validity functions especially in overlapping fuzzy groups.

3.3.6 Probabilistic and Bayesian Segmentation

In Haddon and Boyce (1990) use is being made of a method based on co-occurrences, using information both of regions and edges to improve the efficiency of segmentation of image sequences. The authors approach the image segmentation by integrating information from regions and edges using the so-called co-occurrence matrices.[7] These matrices are used to generate the feature space. In a first phase, an initial segmentation is being done, based on the location of the intensity of each pixel and its adjacent pixels in the matrix. Each pixel is assigned to a vector that determines whether it belongs to a given region or if it is an edge pixel. This temporary segmentation is then improved by characterization (Hansen and Higgins 1997; Rosenfeld et al. 1976) which provides local coherence of the pixels during the segmentation, by minimizing the entropy of the region. If a pixel does not belong to an edge, it is mapped to one of the regions. This categorization is completely unidirectional in the direction of co-occurrence. The regions and edges coherence is therefore satisfied with the assumption that the marginal regions are no more than a pixel in width. The algorithm was tested on synthetic images and infrared natural (continuous tone) images. The authors found that their method gives a satisfactory result for most regions of the images, and the majority of actual edges and boundaries had been properly identified. The algorithm was also tested in series of infrared aerial images. The authors conclude that the method is robust and gives satisfactory results even when images are taken under poor conditions. It is, though, less effective if groups in the co-occurrence field have significant overlaps due to local spatial coherence. Since the technique uses global information into a local context, it was possible to adapt to a multitude of features, such as variability in color and texture. Another conclusion that comes out in this work is that the difference in co-occurrence matrices

---

[7]This corresponds to a family of methodologies based on the analysis of repeated texture structures. In general, a number of matrices is calculated for a texture, and from those various other features are derived. The co-occurrence matrix is defined by a magnitude measure and an angle, and its mathematical expression is: $C_{\theta,d}(x, y) = |\{(m, n) \in (M \times N) \times (M \times N) : d(m, n) = d, tan^{-1}(m - n) = \theta$ or $\pi - \theta$ and $f(m) = x, f(n) = y\}|$ where d(a,b) a distance measure, usually $d[(a, b), (c, d)] = max(|a - c|, |b - d|)$ and $f[(M \times N) \mapsto N(0, 255)]$ the image under analysis. This definition guarantees that the matrix is diagonal.

of a sequence of images is negligible when the circumstances and the content are not significantly variable.

An extension of the above methodology was proposed by Haddon and Boyce (1993). They proposed two similar techniques for segmentation and edge detection based on co-occurrence matrices. The first defines transformations that enhance the difference between typical (and non-typical) features. The second uses the locations of distributions per region in a co-occurrence matrix and defines the position of the corresponding limiting distributions. The techniques were analyzed by labeling in matrices. The matrices can be used to segment regions and simultaneous detect distinct regions. Segmentation and edge detection is carried out simultaneously. The operator used to generate the co-occurrence matrix, effectively creates also the reference positions within the matrix. Convolution is being applied of the operator with a neighborhood of pixels around the pixel to be processed and the pixel is replaced by the label to the co-occurrence matrix on its reference position. The process is repeated for the same operator and various directions. Correlation techniques and procedures of conjugate iterations are used to determine the positions and extent of the distributions in the co-occurrence matrix. The algorithm is applied to Forward Looking InfraRed (FLIR) and multispectral images. Most processes in the algorithm are executed in parallel. The results show a significant improvement over the prior method.

In Haddon and Boyce (1994) the co-occurrence matrix method is being used for the segmentation of FLIR image sequences to their key spatial regions. These regions are subject to classification based on texture. The co-occurrence matrices are normalized in the time axis to achieve consistency in the segmentation of the sequence. The co-occurrence matrices of edges are divided using *Hermite functions*[8] due to their Gaussian structure, and partly due to the presence of Gaussian noise in the images. When the structure is removed, then the texture of the image remains. Then neural networks are being applied (Multi Layer Perceptrons (MLP)[9]) for the classification of the images using the coefficients of Hermite functions. The co-occurrence matrix is a multi-dimensional histogram, in which each element (i, j) corresponds to the frequency with which two events i and j with a specific relationship between them co-occur. The co-occurrence matrices of edges enhance the difference between two pixel intensities. The authors use the Canny edge detection to create the matrices. A sequence of 300 FLIR images taken from low-flying aircraft approaching a bridge were segmented by this method. Images were initially segmented using co-occurrence matrices and normalization in time. Then a co-occurrence matrix was created for each region, which was decomposed by means of orthogonal Hermite

---

[8]Gabor expressed but not proved that the from all real-valued functions, the Hermite functions have the smallest product of implicit uncertainty. It holds that: $g_n(x) = H_n(x)e^{-\frac{1}{2}x^2}$, with $H_n(x) = (-1)^n e^{x^2} \frac{d^n}{dx^n} e^{-x^2}$ the Hermite polynomials.

[9]MLP is the most common type of neural network. It is simple yet with powerful mathematical basis. The input data pass through layers of neurons, with the input layer consisting of as many neurons as the variables of the problem. The output layer has so many neurons as the desired output variables (often just one). The intermediate layers are called hidden.

functions. The high order functions describe the texture of the region and form a vector of 121 features. Every third sample is obtained to generate the validation set. The training and validation sets are uncorrelated. MLP networks were trained until the validation error began to rise (over-training). Additionally, PCA was used to extract the most representative of the 121 features. Very good results were obtained with both a network of one single layer (architecture $25 \times 10 \times 5$) with 96.9 % accuracy in training and 98.3 % in the tests, and with a network of two layers (architecture $25 \times 5 \times 8 \times 5$) with 94.8 % accuracy in training and 88.7 % in the tests. The main regions that were marked were the 'grass', the 'trees', the 'sky', the 'river water that reflects the trees' and the 'river water that reflects the sky'. Problematic regions were mainly the mirage regions. The results after application of a PCA were not as satisfactory as with the use of all the 121 features. This technique appeared robust in the presence of noise and resulted in an overall 93.4 % success in identifying regions of the images.

Haddon et al. (1997) reports of an algorithm for automatic segmentation and classification of images using co-occurrence matrices and Hermite functions. The process involves four steps, including segmentation, texture analysis, initial classification and characterization, and space-time classification. In the segmentation phase, the key regions are separated by using co-occurrence matrices and edge detection. The texture classification of segmented regions is performed using discrete Hermite functions. The co-occurrence matrices are decomposed to calculate features for the regions. The result of the texture analysis is a low-dimensional feature vector describing the texture of each region. Features are selected that contain information about the separation between clusters and are used as inputs to the multi-layered neural network classifier. Local consistency is achieved in the characterization within a region, as well as within the sequence of images. A characterization method is then used to characterize the pixels to correspond to specific coefficients. These compatibility coefficients take into account the current and the previous image in the sequence and the result of the characterization is applied to the previous image for classification. The authors tested the method on a sequence of FLIR images, as well as in snow profile analysis. The authors concluded that the method can be applied to a wide range of applications.

In Haddon and Boyce (1998) is presented a method that ensures consistency in the characterization in classifying the segmented images, both spatially within an image and temporally in a sequence of FLIR images. It is very important to ensure temporal consistency when analyzing a video. Inconsistencies in segmentation between successive images are unacceptable for example to vehicle automatic navigation applications. The image is initially segmented using a technique based on co-occurrence matrices. After segmentation the regions are classified according to texture. Co-occurrence matrices are also used to describe the texture of each region. Then the matrices are decomposed using discrete Hermite functions. The features identified by using linear technical of analysis or PCA, are fed to a MLP neural network for classification. To ensure the spatio-temporal consistency the outputs of the neural network are being taken into account. Two or more outputs of similar values indicate that the outputs are equally likely. Consistency applies only when there is no adequate criteria for a region class, using soft approaches. Neighboring are characterized the regions both in space and time. These neighborhoods affect the likelihood of a region to belong to a class when the result of the neural classifier is unsatisfactory. Tests were conducted in a sequence of 300 (12 s) infrared images. Five groups were found: 'trees', 'grass', 'sky', 'river water reflecting trees' and 'river water that reflects the sky'. The results were very satisfactory after the initial segmentation. After imposing the consistency all regions have been classified and only minimal classification errors were observed.

In Comer and Delp (1999) an algorithm for the segmentation of texture images using a Bayesian method of multiple levels is proposed. This algorithm makes use of a Multi-level Gaussian Auto-Regressive (MGAR) model for the pyramidal representation of the observed image. A larger neighborhood is begin used for the segmentation with respect to single-level algorithms. The samples, which are multi-level representations of the original image, are generated by Gaussian pyramidal decomposition. The nodes in a binary tree are then used for the classification of data and a MGAR model is applied. The value of a random variable is predicted as the linear combination of values of the random variables in previous nodes. Thus a Gaussian pyramidal representation is created. Then a multi-level Markov Random Field model is applied for the pyramid of classes. An optimization criterion is then applied for segmentation. This criterion minimizes the expected value of a number of incorrectly classified nodes in the field of multiple levels. Ultimately, the Expectation Maximization algorithm is applied to estimate the parameters. The method assigns a cost to an erroneous segmentation, based on the number of wrongly classified pixels in the process. The method was tested on two different images with a three-level pyramidal decomposition. The results were satisfactory for both images resulting in a segmentation into regions of homogeneous texture.

3.3.7   Segmentation with Neural Networks

Campbell et al. (1997) proposed a method of automatic segmentation and classification of natural images using neural networks. Initially, the images are segmented using Self-Organizing Feature Maps (SOFM) that use color and texture information. The SOFM used are of 64×64 nodes for better segmentation results. A set of 28 features is extracted from each region. These features include the average color value, position, size, orientation, texture (Gabor filters) and shape (using PCA). Classification is then performed using a MLP with 28 input neurons and 11 output neurons. The training made use of 7,000 regions, whereas the tests were carried out on an independent set of 3,000 samples. Over 80 % of the regions were correctly classified using Learning Vector Quantization (LVQ) while the MLP classified correctly 91.9 % of the regions.

Papamarkos and Gatos (1994) suggested a multi-thresholding method for graylevel images. The process comprises three processing stages. Initially a hill clustering technique is applied in the image histogram to determine the peaks. Then, the areas between the peaks are approximated by real functions using a linear minimax approach. Finally, by applying the one-dimensional Golden Search Minimization (GSM) the global minimum of each of the functions is approximated, giving a threshold value for each function. The method was tested on images with a histogram of two or more peaks in comparison with other known thresholding methods and gave satisfactory results, better than thresholding methods that use probabilistic or global thresholding.

Papamarkos (1999) proposed a new method for reducing the number of colors in color images. The novelty of the method lies in the fact that for the quantization of the colors both the color information of each pixel and local features are taken into account, which are the inputs to a neural network (SOFM). After training the network, the output neurons provide the final image colors (the color palette). The number of output neurons is predefined and corresponds to the number of colors of the output. The output image is formed by the representative colors of the original image while local texture characteristics are preserved. The advantage of the method is that it can use any local features as criteria, leading to different desired results.

In Papamarkos et al. (2000) a segmentation process using SOFM for color images is presented. The particular application of these neural networks is considered as similar to a multi-thresholding method, where the output of the network provides a number of homogeneous clusters. Interesting point in this work is the technique used to identify the optimal number of thresholds, or in other words, the number of segmentation regions. Assuming that the distributions within the regions are Gaussian, it is proposed that the linear combination of probability density functions for each region should be close to the overall density, or the global histogram of the original image. For different numbers of segmentation regions it can assess whether or not an error criterion is minimized.

A special case (and expansion) of Papamarkos et al. (2000), is discussed in Papamarkos and Strouthopoulos (2000) where a multi-thresholding method is proposed for graylevel MRC. The method is based on Page Layout Analysis (PLA) and multi-thresholding through a neural network (SOFM). Initially, through PLA, parts of the image are recognized and classified as text, lines and graphics. During the second phase, the multi-thresholding method is applied to each of these parts; in the parts consisting of text and lines a single threshold is selected, whereas in parts consisting of graphics the number of required thresholds is selected adaptively (Papamarkos et al. 2000). The authors report that they conducted extensive tests on MRC, getting better results in comparison with other methods. They also stated that the method can be applied in all cases of MRC multi-thresholding.

In Papamarkos and Atsalakis (2000) is proposed a method of reducing the number of gray levels in intensity images. The method (which is essentially based on Papamarkos (1999)) describes a procedure that is performed in five steps, including the definition of the features to be used in the classification, PCA on the data, downsampling for training, SOFM neural network training, and final classification through the neural classifier. Furthermore, an adaptive extension of this method is proposed, in which by means of a tree-shaped approach the iterative color reduction method is applied with an increasing number of features being used. The authors report that tests performed on various images gave satisfactory results.

Another extension of the method presented in Papamarkos and Atsalakis (2000) is described in Papamarkos et al. (2002), where an effort is being made to improve the efficiency of the method in Papamarkos and Atsalakis (2000). The process of color reduction is performed essentially as in Papamarkos and Atsalakis (2000), where at the level of the adaptive operations a split and merge process has been added to combine similar classes and to support the decision for the termination of the iterative process. The authors report that the method is particularly important for cases of MRC color reduction with many significant colors.

Atsalakis et al. (2002a) is a different approach to the same idea in the previous works. The problem of determining the optimal number of colors in an image is made here using three proposed methods and their results creates a palette of the image. The number of palette colors is then the number of output neurons of the neural classifier applied as in the above works. Creation of the palette is done by analyzing the one-dimensional histogram per color channel by three methods, including histogram approximation, histogram normalization, and histogram approximation through SOFM. In the experiments carried out, the authors conclude that the method of determining the number of colors using a linear approximation of the histogram gives the best results.

In Atsalakis et al. (2002b) an extension of Papamarkos et al. (2002) is presented, in which the adaptive reduction of colors occurs in parts of the original image and, during the final phase, the colors of all the segments are input to a neural classifier that gives the final representative colors of the image. The aim of this approach was to make possible a Very Large Scale Integration (VLSI) implementation of the algorithm (parallel implementation), which is also proposed in this work.

3.3.8 Other Methods for Segmentation

In Medioni and Yasumoto (1984) a segmentation method that uses the fractal dimension[10] is proposed. The authors calculated the fractal dimension of texture in the frequency domain by approximating a straight line in the logarithmic power spectrum. The difference of the logarithm of the expected values of the two bipolar statistics gives one single texture feature. The process approximates the fractal dimension, which correlates with the expected statistics of the differences compared

---

[10]Assuming S a closed subspace, then for any $\epsilon > 0$ and $N(\epsilon)$ the minimum number of spheres of radius $\leq \epsilon$ necessary to cover all of S, if there exists a δ such that: $\delta = -\lim_{\epsilon \to 0+} \frac{logN(\epsilon)}{log\epsilon}$, then δ is called the fractal dimension of S. In other words, the fractal dimension can be calculated by the limit of the ratio of the logarithm of change in the size of an object to the logarithm of the change in the measurement scale, as this scale tends to zero. In practice the following relation is used: $\delta = log(\text{number of self} - \text{similar pieces})/log(\text{magnification factor})$.

to the statistics of the distance vectors. The main disadvantage of this method is that the segmentation based on a single feature that is extracted from the image is not always feasible.

In Pentland (1984) the problem of fractal description of natural images is addressed. The authors were concerned with the problems of (i) the representation of natural forms like mountains, trees and clouds and, (ii) the calculation of such descriptions from data. It was observed that many images of natural objects and features such as leaves, snowflakes, etc. exhibit fractal characteristics. This makes it impossible to accurately measure them in an image. For example, in measuring the length of a shoreline, whatever the size of the measurement unit, all curves smaller than the unit of measurement are lost. The classical concepts of length and area do not produce consistent measurements for many natural shapes; the basic measurement properties of these shapes vary as a function of the fractal dimension. The fractal dimension is therefore an essential part of a coherent description of such structures. The characterization of a texture based on a fractal surface model makes it impossible to describe the image in a manner robust to the conversion of the scale and a linear transformation of the intensity. The authors showed how continuous tone images can be segmented based on this method. It is implied that in this method the segmentation is more stable in changes in scale compared with thresholding techniques in the intensity domain. Furthermore, the method was tested against correlation techniques and co-occurrence matrix techniques using a tiling of 8 natural texture images created by Laws (1980). In these images, the authors showed that their approach provides better results.

In Xu et al. (1998) a segmentation algorithm was presented that was based on partitioning the image to linked regions of various shapes with the criterion of minimizing the sum of the pixel intensity variations in all regions, provided that each region has a specific number of pixels and neighboring regions show significant differences in the average luminance.

In other works, topological maps have been used as structural methods for segmentation. In Braquelaire and Brun (1998) a technique that uses topological maps and inter-pixel representation was proposed. The authors describe a data structure that allows to store and process regions of every size and form. A model of discrete maps is used, which helps in effective parameter estimation required by the segmentation algorithm. There are two important aspects in a segmented representation: the geometric aspect that describes the shape of the region and the topological side describing neighborhoods and included regions. A topological map is defined as a partition of an oriented surface on a finite set of points. A discrete map is based on both geometric and topological representation levels that act together. The discrete data structure map is used to design the segmentation algorithms. In this study, the authors describe an iterative split-and-merge algorithm using a discrete data structure. During the first phase segmentation is carried out and during the second its refinement, through an iterative way to gradually improve the quality of segmentation.

In Ojala and Pietikäinen (1999) a method of unsupervised texture segmentation is being presented using distributions of features. The proposed algorithm uses local binary pattern distributions and contrast patterns for measuring the similarity between adjacent regions during the segmentation process. Texture information is measured by a method based on Local Binary Patterns and Contrast (LBP/C). The method consists of three phases, including hierarchical partitioning, cumulative merging and classification at pixel level. In the first stage, the image is divided into regions of approximately uniform texture. Then a cumulative merging takes place to merge neighboring regions with similarities using a merge stopping criterion. Finally, classification is performed in pixel level to improve the local character. The authors use a criterion of non-parametric probability as a pseudo-criterion to compare distribution characteristics. The method was tested on four 'mosaic' texture images and two graylevel continuous tone images, giving satisfactory results. The authors conclude that this method is not sensitive to the choice of parameters and does not require a priori knowledge of the number of segments in the image. The method can easily be generalized to other texture features, multi-level information or color features.

In Yoshimura and Oe (1999) a segmentation algorithm is proposed for texture images using Genetic Algorithm (GA) that automatically determine the optimal number of segmentation regions. The authors divide the original image into small rectangular regions and extract texture features by a two-dimensional self-recursive model, and other features such as fractal dimension, the mean value and the variance. Use is made of three types of methods of evolutionary segmentation. The first method uses GA, the second GAs and SOFM and the third GAs and SOFM while considering the optimum number of segmentation regions in the image. Various experiments were performed in a set of images with three or four different texture regions. It was shown that the third method gave the best results, which were visually more accurate than those of other conventional methods. The first method gave good results but it was necessary to define the number of regions a priori. The second method selects the number of regions automatically, while the third method finds the optimal number of regions with homogeneous texture.

In Perner (1999) a methodology for segmentation based on a case-based reasoning is presented. Segmentation is done by mapping to corresponding known cases and using parameters associated with the corresponding case. The images are compared on the basis of features such as the moments. The similarity is estimated as defined in Tversky (1977). When the optimum segmentation parameters are known for a reference image, by matching new images to the reference image, the same features can be used. The system was proven to be working well for computed tomography (CT) brain image segmentation.

3.4 Segmentation of MRC to Detect Rigid Objects

MRC images constitute a distinctive challenge for segmentation applications, as in this case the aim is to segment into specific semantically grouped items within the image structure, so that it would be possible to properly apply different compression methods to optimize compression and transmission. *The important in the case of MRC compression is that a suitable method of segmentation is one that can successfully separate the text and the graphics from the background, which may be complicated and contain regions with strong texture and color.* This section presents a brief description of methods of MRC segmentation, focusing mainly on the identification of text regions. Although numerous works have been published on this subject that ultimately target OCR applications, the main interest in this treatise is in the segmentation that leads to good separation of textual (and similar) regions. There are two basic categories, including the general segmentation methods for tracking objects with compact colors and special methods linked to a compression standard.

3.4.1 General MRC Segmentation Methods

In Perlmutter et al. (1996) a method of text segmentation using Classification Trees (CT) and Tree-Structured Vector Quantization (TSVQ) is presented. A presentation and comparison of similar methods of block-based classifiers is given and it is shown that the use of characteristics of a linear transformation along with either CT or TSVQ may lead to an accurate recognition of text regions. In TSVQ seven features are being used (minimum, maximum, mean, standard deviation, mean to standard deviation ratio, range and median). Each feature vector is transformed using DCT or DWT. The use of a transformation is due to the observation that image regions with text include significantly large amounts of energy distributed among certain transform coefficients, and thus, the existence of text becomes evident in the transform domain. Alongside, a dimension minimization is being done for the features vector by evaluating the segmentation capacity of each transform coefficient. The authors conclude that the use of either CT or TSVQ combined with features in the DCT domain results in a more accurate identification of text regions.

In Chen and Chen (1998) an adaptive segmentation method is proposed that is applicable to color images from the covers of technical magazines. In this method, an initial separation is done identifying parts of the image as 'main' or not, based on their color diversity within each region. The whole process that follows applies only to these main parts of the image and consists of four basic steps, including color quantization, adaptive segmentation based on edges and color, classification of regions, and finally, post-processing. The authors emphasize that in this method assumptions have been made, including that (a) the rotation angle of the image is limited, (b) text characters are uniform in color and (c) the text is written horizontally. The experimental results, in a total of 100 images, showed 98 % success in the segmentation of simple pages and 95 % in complex pages.

In Todoran and Worring (1999) a method of global segmentation of color documents is presented, based on extensive analysis of the expected shape of groups in the RGB color space, taking into account prior knowledge about the creation process of the color document. The authors assume that the document consists of a set of building blocks in a uniform background color and a limited number of colors for each region, and that the texts are clearly separable from the background having distinct colors. According to this model, and knowing that the printing process and digitization of the document leads to a gradual diffusion of the color on the edges, it is possible to make a prediction of the shape of the structures formed in the RGB color cube of the distribution of colors in the image, where point clouds are expected to form around the dominant colors. Based on this model initial line segments are recognized and then classification of each pixel in the corresponding segment is applied. The authors tested their method on a set of synthetic and digitized images and obtained 90 % recognition of text regions in synthetic images and 53 % in digitized image.

In Sobokkta et al. (2000) a text extraction method is presented applicable to color covers of books and periodicals. The method was tested on documents digitized at 200 dpi. The whole process starts with a color reduction through histogram clustering in the RGB color space. Then two parallel processes for text regions recognition follow; at first, the image is divided gradually in the horizontal and vertical direction forming rectangular sections of at least two colors. In the second process, region growing is applied through a search for regions with homogeneous characteristics. By considering the fact that the text is in a horizontal arrangement, both methods aim at a final clustering of regions in a horizontal direction. The results of both methods are compared and combined. The objects identified as text are converted to black and white. The authors indicate that the method can be applied to cases other than the book covers. The results presented give a 98 % success in recognizing characters that belong to text. In the examples shown it is however apparent that the method fails to recognize the text on a complex textured or multi-colored background.

In Kalman et al. (2001) an MRC segmentation method is proposed based on the likelihood of transform coefficients. DCT is used in blocks of 8×8 pixels and estimation follows of the likelihood that the coefficients correspond to text or graphics. The training for the characterization as text or graphics is done with a set of known images. The assessment of the likelihood of coefficients is done based on various features in the transform domain, such as energy, difference, data rate estimation, or differences between probabilities given number of coefficients (eg. 18 coefficients). The authors report that their proposed methods, based on the relative frequency of occurrence of transform coefficients gave better results than previous similar methods.

In Hase et al. (2001) is proposed a text extraction method from color documents, which is based on reducing the colors of the image in the CIELAB color space, which is performed in the histogram domain to identify the representative colors by inductive logic. After the reduction of colors follows the creation of binary images; the number of binary images equals the number of colors detected. An analysis based on the consistency of 8-neighborhoods follows, and then recognition and labeling of objects and final selection of blocks of text. The authors report that the only assumption of their method is the monochrome nature or color uniformity of texts. They also say that the results of the experiments were very satisfactory, setting the limit of their success at the level at which a psycho-visual recognition would be required.

In Pietikäinen and Okun (2001) is presented a method of text detection in MRC, based on the extraction of edges and usage of texture features. In this method, the color images is initially converted to graylevel (intensity). Next, edges are detected using the Sobel method and a subsequent thresholding. Then the calculation of a feature is done in typical non-overlapping tiles of the image, and classification of segments as blocks of text (or not) based on the calculated feature. The authors report that experiments were made with 25 test images and that they were able to achieve satisfactory results, with robustness against changes in both the intensity and the color. They conclude that the most important information for the determination of text is given by the average width and the density of the edges of characters in each region.

In Hase et al. (2003) is proposed a method for MRC segmentation for text extraction. In this method the assumption is that within the image the texts are in a monochromatic representation. To satisfy this assumption resolution reduction takes place initially, reducing the analysis of the digitized document to 50–80 dpi followed by a transformation in the CIELAB color space. Clustering and classification in a few representative colors takes place. Then, criteria for the treatment of super-segmentation are imposed and the merging with the background using characteristics such as minimum color difference or the assessment of color uniformity. The end result of the process is to identify text areas. The authors conclude that the use of 6–10 colors or 4–7 color differences are satisfactory for proper segmentation in the CIELAB color space.

3.4.2 Segmentation Methods Related to Compression

In Rosenholtz and Watson (1996) a method is proposed for the perceptually adaptive JPEG coding. The aim of the authors was to exploit psychovisual characteristics, such as the contrast sensitivity and the luminance/contrast masking, to determine the perceived quantization error as a smooth quantity over the whole surface of the image. The ultimate goal was to determine a quantization table of weights for each 8×8 block of pixels. The authors present results from the application of the method on a typical image where the improvement in compression ratio reaches 22 % while maintaining a similar perceptual image quality.

In Konstantinides and Tretter (2000), one of the most cited works in this category, Konstantinidis and Tretter suggest a JPEG compression method that uses variable quantization for MRC. The method is based essentially on the estimation of an 'activity' feature on the coefficients of DCT. This estimate is based on the performance in compression, i.e. the number of bits per symbol in each 8×8 block of pixels and can be considered optimal for that encoder. Then a scaling factor is extracted for the quantization table that is coherent with the JPEG-3 extension of the compression standard (ISO-IEC-ITU 1996), which supports the use of variable quantization for each block of the image. The comparison made between the two standards (JPEG and JPEG-3) show the clear superiority of the proposed method in the image quality.

In Memon and Tretter (2000) a variable quantization method in JPEG is proposed to improve the perceived image quality. This method, although targeted to the same extension of the JPEG standard (the JPEG-3), approaches the problem of optimization from a different viewpoint with respect to the approach in Konstantinides and Tretter (2000). The main variation stems from the observation that the previous method treats text areas like textured regions. In this study the classification of blocks is made in the image domain by extracting four characteristics. The authors indicate that the method has the advantage of simplicity in implementation and low computational and memory costs, providing the possibility of parallel processing. The presented results on separating blocks of text are satisfactory.

In Jung and Seiler (2003) a segmentation method is proposed to optimize the compression based on JPEG2000. Section 6 of the standard supports multilayer variable coding per image to optimize the compression in MRC. To create these layers a segmentation method is proposed based on measuring the dynamic range of the intensity per block of the image. The median of the dynamic range is used as threshold for the segmentation of the block. Merging of adjacent blocks of similar characteristics follows along with normalization with appropriate filters. The result of the segmentation is a foreground layer with the text and a background layer with the remaining image pixels.

References

Adams, R., & Bischof, L. (1994). Seeded region growing. *IEEE Transactions on Pattern Analysis and Machine Intelligence, 16*(6), 641–647.

Ahuja, N., Rosenfeld, A., & Haralick, R. M. (1980). Neighbor gray levels as features in pixel classification. *Pattern Recognition, 12*(4), 251–260.

Atsalakis, A., Papamarkos, N., & Andreadis, I. (2002a). On estimation of the number of image principal colors and color reduction through self-organized neural networks. *International Journal of Imaging Systems and Technology, 12*(3), 117–127.

Atsalakis, A., Kroupis, N., Soudris, D., & Papamarkos, N. (2002b). A window-based color quantization technique and its embedded implementation. *IEEE International Conference on Image Processing ICIP 2002*, Rochester, USA.

Basu, S. (1987). Image segmentation by semantic method. *Pattern Recognition, 20*(5), 497–511.

Beaulieu, J. M., & Goldberg, M. (1989). Hierarchy in picture segmentation: A stepwise optimization approach. *IEEE Transactions on Pattern Analysis and Machine Intelligence, 11*(2), 150–163.

Bhanu, B., & Faugeras, O. D. (1982). Segmentation of images having unimodal distributions. *IEEE Transactions on Pattern Recognition and Machine Intelligence, 4*(4), 408–419.

Bhanu, B., & Parvin, B. A. (1987). Segmentation of natural scenes. *Pattern Recognition, 20*(5), 487–496.

Bottou, L., Haffner, P., Howard, P., Simard, P., Bengio, Y., & LeCunn, Yann. (1998). High quality document image compression with DjVu. *Journal of Electronic Imaging, 7*(3), 410–425.

Braquelaire, J. P., & Brun, L. (1998). Image segmentation with topological maps and inter-pixel representation. *Journal of Visual Communication and Image Representation, 9*(1), 62–79.

Brodatz, P. (1966). *Textures: A photographic album for artists and designers.* 1 edn. Dover Publications, Inc., ASIN: B000ZGO6XQ.

Brown, M. (1999). *Fisher's Linear Discriminant.*

Buckley, R., Venable, D., & McIntyre, L. (1997) (November 17–20). New developments in color facsimile and Internet fax. *Proceedings of the Fifth Color Imaging Conference: Color Science, Systems, and Applications* (pp. 296–300).

Campbell, N. W., Thomas, B. T., & Troscianko, T. (1997). Automatic segmentation and classification of outdoor images using neural networks. *International Journal of Neural Systems, 8*(1), 137–144.

Canny, J., & Computational, A. (1986). Approach to edge detection. *IEEE Transactions on Pattern Analysis and Machine Intelligence, 8*(6), 679–698.

Chan, F. H. Y., Lam, F. K., & Zhu, H. (1998). Adaptive thresholding by variational method. *IEEE Transactions on Image Processing, 2*(3), 168–174.

Chang, Y. L., & Li, X. (1994). Adaptive image region-growing. *IEEE Transactions on Image Processing, 3*(6), 868–873.

Chen, W., & Chen, S. (1998). Adaptive page segmentation for color technical journals' cover images. *Elsevier Image and Vision Computing, 16*, 855–877.

Cheng, H., & Bouman, C. A. (1998). Trainable context model for multiscale segmentation. *Proceedings of IEEE International Conference on Image Processing (ICIP 98)* (vol. 1(October 4–7), pp. 610–614).

Cheng, H., Bouman, C. A., & Allebach, J. (1997) (May 18–23). Multiscale document segmentation. *Proceedings of IS&T's 50th Annual Conference* (pp. 417–425).

Cheng, H., Bouman, C. A., & Bouman, A. (2001). Document compression using rate-distortion optimized segmentation. *Journal of Electronic Imaging, 10*(2), 460–474.

Cheriet, M., Said, J. N., & Suen, C. Y. (1998). Recursive thresholding technique for image segmentation. *IEEE Transactions on Image Processing, 7*(6), 918–920.

Cho, K., & Meer, P. (1997). Image segmentation from consensus information. *Computer Vision and Image Understanding, 68*(1), 72–89.

Christopoulos, C., Ebrahimi, T., & Lee, S. U. (2002). JPEG2000 Special Issue. *Elsevier signal processing: Image communication* (vol. 17). Elsevier.

Comer, M. L., & Delp, E. J. (1999). Segmentation of textured images using a multi-resolution Gaussian autoregressive model. *IEEE Transactions on Image Processing, 8*(3), 408–420.

Davies, E. R. (1990). *Machine Vision: Theory, Algorithms*. Practicalities: Academic Press. ISBN 978-0122060908.

DeQueiroz, R. L., Buckley, R., & Xu, M. 1999 (February). Mixed raster content (MRC) model for compound image compression. *Proceedings IS&T/SPIE Symposium on Electronic Imaging, Visual Communications and Image Processing* (vol. 3653, pp. 1106–1117).

Dubes, R. C. (1987). How many clusters are best?-an experiment. *Pattern Recognition, 20*(6), 645–663.

Dubes, R. C., & Jain, A. K. (1976). Clustering techniques: The user's dilemma. *Pattern Recognition, 8*, 247–260.

Frigui, H., & Krishnapuram, R. (1999). A robust competitive clustering algorithm with applications in computer vision. *IEEE Transactions on Pattern Analysis and Machine Intelligence, 21*(5), 450–465.

Gambotto, J. P. (1993). A new approach to combining region growing and edge detection. *Pattern Recognition Letters, 14*(11), 869–875.

Gonzalez, R. C., & Woods, R. E. (1992). *Digital image processing*. 3 edn. Prentice Hall, ISBN: 978-0201508031.

Haddon, J. F., & Boyce, J. F. (1990). Image segmentation by unifying region and boundary information. *IEEE Transactions on Pattern Analysis and Machine Intelligence, 12*(10), 929–948.

Haddon, J. F., & Boyce, J. F. (1993). Co-occurrence matrices for image analysis. *Electronics and Communication Engineering Journal*, 71–83.

Haddon, J. F., & Boyce, J. F. (1994). Texture classification of segmented regions of FLIR images using neural networks. *Proceedings of the International Conference on Image Processing*.

Haddon, J. F., & Boyce, J. F. (1998). Integrating spatio-temporal information in image sequence analysis for the enforcement of consistency of interpretation. *Digital Signal Processing, special issue on image analysis and information fusion*.

Haddon, J. F., Schneebeli, M., & Buser, O. (1997) (May 25–30). Automatic segmentation and classification using a co-occurrence based approach. *Proceedings of the 2nd International Conference on Imaging Technologies: Techniques and Applications in Civil Engineering.*

Hansen, M. W., & Higgins, W. E. (1997). Relaxation methods for supervised image segmentation. *IEEE Transactions on Pattern Recognition and Machine Intelligence, 19*(9), 949–961.

Haralick, R., & Shapiro, L. (1991). *Computer and robot vision* (vol. 1). Addison-Wesley, ISBN: 978-0201108774.

Harrington, S. J., & Klassen, R. V. (1997) (October). *Method of encoding an image at full resolution for storing in a reduced image buffer.* Technical report. US Patent 5,682,249.

Hase, H., Shinokawa, T., Yoneda, M., & Suen, C. (2001). Character string extraction from color documents. *Elsevier Pattern Recognition, 34*(7), 1349–1365.

Hase, H., Yoneda, M., Tokai, S., Kato, J., & Suen, C. (2003). Color segmentation for text extraction. *International Journal on Document Analysis and Recognition, 6*(4), 271–284.

He, H., & Chen, Y. Q. (2000). Unsupervised texture segmentation using resonance algorithm for natural scenes. *Pattern Recognition Letters, 21*, 741–757.

Hojjatoleslami, S. A., & Kittler, J. (1998). Region growing: A new approach. *IEEE Transactions on Image Processing, 7*(7), 1079–1084.

Hough, P. V. C. (1959). Machine analysis of bubble chamber pictures. *International Conference on High Energy Accelerators and Instrumentation.*

Huang, J., Wang, Y., & Wong, E. K. (1998). Check image compression using a layered coding method. *Journal of Electronic Imaging, 7*(3), 426–442.

ISO-IEC. (2000a) (December). Information technology—JPEG. (2000). *image coding system—Part 1: Core coding system, ISO/IEC International Standard 15444-1.* ISO/IEC: Technical report.

ISO-IEC-CCITT. (1993a). *Information Technology—Digital Compression and Coding of Continuous-Tone Still Images—Requirements and Guidelines, ISO/IEC International Standard 10918-1, CCITT Recommendation T.81.* Technical report ISO/IEC/CCITT.

ISO-IEC-ITU. (1993). *JBIG, Progressive bi-level image compression, ISO/IEC International Standard 11544 and ITU Recommendation T.82.* Technical report. ISO/IEC/ITU.

ISO-IEC-ITU. (1996). *JPEG-3, Information Technology-Digital Compression and Coding of Continuous-Tone Still Images: Extensions, ISO/IEC 10 918-3, ITU-T Recom.T84.* Technical report. ISO/IEC/ITU.

ISO-IEC-ITU. (2000). *JBIG2, ISO/CEI International Standard 14492 and ITU-T Recommendation T.88.* Technical report. ISO/IEC/ITU.

Jain, A. K., & Dubes, R. C. (1988). *Algorithms for clustering data.* Englewood Cliffs, N.J.: Prentice Hall.

Jung, K., & Seiler, R. (2003). Segmentation and compression of documents with JPEG2000. *IEEE Transactions on Consumer Electronics, 49*(4), 802–807.

Kalman, M., Keslassy, I., Wang, D., & Girod, B. (2001) (October 7–10). Classification of compound images based on transform coefficient likelihood. *Proceedings of the IEEE International Conference on Image Processing* (vol. 1, pp. 750–753).

Konstantinides, K., & Tretter, D. (1998) (October 4–7). A method for variable quantization in JPEG for improved text quality in compound documents. *Proceedings of IEEE International Conference on Image Processing (ICIP 98)* (vol. 2, pp. 565–568).

Konstantinides, K., & Tretter, D. (2000). A JPEG variable quantization method for compound documents. *IEEE Transactions on Image Processing, Correspondence, 9*(7), 1282–1287.

Kurita, T. (1991). An efficient agglomerative clustering algorithm using a heap. *Pattern Recognition, 24*(3), 205–209.

Laws, K. I. (1980) (January). *Textured image segmentation.* Ph.D. thesis, University of Southern California.

Li, L., Gong, J., & Chen, W. (1997). Gray-level image thresholding based on Fisher linear projection of two-dimensional histogram. *Pattern Recognition, 30*(5), 743–749.

Lu, S. W., & Xu, H. (1995). Textured image segmentation using autoregressive model and artificial neural network. *Pattern Recognition, 28*(12), 1807–1817.

Marr, D. (1982). *Vision: A computational investigation into the human representation and processing of visual information*. Henry Holt and Co., ISBN: 0716715678.

Medioni, G. G., & Yasumoto, Y. (1984). A note on using the fractal dimension for segmentation. *Proceedings of 2nd IEEE Computer Vision Workshop* (pp. 25–30).

Mehnert, A., & Jackway, P. (1997). An improved seeded region growing algorithm. *Pattern Recognition Letters, 18*, 1065–1071.

Memon, N., & Tretter, D. (2000) (February). A method for variable quantization in JPEG for improved perceptual quality. *International Conference on Visual Communications and Image Processing*.

Murata, K. (1996) (July). *Image data compression and expansion apparatus, and image area discrimination processing apparatus therefore*. Technical report. US Patent 5,535,013.

Ng, M. K. (2000). A note on constrained k-means algorithm. *Pattern Recognition, 33*, 515–519.

Ohlander, R. B. (1975). *Analysis of natural scenes*. Ph.D. thesis, Carnegie Institute of Technology, Department of Computer Science, Carnegie-Mellon University, Pittsburgh, PA, USA.

Ohm, J. R., & Ma, P. (1997). Feature-based cluster segmentation of image sequences. *Proceedings of the IEEE International Conference on Image Processing* (pp. 178–181).

Ojala, T., & Pietikäinen, M. (1999). Unsupervised texture segmentation using feature distributions. *Pattern Recognition, 32*, 477–486.

Otsu, N. (1979). A threshold selection method from grey level histograms. *IEEE Transactions on Systems, Man and Cybernetics, 9*(1), 62–66.

Papamarkos, N. (1999). Color reduction using local features and a Kohonen Self-Organized Feature Map neural network. *International Journal of Imaging Systems and Technology, 10*(5), 404–409.

Papamarkos, N., & Atsalakis, A. (2000). Gray-level reduction using local spatial features. *Computer Vision and Image Understanding, 78*(3), 336–350.

Papamarkos, N., & Gatos, B. (1994). A new approach for multilevel threshold selection. *Computer Vision, Graphics, and Image Processing-Graphical Models and Image Processing, 56*(5), 357–370.

Papamarkos, N., & Strouthopoulos, C. (2000). Multithresholding of mixed type documents. *Engineering Applications of Artificial Intelligence, 13*, 323–343.

Papamarkos, N., Strouthopoulos, C., & Andreadis, I. (2000). Multithresholding of color and gray-level images through a neural network technique. *Image and Vision Computing, 18*, 213–222.

Papamarkos, N., Atsalakis, A., & Strouthopoulos, C. (2002). Adaptive color reduction. *IEEE Transactions on Systems, Man, and Cybernetics-Part B, 32*(1), 44–56.

Pauwels, J., & Frederix, G. (1999). Finding salient regions in images: Nonparametric clustering for image segmentation and grouping. *Computer Vision and Image Understanding, 75*, 73–85.

Pavlidis, T. (1982). *Algorithms for graphics and image processing*. Berlin Heidelberg: Springer. ISBN 978-3642932106.

Pennebaker, W. B., & Mitchell, J. L. (1993). *JPEG Still Image Compression Standard*. New York: Springer.

Pentland, A. (1984). Fractal-based description of natural scenes. *IEEE Transactions on Pattern Analysis and Machine Intelligence, 6*(6), 661–674.

Perkins, W. A. (1980). Area segmentation of images using edge points. *IEEE Transactions on Pattern Recognition and Machine Intelligence, 2*(1), 8–15.

Perlmutter, K., Chaddha, N., Buckheit, J., Gray, R., & Olshen, R. (1996) (May 7–10). Text segmentation in mixed-mode images using classification trees and transform tree-structured vector quantization. *ICASSP 1996* (vol. 4, pp. 2231–2234).

Perner, P. (1999). An architecture for a CBR image segmentation system. *Engineering Applications of Artificial Intelligence, 12*, 749–759.

Pietikäinen, M., & Okun, O. (2001) (September 10–13). Edge-based method for text detection from complex document images. *Proceedings of the 6th IEEE International Conference on Document Analysis and Recognition* (pp. 286–291).

Prager, J. M. (1980). Extracting and labeling boundary segments in natural scenes. *IEEE Transactions on Pattern Analysis and Machine Intelligence, 2*(1), 16–27.

Ramos, M., & DeQueiroz, R. L. (1999) (October). Adaptive rate-distortion-based thresholding: Application in JPEG compression of mixed images for printing. *Proceedings of IEEE International Conference on Image Processing (ICIP 99)* (pp. 25–28).

Rosenfeld, A., Hummel, R., & Zucker, S. (1976). Scene labeling by relaxation operations. *IEEE Transactions on Systems, Man and Cybernetics*, *6*(6), 420–433.

Rosenholtz, R., & Watson, A. (1996) (September 16–19). Perceptual adaptive JPEG coding. *IEEE International Conference on Image Processing* (vol. 1, pp. 901–904).

Said, A., & Pearlman, W. A. (1996). A new fast and efficient image codec based on set partitioning in hierarchical trees. *IEEE Transaction on Circuits Systems and Video Technology*, *6*(3), 243–250.

Singh, S., & Al-Mansoori, R. (2000). Identification of regions of interest in digital mammograms. *Journal of Intelligent Systems*, *10*(2), 183–217.

Sobokkta, K., Kronenberg, H., Perroud, T., & Bunke, H. (2000). Text extraction from colored book and journal covers. *International Journal on Document Analysis and Recognition*, *2*, 163–176.

Taubman, D. S., & Marcellin, M. W. (2002). *JPEG2000 Image Compression Fundamentals, Standards and Practice*. Kluwer Academic Publishers, ASIN: B011DB6NGY.

Todoran, L., & Worring, M. (1999). Segmentation of color documents by line oriented clustering using spatial information. *International Conference on Document Analysis and Recognition ICDAR'99* (pp. 67–70).

Tversky, A. (1977). Feature of similarity. *Psychological Review*, *84*(4), 327–350.

Vernon, D. (1991). *Machine vision: Automated visual inspection and robot vision*. Prentice-Hall, ISBN: 978-0135433980.

Wallace, G. (1991). The JPEG still picture compression standard. *Communications of the ACM*, *34*(4), 30–44.

Xu, Y., Olman, V., & Uberbacher, E. C. (1998). A segmentation algorithm for noisy images: Design and evaluation. *Pattern Recognition Letters*, *19*, 1213–1224.

Yarman-Vural, F., & Ataman, E. (1987). Noise, histogram and cluster validity for Gaussian-mixtured data. *Pattern Recognition*, *20*(4), 385–401.

Yeung, M., Yeo, B. L., & Liu, B. (1998). Segmentation of video by clustering and graph analysis. *Computer Vision and Image Processing*, *71*(1), 94–109.

Yoshimura, M., & Oe, S. (1999). Evolutionary segmentation of texture image using genetic algorithms towards automatic decision of optimum number of segmentation areas. *Pattern Recognition*, *32*, 2041–2054.

Zahid, N., Limouri, M., & Essaid, A. (1999). A new cluster validity for fuzzy clustering. *Pattern Recognition*, *32*, 1089–1097.

Chapter 4
Compression Optimization

—Διαίρει καὶ βασίλευε *[diaírei kài basíleue], in ancient
greek, 'divide and rule'.*

Philip of Macedonia

4.1 Introduction

Any method for image compression is directly related to the content of the image
being compressed. Since continuous tone images have quite different content characteristics when compared with artificial images (graphics) and digitized documents,
there is no generic compression method efficiently and readily applicable to all cases.
This means that in the case of color MRC, it is expected (at least in the general case)
that none of the conventional compression methods will yield satisfactory results,
both in terms of the compression ratio and the quality of the compressed image. The
ways to tackle with this problem may be based,

- either on employing an *image analysis approach* that consists basically on the
  development of a new compression method that will eloit new features of the
  images, so as to enable efficient compression of MRC

  – this approach leads, typically, to inductive tactics (heuristics), in which the
    composite image is seen as a set of homogeneous regions, which are recognized,
    divided, grouped and properly compressed

- or on employing a *segmentation approach* that consists on the 'transformation' of
  images in such a way to allow the efficient application of one or more conventional
  compression methods

  – in this approach, the composite image is segmented into two or more parts (sub-
    images) of known statistical characteristics (eg. continuous tone sub-images
    or textual sub-images), and a different and appropriate compression method is
    applied in each sub-image

© Springer Nature Singapore Pte Ltd. 2017
G. Pavlidis, *Mixed Raster Content*, Signals and Communication Technology,
DOI 10.1007/978-981-10-2830-4_4

The *segmentation approach* was the selected choice upon which to build in the part concerned with the optimized MRC compression. Nevertheless, both approaches were carefully studied in order to select the most efficient method. Most of the methods in the *image analysis approach* are usually intended to identify and distinguish the text regions for OCR and conversion into text, making several assumptions about the color depth and characteristics of the concerned regions. In contrast, the methods categorized in the *segmentation approach* are not targeting to a clear separation of the text, but to distinguish regions with similar statistical characteristics and create 'layers' or subsets of the initial images, which can then be encoded with a different and appropriate method.

In this light, this chapter presents results of a study on MRC image compression, in which images are represented by homogeneous information layers. Initially, MRC compression based on generic methods is presented as compared with methods in which the images are segmented into layers. The improved results attained by using methods that exploit segmented images (layers) both in compression ratio and in compressed image quality, lead to an unforced, natural selection of the layered image representation methods. It becomes apparent that a composite image is preferable to be decomposed in order to achieve better compression results. The chapter also includes results of a study on the appropriate methods of filling gaps in the sparse layers of decomposed images, and their final adaptive encoding with typical compression methods. In addition, a comprehensive MRC compression method is presented that uses a layered representation to optimize the compression ratio and the quality of the encoded image.

4.2 Global or Layered Compression?

This section reports the results of a comparative study on MRC compression; the comparison considered the cases in which typical compression methods are applied either globally or on a layered manner. The layered representation should be understood as a decomposition of the composite image into two complementary images: the *foreground image* and the *background image*. The foreground image comprises all high frequency content (text and line art) while the background contains continuous tone images (or, in the simplest case, a background color). A first important observation in this layered representation of MRC is that *the background layer requires a higher resolution for color information with respect to texture and structure, unlike the foreground, in which important role plays the structure and not the color fidelity*. In order to validate this claim, the following experiment was conducted: a set of digitized MRC images was manually segmented into the two layers.

Since the foreground layer is actually an image extracted from the initial image, this extraction creates *holes* or *gaps* of missing data in the remaining background layer. These gaps in the background layer need to be filled with *appropriate* data. On the other hand, the foreground can be simply filled with a single color, like white. Three typical compression methods were selected for the encoding of the

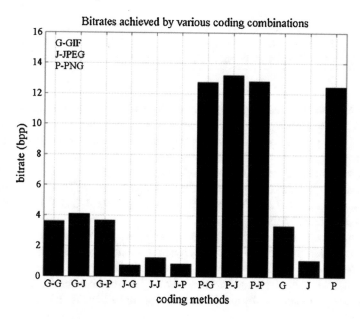

Fig. 4.1 Average data rates obtained by encoding MRC with various methods and combinations

layers, namely JPEG with quality factor 20, Portable Network Graphics (PNG)-24 bpp lossless, and Graphics Interchange Format (GIF)-8 bpp lossless, which encoded the layers using all possible combinations. It should be noted that the GIF method employs lossless Lempel-Ziv-Welch (LZW)-based compression using a color palette of 256 colors, applying a typical color reduction and color diffusion. The JPEG and PNG methods operate on all three color channels. The PNG algorithm performs lossless compression and is based on the use of a dictionary (such as in LZW).

The results confirm the claim, indicating higher compression ratios and compressed image quality for the case in which the background is encoded using lossy compression and the foreground is losslessly encoded after a color reduction.

Figure 4.1 shows a graph with the data rate achieved in this experiment. The horizontal axis consists of the combinations of compression methods applied. In positions in which two methods are shown (denoted by two initials), the first method relates to the compression of the background and the second to the foreground layer respectively. Purely for comparison purposes, the right-hand graph bars represent the application of a single compression method without prior segmentation into layers. The data rate refers to the amount of information obtained after encoding either using a global or a layered approach. For a more accurate presentation of the results, Table 4.1 reports the numerical values that are graphically depicted in Fig. 4.1. The graph of Fig. 4.1 and the results in Table 4.1 reveal a number of interesting facts:

- *The compressing of the background layer plays a major role in the overall data rate to be achieved.* This is confirmed by comparing the results of compression

Table 4.1 Average data rates obtained by encoding MRC with various methods and combinations

Method	G-G	G-J	G-P	J-G	J-J	J-P	P-G	P-J	P-P	GIF	JPG	PNG
Rate (bpp)	3.62	4.08	3.69	0.76	1.23	0.83	12.75	13.22	12.82	3.36	1.08	12.46

using the PNG method (lossless) and the GIF (with color reduction) with the lossy JPEG for the background layer. As seen from the results, application of JPEG on the background layer greatly improves the final compression ratio that can be achieved.

- The use of any layered compression scheme with JPEG compression for the background layer, results an improved data rate which is more efficient than that of the global application of JPEG on the composite image (see J-G, J-J, J-P relative to J on the Figure and Table).
- *It is preferable in terms of data rate efficiency, to use a layered coding scheme with a different method for the background and the foreground (see for example J-G, J-P relative to J-J in the graph).* For an optimized data rate, the compression method of each layer should be selected based on the characteristics of the layer and the requirements set by the HVS modeling applied on the specific information content.
- It is evident that the best compression efficiency is achieved in the cases J-G (0.76 bpp), which corresponds to JPEG compression on the background and GIF-8 on the foreground layers, and J-P (0.83 bpp), which corresponds again to JPEG compression on the background and PNG-24 on the foreground.

In addition and complementing the data rate results, Fig. 4.2 shows the corresponding results of obtained compressed image quality. Apparently, to assess the compressed image quality, decoding and reassembly of the two layers into a single MRC image had to take place, using pixels of the background and foreground layer depending on their original locations. The quality assessment is based on PSNR measurements in dBs, and it is worth noting that in this case the measure does not reflect in the best way the perceived quality. *In fact, PSNR measurements are not suitable for image quality comparisons when color reduction (or generally a color quantization) takes place, since it can only capture in an absolute mathematical way the mean squared error between the images being compared but may not reflect the true similarity or difference between them as perceived by the HVS.* Thus, the use of PSNR here is due to its wide usage for compatibility reasons and understanding of the process. Table 4.2 reports numerically the results plotted in Fig. 4.2. From the graph in Fig. 4.2 and the results in Table 4.2 a number of conclusions can be drawn:

- All combinations of layered coding methods used in the experiment yielded quality estimates (PSNR values) with very close values (around 35 dB). Noting also that the JPEG encoding used a low quality factor, the perspective that a strong compression may be applied on the background layer without a substantial loss in the final image quality can be definitely supported. A further optimization in the performance of a layered compression method should also include an effect on the foreground layer.

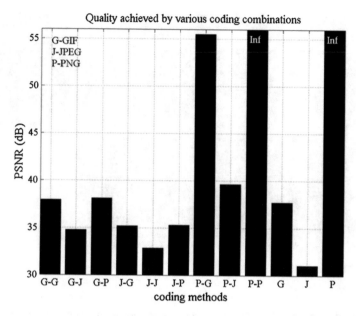

Fig. 4.2   Average quality achieved by encoding MRC with various methods and combinations

Table 4.2   Average quality achieved by encoding MRC with various methods and combinations

Method	G-G	G-J	G-P	J-G	J-J	J-P	P-G	P-J	P-P	GIF	JPG	PNG
PSNR (dB)	37.93	34.78	38.11	35.18	32.84	35.30	55.49	39.64	∞	37.70	31.04	∞

- It is apparent that the PSNR measure is not adequate for assessing the quality of layered MRC encoding. This fact is demonstrated in Fig. 4.3, which gives an example where a magnified portion of one of the experimental images is shown, and it is evident that a high PSNR value does not agree with the perceived image quality.

  Some general conclusions may be drawn from the example in Fig. 4.3:

- Using a color reduction method in the background layer results in a significant change in a large proportion of the initial image and while it can yield high PSNR values, deformations are evident and easily perceived (Fig. 4.3a). In addition, it is not possible to achieve a significant reduction in the data rate.
- The application of global lossy coding should be avoided, because even though significant compression ratios can be achieved, largely noticeable distortions also appear that are particularly evident to a human observer (Fig. 4.3b).
- The application of a global lossless coding is to be considered the worst choice since it typically yields low compression ratios of the order of 2:1 (Fig. 4.3g).
- The use of layered coding, in which lossy coding is applied to the background layer and encoding by reducing colors is applied to the foreground layer, gives the best results in terms of compression ratio and objective image quality. The most

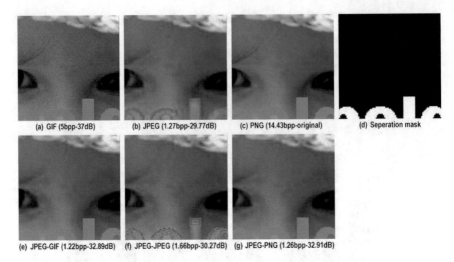

(a) GIF (5bpp-37dB)   (b) JPEG (1.27bpp-29.77dB)   (c) PNG (14.43bpp-original)   (d) Seperation mask

(e) JPEG-GIF (1.22bpp-32.89dB)   (f) JPEG-JPEG (1.66bpp-30.27dB)   (g) JPEG-PNG (1.26bpp-32.91dB)

Fig. 4.3 Portion of an MRC test image, in which different coding combinations were applied

important though is the fact that the method is appropriate because it gives the best results in terms of subjective (perceived) image quality (Fig. 4.3e).

- The use of layered coding, in which lossy coding is applied to the foreground layer is to be avoided, because it leads both to strong deformations in this layer and in an unsatisfactory compression ratio (Fig. 4.3z).
- It should also be noted that using a lossless compression method on the foreground layer, although resulting satisfactory image quality, it can not guarantee a positive contribution to the compression ratio. Specifically, as shown in Fig. 4.3g, the data rate performance is competitive to the case in Fig. 4.3e, mainly due to the fact that the foreground layer consists of uniformly colored regions. If there is no uniformity (or homogeneity), or if special color complexity is present in the foreground, the method loses its competitiveness. This claim was confirmed by an extension of the experiment described as follows: the foreground was deformed by applying monochromatic uniform and Gaussian noise at a rate of 10 % (as shown in Fig. 4.4). Then the experiment was repeated and the obtained results confirmed precisely this claim, as shown in the graphs of Fig. 4.5.

In both cases of noise corrupted foreground, the PNG method requires a higher data rate in the final representation of the data. The data rate difference with the GIF method increased to 1.5 bpp, while it was initially very close to null (0). At the same time, the quality of the final image remained at a similar level in both methods, as shown in the graphs of Fig. 4.6. It should, of course, be noted that it is usually expected to find objects with solid and uniform colors in the foreground of typical MRC images.

The main outcome of these experiments is that a layered representation of MRC images and the application of a different compression method on each of the layers should be expected to yield better MRC compression results, both in data rate and

Fig. 4.4 Portion of an
experimental image with
monochromatic uniform
noise in the foreground layer
(text)

in compressed image quality. Thus, when an MRC image is given, a method of segmentation and decomposition into layers should be applied in order to capture, distinguish and group the regions with similar characteristics, so that a combination of coding[1] methods can be naturally adapted to achieve optimal performance in both compression ratio and in objective and subjective quality of the reconstructed image.

4.3   Layered Coding

Layered coding of MRC is a very efficient approach to the compression of compound images.[2] The experiments presented thus far involved pre-segmented MRC images, in which the background and foreground layers were given in advance. In the general case of an unknown MRC image in compound form, the decomposition into layers must precede any treatment towards an efficient compression. The decomposition may be accomplished using a segmentation method that would be suitable to identify the structural features of the image, in order to further arrange those structurally similar regions in the respective layers. Following the foreground-background model (i.e. the significant-complementary information separation), it is expected that the segmentation aims to separate regions of continuous tone images and large uniform areas in the background and the text with line drawings in the foreground. Final result of such a segmentation is one binary mask representing the separation of pixels into background and foreground and the two images of the respective layers, as shown

---

[1]Note that the use of the terms 'coding' and 'compression' are being used interchangeably throughout the text in this Chapter. The main aim is to focus on 'compression' so any reference to 'coding' should be regarded as clearly connected to compression.

[2]Another term for MRC images.

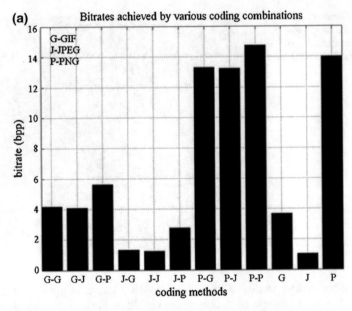

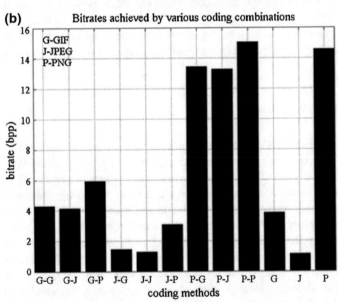

Fig. 4.5 Average data rates in images with corrupted foreground. **a** Uniform, and **b** Gaussian noise

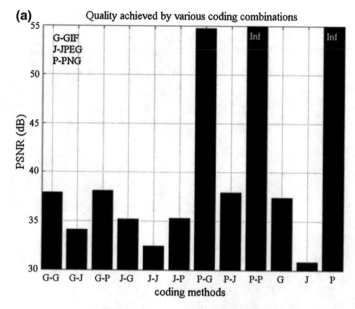

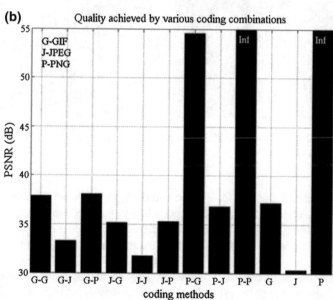

Fig. 4.6 Average quality in images with corrupted foreground. **a** Uniform, and **b** Gaussian noise

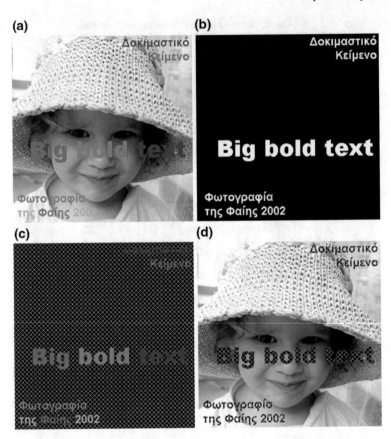

Fig. 4.7 Layered representation of MRC: **a** compound image, **b** separating mask, **c** foreground layer and **d** background layer

in Fig. 4.7. In this figure, the foreground and background layers are represented by images of a size equal to the initial compound image, and they are supplemented with data from the initial image on the respective pixel locations defined by the separation mask. All other positions in the images are vacant (denoted by the cross-hatching pattern).

An immediate issue caused by this layered representation is the creation of gaps or holes or empty regions in the two layers, as defined by the separation mask. Apparently, there should be a method of filling those gaps so that a typical compression method can be applied. Otherwise a new method for sparse image compression should be developed, a fairly complex problem with few reports in the literature. Further, a significant factor in the efficient use of typical compression methods is the identification of any relevance of the characteristics of the layers with a corresponding compression method. Since the segmentation that has already taken place to produce the layers was selected and applied according to this requirement, this notice here has to do more with the methodology to fill the gaps in the complementary layers in order for the conventional methods of compression to be applicable. Thus, once a 'satisfactory' segmentation method is applied followed by an 'intelligent' method to fill-in the gaps in the layers, it is possible to efficiently apply a typical coding method with guaranteed satisfactory compression ratios and compressed image quality.

A side issue is the selection of the layered representation method, since the MRC standard (ITU-T 1999) suggests two alternatives:

- *Representation using three layers, foreground-background-mask*, in which the binary mask defines the separation of pixels of the original image into the two main layers. All layers are images of the same size, as shown in Fig. 4.7. Ultimately, three distinct images are being created,

  - *a binary mask*, which is the product of segmentation and aims to distinguish the pixels according to their response to the segmentation criterion
  - *a foreground image*, which contains color information for the pixels identified in the segmentation mask, typically of high spatial frequency (including text and graphics)
  - *a background image*, which contains the remaining pixels, and is generally of low spatial frequency (including regions of continuous tone images and regions with uniform color or smooth color transitions)

- *Representation using a background and multiple foreground-mask layers*, in which there is a background layer and as many binary masks as the colors of the foreground. The masks can be of an arbitrary size, depending on the spatial distribution of each color in the initial image. Practically, this representation requires the application of a color reduction method in the foreground to produce as few masks as possible (keep few perceptually representative colors). It is expected that a definition of the color must accompany each mask, along with the coordinates of the top-left corner of the mask pixels relative to the position in the initial image. Figure 4.8 shows an example of generating multiple masks from a foreground image of scattered pixels.[3] The whole process of decomposition into multiple layers is as follows:
  For each and every distinct color in the foreground,

1. Identify all the pixels with this color
2. Identify of the bounding box for these pixels and form a sub-image
3. Convert the sub-image to a binary mask with a TRUE value in locations where the pixels have the selected color
4. Store the mask along with the corresponding color and the coordinates of the upper-left corner of the mask in the initial compound image

---

[3]The checkerboard pattern that is shown in the image is used to denote the transparency or the gaps in the layers.

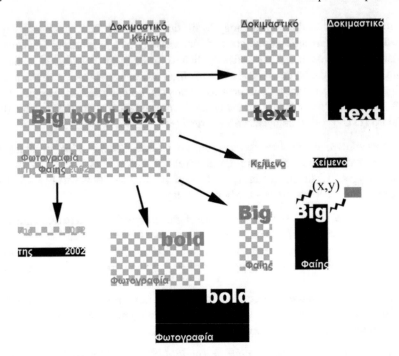

Fig. 4.8 Foreground separation into multiple masks of distinct colors

Within this framework of the layered decomposition of MRC the encoders that may be used herein depend on the characteristics of each layer, and typical selections include the well known JPEG (ISO-IEC-CCITT 1993a; Pennebaker and Mitchell 1993; Wallace 1991) or JPEG2000 (ISO-IEC 2000a; Taubman and Marcellin 2002; Christopoulos et al. 2002) for the layers with true colors and JBIG (ISO-IEC-ITU 1993) or JBIG2 (ISO-IEC-ITU 2000) for the binary layers (such as the masks). Many are the published works (Mukherjee et al. 2001, 2002; Jung and Seiler 2003; DeQueiroz 1999, 2000; DeQueiroz et al. 2000; Yang et al. 1995; Yang and Galatsanos 1997; Ballester et al. 2001; Tony Chan and et al. 2000; Bertalmio et al. 2000, 2001; Oliveira et al. 2001; Masnou and Morel 1998; Masnou 2002; Igehy and Pereira 1997; Efros and Leung 1999; Gorla et al. 2001; Wei and Levoy 2000; Bertalmio et al. 2003; Rane et al. 2002; Shirani et al. 2000; Huang et al. 1998; Chen et al. 1994; Bauschke and Borwein 1996; Jiang and Zhang 2003; DeBrunner et al. 2000; Combettes 1997; Bottou and Pigeon 1998; Stasinski and Konrad 2000) that have demonstrated the validity of this approach.

Considering the three-layer representation of a true color image, it is apparent that it leads essentially to an augmentation of the data rate from 24 to 49 bpp for the same amount of pixels (the representation includes two images of 24 bpp and one of 1 bpp, same size as the initial image). Even though this representation seems to introduce further redundancies, it is essentially the key to a more efficient compression. If the segmentation that takes place is sufficiently intelligent (that is, adapted to the task at hand), then the resulting layers may be compressed with very high efficiency and improved compressed image quality. Furthermore, even if the model appears simplistic, it has proven to be extremely efficient and has been successfully incorporated in various commercial products such as DjVu (Bottou et al. 1998) and Digipaper (Huttenlocher et al. 1999). In these cases, the model, although based on the same concept, it is not compatible with any international standard or definition. In fact, it was only after the final formation of the ITU MRC Rec. T.44 (ITU-T 1999), that the representation of decomposed MRC got a standardized form. Using this standard, L. Sharpe and R. Buckley proposed a layered architecture for JPEG2000 (Sharpe and Buckley 2000) in 2000, Mukherjee et al. proposed a fully-compatible with the standard system related with JPEG in 2001 (Mukherjee et al. 2001) and a fully-compatible with the standard system related with JPEG2000 in 2002 (Mukherjee et al. 2002). In (Jung and Seiler 2003), Jung and Seiler proposed an integrated segmentation and JPEG2000 compression system for MRC.

One of the major issues that arises in any of the layered approaches (three|multi-layer) is that there must be an algorithm for filling the gaps in the various layers so that it is possible to apply a typical encoding scheme. There are various techniques for this goal, although many of them did not initially appear as solutions to this very problem. The important characteristic in this problem is that the filling-in of the missing information should be done in a way that is consistent with the characteristics of the encoder to be used. Apparently, whatever the approach, the data filling is to be done using information from 'existing' regions of the layers, so as to ensure a kind of continuity in the entire layer. Data filling in overall and under this scope is not a simple task. In (DeQueiroz et al. 2000) de Queiroz et al. based on his previous work (DeQueiroz 1999, 2000), focuses on the problem of segmentation and gaps filling with an ultimate goal to create a complete system for MRC compression with optimal characteristics, based on the standard MRC representation.

In the following sections, the difficult problem of gap-filling is being treated. After a review of existing gap-filling methods, an innovative method based on non-linear successive projections is described, which was properly designed to ensure consistency with the features of the standard JPEG encoder. This method is obviously applicable to the background layer and leads to efficient compression of this layer, while improving the subjective and objective quality of the final image. Then a way to apply the second method of the layered representation (background-multiple masks) is described, which in combination with the method of gaps-filling in the background leads to a better compression and quality outcome.

4.4 Methods for Data Filling in Images

In most cases, the foreground layer is filled by horizontal stripes of the same color based on the color of existing pixels in the neighboring regions, thus ensuring continuity on the horizontal axis, while all other gaps can be filled with a single uniform color (usually white). In this manner, the image becomes consistent with all encoders that act in a block-based manner on the image and use causal neighborhoods as prediction domains. Furthermore, if the foreground is not expected to be very dense and crowded in pixels (it is usually expected to be quite sparse), the compression can easily be optimized and it is expected to lead to satisfactory results, as will later be pointed out.

What makes a real difference in the efficiency of the overall layered MRC compression system is the compression of the background layer. This layer is typically well crowder and dense in pixels, and may be filled with a variety of methods. In most of the published works on this subject, two directions can be, basically, identified, including

- methods to achieve data recovery (reconstruction of the missing data)
- methods aimed strictly to improve the compression ratio

4.4.1   Data Filling for Improved Reconstruction

Generally, in the family of methods aiming to reconstruct missing data, this reconstruction is driven by the need to restore an ideal or imaginary image. Many published studies coming from other fields of image processing, such as the reconstruction of a damaged image, address the problem in the general approach of supplementing or replacing of lost or damaged information (as in Yang et al. 1995). Other image processing fields may also contribute significantly to the restoration, such as special compression distortion removal techniques (as in Yang and Galatsanos 1997). Another promising technique made its appearance in the field of reconstruction, inspired by the way art conservators act: it is the technique of *image inpainting*. Other relevant techniques include *disocclusion* and *artificial texture synthesis*.

In the *inpainting* techniques (Ballester et al. 2001; Tony Chan and et al. 2000; Bertalmio et al. 2000, 2001; Oliveira et al. 2001) local characteristics are being used for the diffusion of information from existing parts of the image to the gaps, through an iterative process. In particular, taking advantage of the local gradient revealed by the Laplacian of the image, and the direction of the isophote (orthogonal gradient), it becomes possible to iteratively transfer image data to blank areas (gaps). The diffusion of data is obtained by the solution of the partial differential equation

$$\frac{\partial I}{\partial t} = \nabla(\Delta I) \cdot \nabla^{\perp} I \tag{4.1}$$

where, $I, \nabla, \Delta, \nabla^{\perp}$ are the image, the gradient, the Laplacian operator, and the orthogonal gradient operator, and t is an artificial variable which represents the virtual time in the process.

Disocclusion (Masnou and Morel 1998; Masnou 2002) may, in fact, be defined as a concept opposite to occlusion, which, in turn, is the result of the existence of an object in front of another, or in front of the background, making parts of the scene invisible. Reversing this definition one may say that the idea of disocclusion describes the attempt to remove an imaginary object, which would be responsible for a given gap in the background. Generally, disocclusion takes into account level lines in a neighborhood around the gaps and tries to extend them in the gaps while guaranteeing that various criteria hold, such as the criterion of non-intersection of the level lines using the definition of T-junctions.

Furthermore, in *artificial texture synthesis* (Igehy and Pereira 1997; Efros and Leung 1999; Gorla et al. 2001; Wei and Levoy 2000), the gaps in images are globally replaced by existing parts according to characteristics of adjacent regions. In particular, in most methods, after defining a neighborhood around the gaps, a search algorithm identifies similar regions in the image (based on a distance criterion). When the 'best' regions is identified, then this region is being used to fill the gaps.

In addition, *hybrid image-inpainting-texture-synthesis methods* have been proposed (Bertalmio et al. 2003; Rane et al. 2002) and their effects in restoring missing information are most encouraging. According to these methods, the gaps in the images are sometimes filled with inpainting and sometimes with texture synthesis according to the characteristics of a predefined neighborhood, thus using the advantages of both involved methods.

Finally, there are many works that deal with the problem of the *recovery of information* that was lost during the transmission through noisy channels. In particular, in cases where the images are encoded after being segmented, the errors could lead to damage or loss of the whole of the affected parts of the image. When JPEG2000 coding is used, whole code-blocks may be missing or distorted parts in the image may occur. The loss of parts is more obvious in JPEG encoding, which is designed to operate on blocks (the typical 8×8 pixel blocks). Many works have already contributed to the recovery of lost information in JPEG compression during transmission (Shirani et al. 2000). In such cases the decoder attempts to fill-in the lost information by making use of prior knowledge about the distribution of DC and AC coefficients (Smoot 1996; Reininger and Gibson 1983; Smoot and Rowe 2003) and the encoder characteristics.[4]

4.4.2 Data Filling for Improved Compression

Apart from methods that tackle gaps filling as a reconstruction problem there are other methods that consider it a problem of compression optimization. These methods are the focus of this section.

In (Huang et al. 1998) Huang et al. suggested a layered method for compressing images of scanned bank checks, where gaps in the background layer were filled

---

[4]It should be noted that the use of JPEG implies that the images to be encoded are of particular characteristics (generally smooth) and are consistent with the characteristics of the encoder, and therefore, are compressed sufficiently. This is true in this case, since the images of the background layer are expected to be smooth.

with the average values of existing neighboring pixels. This method is obviously not optimal and is not related to a specific encoder, but it leads to an improved compression efficiency, especially in cases in which there is relative smoothness in the background layer and the gaps are not extensive.

In (Chen et al. 1994) Chen et al. proposed an iterative Projections Onto Convex Sets (POCS) method (Bauschke and Borwein 1996; Jiang and Zhang 2003; DeBrunner et al. 2000; Combettes 1997), for encoding arbitrarily shaped parts of images using a conventional DCT-based encoder. This method was given the name of Linear Successive Projections (LSP). In LSP, an image is sequentially projected from the convex set of images that can be represented through a selected number of transform coefficients into the convex set of images whose pixel values outside the gaps are defined by the original image. What is important is that the process converges, as suggested by the theory of POCS.

In (Bottou and Pigeon 1998) Bottou and Pigeon, suggested an extension of the above algorithm in a DWT-based coding system, where the dimension of the scale is an important factor of acceleration. This algorithm has been incorporated into the coding engine of DjVu IW44 for the background layer of MRC images.

In (Stasinski and Konrad 2000) Stasinski and Konrad, proposed a variation of (Chen et al. 1994), which is again using two projection operators: a bandwidth restriction and a sample replacement operator. The bandwidth restriction operator acts on the frequency domain in an oversampling mesh, providing significant versatility in shaping the spectrum. Additionally, the researchers proposed a specially designed FFT to reduce complexity. The recovery algorithm comprises three steps,

1. Fourier transform of the reconstruction error on an irregular sampling grid
2. Update and restriction of the bandwidth in the frequency domain
3. Inverse Fourier transform and reconstruction

In (DeQueiroz 1999, 2000; DeQueiroz et al. 2000) de Queiroz et al. proposed a gaps filling method targeted especially to the use of the JPEG encoder, in which

- parts belonging to the background are not affected
- parts belonging entirely to the foreground are filled with the mean value of the pixels of the previous part (in consistency with the DPCM coding of the DC coefficients in the JPEG encoder)
- parts that are partially empty are filled through an iterative process that disseminates information from existing pixels into the gaps, using the average values of the immediately adjacent pixels, resulting in smooth transitions, in order to minimize the influence of the AC coefficients.

Obviously this method was designed having a particular encoder in mind and thus led to very good results (despite the simplicity of the approach). In (DeQueiroz 2000) de Queiroz also mentions (without further analysis) the possibility to fill gaps in the transformation domain (similar to the POCS approach (Chen et al. 1994), also taking into account the quantization.

4.4.2.1 Data Filling Using Non-linear Successive Projections

Inspired by the works in data filling methods for improved compression, Pavlidis et al. in (Pavlidis et al. 2004; Pavlidis and Chamzas 2005) proposed a system that implements a Non-Linear Successive Projections (NLSP) to enable data filling of the background layer of decomposed MRC images in a compression-wise efficient manner, as implied in (DeQueiroz 2000; Chen et al. 1994). The aim in these works was to use a general compression standard (JPEG) and to test the idea that the expansion of POCS theory incorporating a non-linear process (quantization) might lead to a further improvement of the compression ratio. If the encoder is general, the rest of the system should of course be consistent with it, in terms of its design and functional characteristics. Only this way it may be expected to improve the compression ratio.

The typical JPEG encoder uses DPCM coding to reduce the magnitude of the quantized DC coefficients and Huffman coding to encode these DPCM differences along with the quantized AC coefficients. In this perspective, it seems valid to claim that the required data filling can be simulated by a prediction of both the DC coefficients and several AC coefficients for each region that includes gaps, in order to transform the image in a format that requires fewer bits during entropy coding. Given a color image, a binary mask,[5] and that the encoder is a standard baseline JPEG, the proposed system:

- performs an initial gap filling (pre-filling) of the background layer to create a 'good' baseline for the NLSP
- compresses the pre-filled image using standard JPEG with a given quality factor, and estimates the data rate after compression
- decompresses the compressed data stream and assesses the quality only in the areas in which background pixels initially existed
- restores the initial background pixel values
- updates the DC coefficients' values of the entirely empty blocks
- repeats the process until convergence; convergence is achieved when no further changes are measured in compression ratio and the estimated quality (PSNR)

A mathematical formulation of the functions performed by the system uses two sets,

1. the set of images represented by a subset of the transform coefficients quantized by the quantizer $q(i, j)$ of JPEG

$$Q = \{\hat{f} \mid \hat{f}(i, j) = f(i, j)/q(i, j), \quad \forall i, j\} \tag{4.2}$$

2. the set of images whose pixel values in the non-empty areas are equal to those of the pixels of the initial image (as defined by the separation mask)

$$P = \{\hat{f} \mid \hat{f}(x, y) = f(x, y), \quad x, y \notin M\} \tag{4.3}$$

---

[5]It is supposed that in these masks a TRUE value represents the foreground.

where f, q, M the image, the quantizer and the mask, i, j the coordinates in the transform domain and x, y the spatial coordinates.

Obviously the second set P is convex, whereas the first set Q is not. Since this holds, the system cannot be described by the POCS theory, and therefore there is no guarantee to converge to a global minimum. Despite the lack of theoretic proof for convergence, the researchers simulated a system that successively projects between sets P and Q (as described above), and found that the system always converges to a sub-optimum. In addition, it was also demonstrated that the convergence speed as well as the final approximation of the global optimum was found to depend significantly on the initial pre-filling method employed. Figure 4.9 shows the block diagram of method for the data filling of the background layer of a decomposed MRC image using non-linear successive projections.

Intuitively, what happens is that during the iterations, diffusion of information from existing pixels to the pre-filled gaps and vice versa occurs. This can be illustrated by a simple example, in which in a horizontal stripe of 8×8 blocks of pixels there are partially empty blocks, as shown Fig. 4.10a. In this case, all empty pixel locations appear in black. The boundaries of the JPEG blocks are marked in blue color. After the pre-filling process, the image takes the form shown in Fig. 4.10b. All empty locations are now filled with initial values. Finally, Fig. 4.10c shows the result of the NLSP after achieving convergence, adapted for JPEG compression with quality factor 50 (which essentially determines the quantizer of the JPEG $q(i, j)$).

As the method is limited to achieve sub-optimality, one may assume that the initial conditions (that is, the pre-filling method) are of particular importance for improving the approximation of the global optimum. For the experimental proof of this hypothesis, extensive tests were conducted with different data filling methods in order to assess the effect of each method in the overall system efficiency. Thirteen (13) were the methods that have been tested and compared, including

(0) no pre-filling—all the gaps are filled with black
(1) pre-filling using the mean DC value in a 3×3 neighborhood of 8×8 blocks
(2) pre-filling with the gradient of DC coefficient values from the left and right adjacent blocks (similarly to a curve fitting between DC values)
(3) pre-filling with the weighted mean of DC coefficient values in a causal neighborhood of 8×8 blocks (as defined in (Shirani et al. 2000))
(4) pre-filling with the global mean pixel value of the image
(5) pre-filling using the most probable background color value (estimated as the color that corresponds to the maximum peak of the Gaussian smoothed histogram of the image—smoothed iteratively until it reaches a bimodal form)
(6) pre-filling of totally empty blocks with the DC coefficient value of the previous block; pre-filling of partially empty blocks with the weighted average of the existing pixels of the previous and current block
(7) pre-filling using the de Queiroz method (as defined in (DeQueiroz et al. 2000))
(8) pre-filling using fast image inpainting (kernel 1) (as defined in Oliveira et al. 2001)

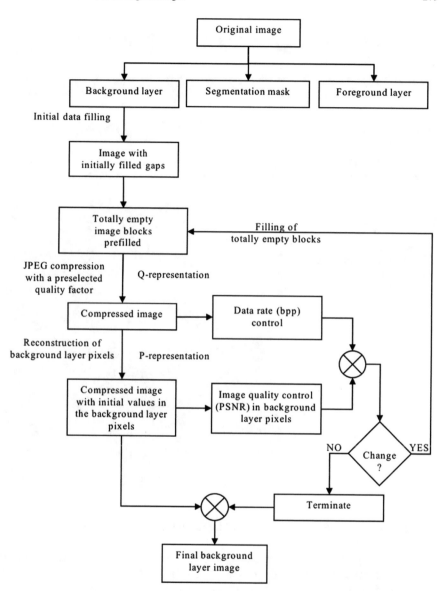

Fig. 4.9 Block diagram of the non-linear successive projections method for data filling of the background layer in decomposed MRC images for sub-optimal JPEG-matched compression

(a)

(b)

(c)

Fig. 4.10 **a** Initial image stripe (height of 8 pixels) with 'problematic' 8 × 8 blocks, **b** the image stripe after the pre-filling, **c** the image stripe after reaching convergence in the NLSP

Fig. 4.11 A test image background layer and the localization of the 8 × 8 blocks with gaps that need data filling so that the image could be efficiently compressed

(9) pre-filling using fast image inpainting (kernel 2) (as defined in Oliveira et al. 2001)

(10) pre-filling using standard image inpainting (as defined in Bertalmio et al. 2000)

(11) pre-filling using LSP (as defined in (Huang et al. 1998))

(12) pre-filling using modified LSP with a one-level DWT (similar to Bottou and Pigeon 1998)

Figures 4.11, 4.12, 4.13 and 4.14 show an example of a background layer with gaps and the result of various data filling methods used in this research (8 selected out of the total 13 methods described above). The methods presented in these images, according to the definition, are (0) no pre-filling, (1) DC mean of neighbors, (2) DC gradient, (3) weighted causal DC mean, (4) global image mean, (5) most probable background, (6) DC-Huffman and (7) DeQueiroz. It is clear how these methods differentiate in providing data filling for the background layer. Figure 4.15 shows a

Fig. 4.12 Result of various pre-filling methods used in this research: (0) no pre-filling, (1) DC mean of neighbors, (2) DC gradient and (3) weighted causal DC mean

selection from the dataset of the MRC color images and accompanying masks that were used in this research. Apparently the images exhibit a wide range of features with multiple colors, even various types of texture, along with texts in various page positions, languages, colors and font sizes, whereas handwritten text is also included in the dataset. The images are mostly scanned compound images of high resolution.

As expected, method (0) yielded the worst results, leading to low compression ratio and compressed image quality. Methods (4) and (5) also led to unsatisfactory results mainly due to the fact that these methods are global and fail to capture local features, thus not adapted to the encoder. Poor results were also attained by method (1) which, even though it is based on the prediction of the DC value in the DCT domain, it is not strictly related to the encoder characteristics. Methods (8), (9) and (10) based on image inpainting are actually reconstruction methods: their effect was to create the image that intuitively simulated the ideal image behind the scenes. The results were really impressive in terms of reconstruction, but both the efficiency of the compression, and the excessive time required for their application, render them

Fig. 4.13 Result of various pre-filling methods used in this research: (4) global image mean, (5) most probable background, (6) DC-Huffman and (7) DeQueiroz

poor choices for this application. Methods (11) and (12) provided, in general, good results. On the other hand, the methods (2), (3), (6) and (7), in which the data filling actually took into consideration the selected encoder, provided the best results. Of these, method (6) yielded, on average, the best combination of compression-quality results. This method was named *DC-Huffman* since it is based on using previous DC coefficient values, following the Huffman coding model of the JPEG encoder.

Figure 4.16 shows a graph of the overall assessment of the NLSP system performance for all the pre-filling methods for a standard MRC image and a preselected JPEG compression quality factor 50. In this figure, the methods are ordered from left to right according to their efficiency in compression. The measurements are expressed as relative to method (0) (i.e. without pre-filling before the application of NLSP). It should be noted that all the quality measurements are in terms of PSNR; in particular, use is made of the PSNR estimation proposed in (Kang and Leou 2003), in which the quality is assessed in the YUV color space instead of RGB, and the

Fig. 4.14 The background
layer after data pre-filling
(using the DC-Huffman
method) with the affected
8×8 blocks visible .

final quality estimate is a linear combination of the estimates in each of the three
channels,

$$PSNR = \frac{1}{6}(4 \cdot PSNR_Y + PSNR_U + PSNR_V) \tag{4.4}$$

where $PSNR_Y$, $PSNR_U$, and $PSNR_V$ are the corresponding PSNR values of the
Y, U, and V color channels of the image.

Figure 4.17 shows the total gain in compression and quality by the DC-Huffman
method in comparison with the worst pre-filling method (method 0) for two representative images and four different quantization tables (for JPEG quality factors 20,
50, 70, and 90). Test image 'fay' is an image with a continuous tone background
and areas of strong texture, while image 'magazine' is a standard digitized magazine
page with basically white background. These images and the corresponding masks
are shown in Fig. 4.18. Figure 4.19 shows the effect of the method after reaching
convergence on pre-filling method (0) and method (6) (DC-Huffman) for these two
test images and a JPEG quality factor 50. If one carefully observes Fig. 4.19a, b in
Fig. 4.20 it can be easily seen that without pre-filling the final image is not expected
to be filled with data stemming from the existing pixels. Additionally, since abrupt
changes and edges are maintained, ringing artifacts are expected to be significant in
the areas around the gaps, leading to a further reduction in overall perceived image
quality. On the other hand, the DC-Huffman pre-filling leads to the creation of a
smooth image with no discontinuities, in consistency with the encoder being used.
Both the objective and the subjective quality obtained is improved significantly.

The algorithm was tested on a set of standard MRC test images, both graylevel
and true-color, and similar positive results were obtained. In the process of these

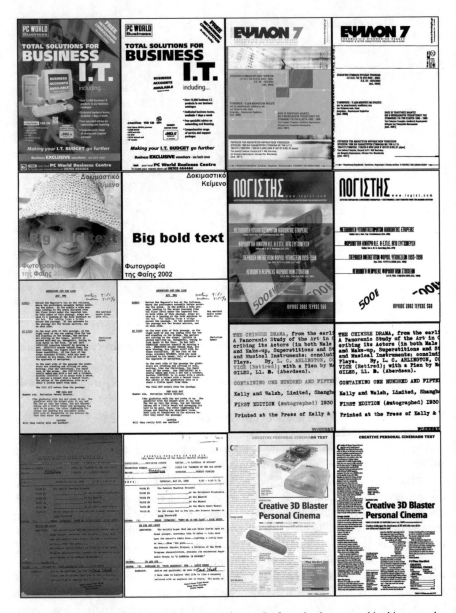

Fig. 4.15 A selection of images and accompanying masks from the dataset used in this research

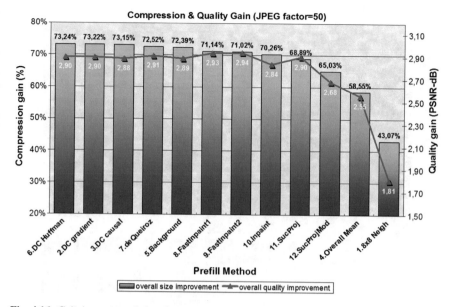

Fig. 4.16 Gain in compression and quality for all methods of pre-filling relative to no pre-filling (0)

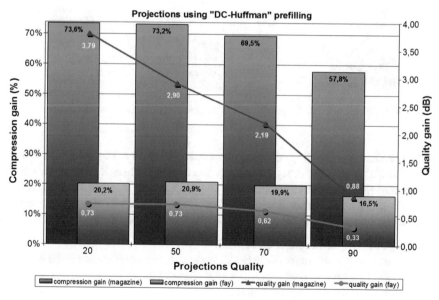

Fig. 4.17 Gain in compression and quality by the DC-Huffman method in comparison to no pre-filling for representative test images 'magazine' and 'fay'

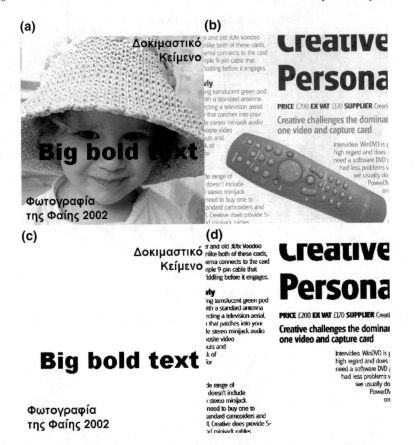

Fig. 4.18 Representative test images. **a** 'fay', **b** 'magazine' and the corresponding masks

tests, the Otsu thresholding technique was used (Otsu 1979) to create the binary
mask (based on the luminance channel of the images). Then, this mask was used
to separate foreground and background. The background was filled using five (5)
methods and was compressed with standard JPEG. Figure 4.21 shows the average
gain in compression and quality for each of the following pre-filling methods,

(1) No pre-filling and application of NLSP
(2) Pre-filling with the most likely background color and application of NLSP
(3) Pre-filling with DC-Huffman and application of NLSP
(4) Filling with the de Queiroz method
(5) Filling with the classic LSP method

The results reflect the performance of each method when the objective is (a)
the achievement of a specified quality or (b) the achievement of a specific data
rate. Apparently, in these experiments the foreground layer was disregarded. Overall
results show that the DC-Huffman method proved more efficient in a rate-distortion

Fig. 4.19 Comparison between no pre-filling and pre-filling with the DC-Huffman method: image 'fay'. **a** at 24,967 bytes (32.85 dB PSNR—ringing artifacts are apparent), **b** at 19,479 bytes (33.58 dB PSNR), image 'magazine', **c** at 54,039 bytes (33.76 dB—ringing artifacts are apparent), **d** at 14,500 bytes (36.66 dB)

sense, giving good compression rates with improved compressed image quality at a fast convergence rate. The average number of iterations during the NLSP process was 22 for method (3), the lowest in all NLSP methods and approximately the same as in methods (4) and (5). The number of iterations showed an increasing trend with the increase of the target data rate, as shown in Fig. 4.22, which depicts the number of iterations as a function of the required JPEG quality factor. In particular, this increasing trend seems to be linear and can be approximated by a line with a slope of $0.1533 \approx \frac{46}{30}$ (f being the JPEG quality factor),

$$\text{iter} = \frac{1}{30} \left(46 \cdot f + 80 \right) \tag{4.5}$$

(a)   **(b)**

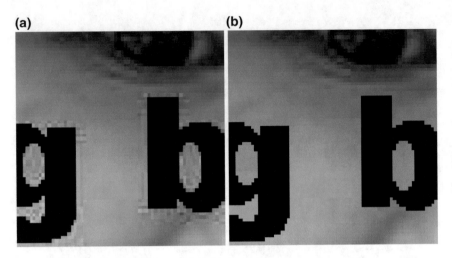

Fig. 4.20 Magnified portion of Fig. 4.19a, b respectively

It should be emphasized that inpainting methods were not included in these experiments, because, due to their iterative nature (iterative data diffusion), they usually require hundreds or thousands of iterations for a typical data filling task for such background images. Thus, a comparison with the other selected methods would not be fair and of no practical importance. As described in previous paragraphs, the purpose of inpainting methods is on an entirely different basis, aiming to a perceptually proper image recovery without taking into account requirements for compression.

To neutralize the effect of a 'bad' segmentation during the creation of the binary mask by the Otsu method, another set of experiments was designed, in which the masks were prefabricated (a ground-truth was formed). During further experiments, using the above five (5) methods throughout the set of test images, and using the full range of values for the JPEG quality factor (10 to 90 in steps of 10), interesting results were obtained, as shown in Figs. 4.23, 4.24 and 4.25[6] and summarized in Table 4.3; while the best compression ratio was achieved by method (4), the corresponding quality was similar to that of method (2), which gave one of the worse compression results. The second best compression ratio achieved was by method (3), which gave better quality in the overall. At the same time, the complexity of different methods have led to varying convergence times, revealing a significant advantage of method (3) over methods (4) and (5) that required significantly more running times.[7]

---

[6]It should be noted that in these figures the methods have been arranged based on their performance. Following the definition of the experiment, the order in which the methods appear is (1), (2), (5), (3), (4) in Figs. 4.23, 4.24, whereas in Fig. 4.25 results are shown relative to the results attained with method (1) and arranged as (2), (3), (5), (4).

[7]The experimentation platform was based on an implementation in MatLab, which can not be considered optimal. In this sense, the computational times referenced may only be taken into account as indicative relative values purely for comparison.

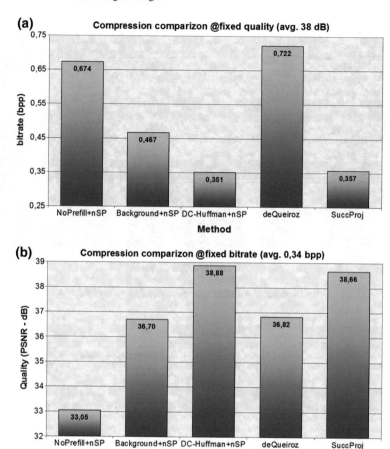

Fig. 4.21 Average gain in compression and quality for five data filling methods: **a** when targeting a specific image quality, **b** when targeting a specific compression ratio

Figures 4.26, 4.27 and 4.28 show three representative results attained in this experiment. In these images the two top-left images are the initial compound image and the background mask, whereas the rest of the images correspond to the results of the compared data filling algorithms: no pre-filling and application of NLSP, prefilling with a background color and application of NLSP, DC-Huffman pre-filling and application of NLSP, application of linear successive projections and the de Queiroz method. In all examples shown, the JPEG quality factor was set to 50 (default value). Tables 4.4, 4.5 and 4.6, list the raw measurements for each of the three examples shown (for SSIM see (Wang et al. 2004)). Apparently in all cases the DC-Huffman + NLSP method outperforms all other methods is terms of rate-quality and also in terms of running times as already pointed out in previous results.

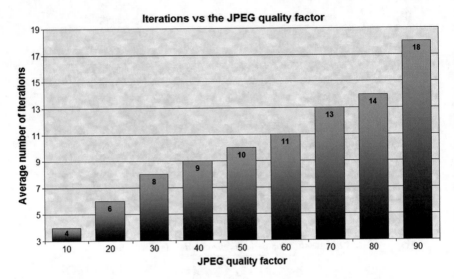

Fig. 4.22 Average number of iterations versus the JPEG quality factor when using DC-Huffman data filling

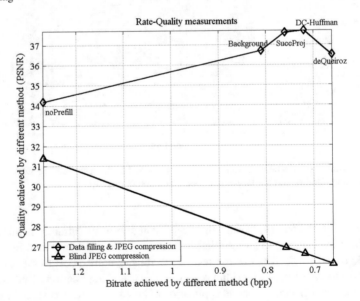

Fig. 4.23 Overall average comparative rate-quality results for data filling and compression

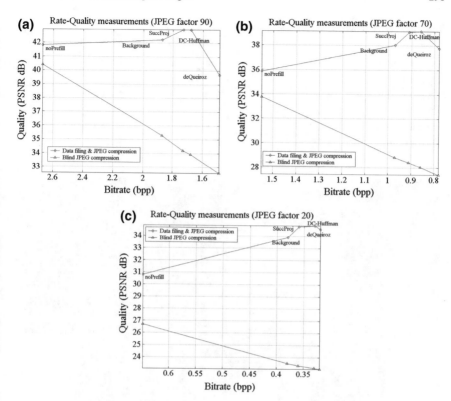

Fig. 4.24 Rate-quality comparative results for data filling and compression using three JPEG quality factors

Fig. 4.25 Relative convergence times for the compared data filling and compression methods

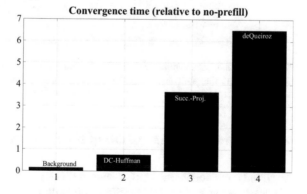

Table 4.3 Overall average data filling and compression results

Method	Bitrate (bpp)	Quality (dB)	Time (s)
(4) DeQueiroz	0.66	36.5	437
(3) DC-Huffman	0.72	37.7	50
(5) Successive Projections	0.76	37.6	248
(2) Background	0.81	36.7	8
(1) No Prefill	1.27	34.2	67

Fig. 4.26 Data filling example #1; from *top-left* compound image and binary mask, no pre-filling and NLSP, pre-filling with the most probable background color and NLSP, DC-Huffman pre-filling and NLSP, linear successive projections, DeQueiroz method

4.4.3 Efficient Compression of the Foreground Layer

As already mentioned, the MRC standard (ITU-T 1999) suggests two alternatives to a layered representation, which differ only with respect to the representation of the foreground:

- Three-layered foreground-background-mask decomposition, in which a binary mask defines the pixel locations of the original image that belong to the foreground. All layers are images of the same size.

Fig. 4.27 Data filling example #2; from *top-left* compound image and binary mask, no pre-filling and NLSP, pre-filling with the most probable background color and NLSP, DC-Huffman pre-filling and NLSP, linear successive projections, DeQueiroz method

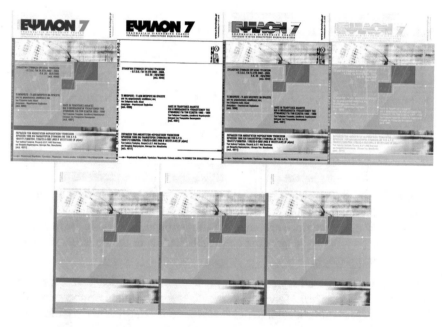

Fig. 4.28 Data filling example #3; from *top-left* compound image and binary mask, no pre-filling and NLSP, pre-filling with the most probable background color and NLSP, DC-Huffman pre-filling and NLSP, linear successive projections, DeQueiroz method

Table 4.4 Raw results for the data filling example in Fig. 4.26

Method	Image size (bytes)	Quality (dB)	Quality-SSIM
No pre-filling + NLSP	210,423	32.47	0.676
Background + NLSP	212,969	32.46	0.675
DC-Huffman + NLSP	151,462	32.84	0.681
LSP	152,647	32.83	0.681
DeQueiroz	147,374	32.69	0.680

Table 4.5 Raw results for the data filling example in Fig. 4.27

Method	Image size (bytes)	Quality (dB)	Quality-SSIM
No pre-filling + NLSP	24,962	32.59	0.905
Background + NLSP	22,262	33.07	0.908
DC-Huffman + NLSP	19,754	33.36	0.913
LSP	20,092	33.51	0.915
DeQueiroz	19,182	33.14	0.912

Table 4.6 Raw results for the data filling example in Fig. 4.28

Method	Image size (bytes)	Quality (dB)	Quality-SSIM
No pre-filling + NLSP	225,859	33.63	0.795
Background + NLSP	148,985	34.67	0.800
DC-Huffman + NLSP	108,208	34.99	0.802
LSP	109,505	34.99	0.803
DeQueiroz	106,099	34.84	0.803

- Multi-layered background-foreground masks representation, in which there is a background layer and a number of binary masks that correspond to the pixel locations of the representative colors of the foreground. The size of the foreground masks can be arbitrary, depending on the distribution of the represented color in the original image.

In order for the second representation model to be of practical value, it requires the reduction of colors in the foreground layer to keep to a limited number of required masks. It is expected that a color definition must accompany each mask, along with the coordinates of its top-left corner relative to its position in the initial image.

For the choice of the representation which would lead to a better compression performance, the following tests were conducted:

(a) The foreground layers of manually generated ground-truth MRC images were compressed using both the representations,

Fig. 4.29 A sparse foreground image (*left*) and its filled-in counterpart (*right*); PNG compression leads to 0.11 bpp; JBIG compression of the corresponding arbitrarily-sized binary masks leads to 0.09 bpp

- by reducing the colors to 16 (4 bpp), filling the gaps with the most prominent color (see Fig. 4.29 that corresponds to Fig. 4.7), and applying PNG lossless compression, and
- by reducing the colors to 16 (4 bpp), creating multiple binary masks of arbitrary sizes (one for each color) (see Fig. 4.8 that again corresponds to Fig. 4.7) and applying JBIG lossless compression on each binary foreground mask; in the experiments both the number of bytes required to store the color of each mask and those needed for the position in the initial image were also accounted for (i.e. 5 extra bytes were summed to the results).

(b) The same experimentation was applied to images in which automated segmentation was imposed using a simple global thresholding right after contrast enhancement. Obviously, this method is not suitable for the best outcome in MRC, but was selected as an extreme case for demonstration. Figure 4.30 shows an example of such a case with a rather dispersed foreground mask with gaps, where part of one of the binary masks is shown to highlight how a poor segmentation and an inexact color reduction has resulted in groupings of dispersed pixels, and the same part of the foreground image is shown after the color reduction and before the final compression. The result for the given image was 0.62 bpp for the case of multiple masks and 0.52 bpp for the case of a global compression.

In the first group of tests, the multiple binary masks approach yielded slightly better average compression results with a bitrate of 0.09 bpp versus 0.10 bpp yielded by a direct compression of the foreground. On the contrary, during the second group of tests the efficiency reversed yielding 0.80 bpp versus 0.63 bpp. In overall, the average data rate was 0.6 bpp for multiple masks against 0.5 bpp for the direct foreground compression. It is worth noting that the multiple masks method gives

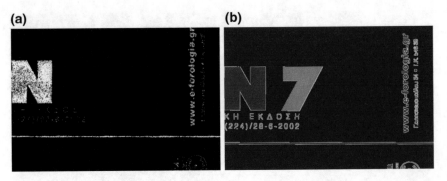

Fig. 4.30 Part of a test MRC image used during the experiments: **a** part of a mask showing the dispersion of pixels due to poor segmentation, **b** the same part of the foreground layer after the color reduction

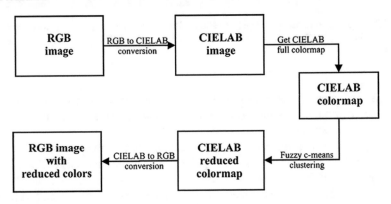

Fig. 4.31 General framework of color reduction based on fuzzy c-means clustering

better compression results when the segmentation preceding is adapted to the image content. Extensive tests have shown that the segmentation is very important in the efficiency of compression of the foreground (as also of the background). It should finally be noted that the color reduction in the experiments was implemented by the well known *fuzzy c-means* clustering method on the palette of all colors of each image in the CIELAB color space. The overall process is shown in the Fig. 4.31.

Concluding this chapter, it is worth summarizing some interesting findings of the research presented on MRC compression when using a segmentation to multiple layers. Most importantly, *compression of MRC images is preferable to be tackled within a framework of a layered representation, selecting a different compression method for each layer*. Specifically,

- based on the observation that the foreground layer is expected to contain objects with solid colors, a color reduction method accompanied by a lossless compression method is expected to yield better results in a rate-distortion sense,[8]
- based on the observation that the background layer is expected to exhibit substantial color variation but reduced structural information, a method propel data filling of the gaps of the layer accompanied by a lossy compression method is expected to yield better results in a rate-distortion sense.

Based on these observations, a method was presented for the optimized data filling of the background layer, which is based on non-linear successive projections (NLSP) inspired by the general theory of projections onto convex sets (POCS). In this method, in order to ensure coherence with the selected encoder (JPEG in this case), a specific pre-filling method was developed for the efficient generation of initial data, which was tested over many other available techniques and gave successful results. In the experiments conducted, an improvement of up to 70 % in compression ratio was observed in comparison with a non-layered approach. At the same time it was observed that the technique depends heavily on the initial pre-filling of the image. The DC-Huffman pre-filling method that was developed led to very good results in a rate-distortion sense. The image quality was improved up to 3 dBs in PSNR, along with a substantial reduction of ringing artifacts in regions with discontinuities. Finally the problem of filling and compressing the foreground layer was addressed, which, by its nature, does not include the problems of the background and can be treated with standard methods that are able to yield satisfactory results. Nevertheless, experiments were conducted with two different approaches and it is worth noting that the multiple masks method yielded better results when the prior segmentation into layers is better adapted to the image content.

References

Ballester, C., Bertalmio, M., Caselles, V., Sapiro, G., & Verdera, J. (2001). Filling in by joint interpolation of vector fields and gray levels. *IEEE Transactions on Image Processing, 10,* 1200–1211.

Bauschke, H., & Borwein, J. (1996). On projection algorithms for solving convex feasibility problems. *SIAM Review, 38,* 367–426.

Bertalmio, M., Bertozzi, A., & Sapiro, G. (2001). Navier-Stokes, fluid dynamics and image and video inpainting. In *Proceedings of the IEEE Conference on Computer Vision and Pattern Recognition.*

Bertalmio, M., Sapiro, G., Ballester, C., & Caselles, V. (2000). Image inpainting. In *Proceedings of the 27th International Conference on Computer Graphics and Interactive Techniques (SIGGRAPH 2000)* (pp. 417–424).

Bertalmio, M., Vese, L., Sapiro, G., & Osher, S. (2003). Simultaneous structure and texture image inpainting. *IEEE Transactions on Image Processing, 12(8).*

Bottou, L., & Pigeon, S. (1998). Lossy compression of partially masked images. In *Proceedings of the IEEE Data Compression Conference.*

---

[8]Better results in a rate-distortion sense signify that the compression under study is expected to yield a low data rate and a high objective and subjective reconstructed image quality.

Bottou, L., Haffner, P., Howard, P., Simard, P., Bengio, Y., & LeCunn, Y. (1998). High quality document image compression with DjVu. *Journal of Electronic Imaging, 7*(3), 410–425.

Chan, T., Osher, S., Shen, J., Strong, D., Blomgren, P., Wong, C. K., et al. (2000). *Homepage for Digital Image Inpainting.*

Chen, H., Civanlar, M., & Haskell, B. (1994). A block transform coder for arbitrary shaped image segments. In *Proceedings of the IEEE ICIP 1994* (pp. 85–89).

Christopoulos, C., Ebrahimi, T., & Lee, S. U. (2002). JPEG2000 special issue. In *Elsevier signal processing: Image communication* (Vol. 17). Elsevier.

Combettes, P. (1997). Generalized convex set theoretic image recovery. *IEEE Transactions on Image Processing, 6*, 493–506.

DeBrunner, L., DeBrunner, V., & Yao, M. (2000). Edge-retaining asymptotic projections onto convex sets for image interpolation. In *4th IEEE Southwest Symposium on Image Analysis and Interpretation.*

DeQueiroz, R. (1999). Compression of compound documents. In *Proceedings of the IEEE ICIP 1999.*

DeQueiroz, R. (2000). On data filling algorithms for MRC layers. In *Proceedings of the IEEE ICIP 2000* (pp. 10–13)

DeQueiroz, R., Fan, Z., & Tran, T. (2000). Optimizing block-thresholding segmentation for multilayer compression of compound images. *IEEE Transactions on Image Processing, 9*(9), 1461–1471.

Efros, A., & Leung, T. (1999). Texture synthesis by non-parametric sampling. In *Proceedings of the IEEE ICCV* (pp. 1033–1038).

Gorla, G., Interrante, V., & Sapiro, G. (2001). Growing fitted textures. In *Proceedings of the SIGGRAPH 2001.*

Huang, J., Wang, Y., & Wong, E. K. (1998). Check image compression using a layered coding method. *Journal of Electronic Imaging, 7*(3), 426–442.

Huttenlocher, D., Felzenszwalb, P., & Rucklidge, W. (1999). Digipaper: A versatile color document image representation. In *Proceedings of the IEEE ICIP 1999.*

Igehy, H., & Pereira, L. (1997). Image replacement through texture synthesis. In *Proceedings of the IEEE ICIP 1997* (pp. 26–29).

ISO-IEC. (2000a). Information technology—JPEG. (2000). *image coding system—Part 1: Core coding system, ISO/IEC International Standard 15444–1.* ISO/IEC: Technical report.

ISO-IEC-CCITT. (1993a). *Information Technology—Digital Compression and Coding of Continuous-Tone Still Images—Requirements and Guidelines, ISO/IEC International Standard 10918-1, CCITT Recommendation T.81.* Technical report ISO/IEC/CCITT.

ISO-IEC-ITU. (1993). *JBIG, Progressive bi-level image compression, ISO/IEC International Standard 11544 and ITU Recommendation T.82.* Technical report ISO/IEC/ITU.

ISO-IEC-ITU. (2000). *JBIG2, ISO/CEI International Standard 14492 and ITU-T Recommendation T.88.* Technical report ISO/IEC/ITU.

ITU-T. (1999). *Mixed Raster Content (MRC ITU-T Recommendation T.44).* Technical report ITU-T.

Jiang, M., & Zhang, Z. (2003). Review on POCS algorithms for image reconstruction. *Computerized Tomography Theory and Applications, 12*(1), 51–55.

Jung, K., & Seiler, R. (2003). Segmentation and compression of documents with JPEG2000. *IEEE Transactions on Consumer Electronics, 49*(4), 802–807.

Kang, L. W., & Leou, J. J. 2003. A new error resilient coding scheme for JPEG image transmission based on data embedding and vector quantization. In *Proceedings of the IEEE International Symposium on Circuits and Systems—ISCAS2003* (Vol. 2, pp. 532–535)

Masnou, S. (2002). Disocclusion: A variational approach using level lines. *IEEE Transactions on Image Processing, 11*(2).

Masnou, S., & Morel, J. (1998). Level-lines based disocclusion. In *Proceedings of the IEEE ICIP 98.* IL, USA.

Mukherjee, D., Chrysafis, C., & Said, A. (2002). JPEG2000-matched MRC compression of compound documents. In *Proceedings of the IEEE ICIP2002* (pp. 22–25).

Mukherjee, D., Memon, N., & Said, A. (2001). JPEG-matched MRC compression of compound documents. In *Proceedings of the IEEE ICIP 2001* (pp. 434–437).

Oliveira, M., Bowen, B., McKenna, R., & Chang, Y. (2001). Fast digital image inpainting. In *Proceedings of the International Conference on Visualization, Imaging and Image Processing (VIIP 2001)* (pp. 417–424).

Otsu, N. (1979). A threshold selection method from grey level histograms. *IEEE Transactions on Systems, Man and Cybernetics, 9*(1), 62–66.

Pavlidis, G., Tsekeridou, S., & Chamzas, C. (2004). JPEG-matched data filling of sparse images. *IEEE International Conference on Image Processing ICIP 2004* 24–27 October.

Pavlidis, G., & Chamzas, C. (2005). Compressing the background layer in compound images, using JPEG and data filling. *Elsevier Signal Processing: Image Communication, 20*, 487–502.

Pennebaker, W. B., & Mitchell, J. L. (1993). *JPEG still image compression standard.* New York: Springer

Rane, S., Sapiro, G., & Bertalmio, M. (2002). Structure and texture filling-in of missing image blocks in wireless transmission and compression applications. In *Proceedings of the IEEE ICIP 2002* (pp. 22–25)

Reininger, R., & Gibson, J. (1983). Distributions of the two-dimensional DCT coefficients of images. *IEEE Transactions on Communications, 6*(31), 835–839.

Sharpe, L. H., & Buckley, R. (2000). JPEG jpm file format: A layered imaging architecture for document imaging and basic animation on the Web. In *Proceedings of the SPIE, Applications of Digital Image Processing*, (Vol. 4115) (XXIII) (pp. 464–475).

Shirani, S., Kossentini, F., & Ward, R. (2000). Reconstruction of baseline JPEG coded images in error prone environments. *IEEE Transactions on Image Processing, 9*(July), 1292–1299.

Smoot, S. (1996). Study of DCT coefficient distributions. In *Proceedings of the SPIE Symposium on Electronic Imaging* (Vol. 2657).

Smoot, S., & Rowe, L. (2003). Laplacian model for AC DCT terms in image and video coding. In *Proceedings of the 13th International Workshop on Network and Operating Systems Support for Digital Audio and Video Table of Contents* (pp. 60–69).

Stasinski, R., & Konrad, J. (2000). POCS-based image reconstruction from irregularly-spaced samples. In *Proceedings of the IEEE ICIP 2000* (pp. 10–13).

Taubman, D. S., & Marcellin, M. W. (2002). *JPEG2000 image compression fundamentals, standards and practice.* Kluwer Academic Publishers, ASIN: B011DB6NGY.

Wallace, G. (1991). The JPEG still picture compression standard. *Communications of the ACM, 34*(4), 30–44.

Wang, Z., Bovik, A. C., Sheikh, H. R., & Simoncelli, E. P. (2004). Image quality assessment: From error visibility to structural similarity. *IEEE Transactions on Image Processing, 13*(4), 600–612.

Wei, L., & Levoy, M. (2000). Fast texture synthesis using tree-structured vector quantization. In *Proceedings of the SIGGRAPH 2000.*

Yang, Y., Galatsanos, N., & Katsaggelos, A. (1995). Projection-based spatially adaptive reconstruction of block-transform compressed images. *IEEE Transactions on Image Processing, 4*(7), 896–908.

Yang, Y., & Galatsanos, N. (1997). Removal of compression artifacts using projections onto convex sets and line process modeling. *IEEE Transactions on Image Processing, 6*(10), 1345–1357.

Chapter 5
Transmission Optimization

—*Data transmission is no longer something scary you don't want in your backyard. Now you want it directly in front of your house.*

Douglas Coupland

—*Structure is more important than content in the transmission of information.*

Abbie Hoffman

5.1 Introduction

The ever increasing need to transmit multimedia data through high speed communication systems, in which various restrictions and broadcasting policies may apply, paves the way for research targeting new algorithms and new information management procedures even before transmission takes place. One of the main challenges in the design and operation of networks that manage the transmission of images, either sensitive or not to delays, is network congestion (Chamzas and Duttweiler 1989). In addition, the high demand for multimedia data and the increasing volume of information being transmitted, along with extraneous factors (like electromagnetic interference) can create conditions likely to lead to the occurrence of errors that result in information distortion or to transmission outages that correspond to the loss of parts of the transmitted information.

The digital representations of images correspond to a large amount of data bits, and therefore, compression techniques are essential for an efficient transmission. Typical image compression standards, such as lossless JPEG (L-JPEG) and JPEG-LS, can yield compression ratios around 2:1 on the average. A key issue is that the algorithms, on which these methods are based upon, can not control or allow a controlled packet loss during transmission over a network. Therefore, when a network congestion occurs during the transmission of a compressed image, the rest of the image (or at least part of it) is negatively affected, unless the network user is provided with a *control-confirm-repeat* transmission mechanism. Such a communication protocol can on one hand help to recover lost information, with the negative effect of adding-

© Springer Nature Singapore Pte Ltd. 2017
G. Pavlidis, *Mixed Raster Content*, Signals and Communication Technology,
DOI 10.1007/978-981-10-2830-4_5

up to the congestion of the network, without taking into account the inherent increase in transmission delay. Therefore, for an effective mechanism to deal with congestion, the loss of information should be known or possibly in a way controlled by the user. In such a case, specific cost advantages could be provided to the user through different treatment of parts of information sent or received over the network. It might, for example, be an agreement, on which information packets are considered important or primary and which are complementary or secondary, so costs may be estimated differently in each case, depending on the situation in the network. In separating the transmitted information into significant and complementary, the provider of the network services could guarantee the safe transmission of important packets and the under-conditions transmission or even loss to an estimated pre-determined degree, of complementary packets.

Throughout this chapter, the idea of information segmentation into important and complementary permeates the methods developed and described. Initially, a method of image segmentation is described for the separation of the information and sequential transmission, which faces the overall challenge of improving both the compression results and the transmission efficiency using conventional compression methods in a lossy+lossless compression scheme and the application of vector quantization using neural networks. Then an extensive analysis of the behavior of the JPEG2000 image compression standard follows, with simulations of different noise scenarios and rates, focusing again on the segmentation of the information of the image into significant and complementary, which leads to conclusions about the possibilities of applying transmission priorities, contingency plans and cost policies.

5.2 MSICT: Optimizing Compression and Transmission

This section describes a method to comprehensively tackle with optimized compression and image transmission using standard compression techniques that has been proposed in (Pavlidis et al. 2001) and extended in (Pavlidis et al. 2005). The method is based on image segmentation into significant and complementary parts and on a lossy+lossless compression scheme, wherein vector quantization is applied using a neural network for the classification of image information and progressive transmission.

One of the basic techniques in image compression is the reduction of the number of colors with minimum optical distortion (Carpentieri et al. 2000). One of the most famous and widely known algorithms to achieve compression by reducing the information in the color space of the image and is based on the minimum square error estimate is vector quantization (VQ). In addition, it has been shown that compression is further improved by using an entropy coder (EC) just after the application of VQ, as shown in Fig. 5.1. The method presented in this section is based on this principle and forms a VQ-EC compression scheme based on the use of unsupervised neural networks. Code-vectors of the VQ are generated through the use of a neural network (NN) that is fed with the probability distributions of parts of the images on which it

Fig. 5.1 Typical diagram of
a vector quantization-entropy
coding compression system

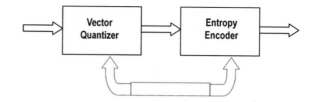

applies. This quantizer is essentially a simplified and fast implementation of a typical
VQ.

The method presented was named Multi-Segment Image Coding and Transmission (MSICT) scheme and is suitable for the encoding and progressive transmission of images in telecommunication networks with noise and/or congestion. It can be regarded as *partially adaptive context selection method* that makes use of a simple classification process for separating the information according to local statistical characteristics so that the compression with a typical entropy encoder becomes more efficient, while supporting progressive transmission in networks with noise and congestion. Potential information loss in case of transmission errors will result in less noticeable distortion, as the parts that are possibly lost are scattered throughout the image surface rather than being concentrated to form large gaps. The method has been tested extensively and was compared with the most common and efficient compression schemes (L-JPEG, JPEG-LS, Context-based Adaptive Lossless Image Coder (CALIC), Fast, Efficient Lossless Image Compression System (FELICS) and JPEG2000), and turned out to yield comparable results in compression efficiency, while allowing progressive transmission and protection against the occurrence of errors, through a substantially low complexity scheme.

In general, as already noted, a compression is called lossless (reversible) when the reconstructed image is an exact copy of the original. Lossless compression can be achieved due to the existence of redundancy in the data, in the sense that the information that is encoded can be modeled by non-uniform distributions. It is worth noting that there are many applications that require lossless compression (such as medical applications, digital archival and processing of satellite images) (Carpentieri et al. 2000; Pennebaker and Mitchell 1993; Wu and Memon 1997, 2000; Wu 1996; Langdon 2001; Christopoulos et al. 2002; Weinberger et al. 1995; Howard and Vitter 1993; Rabbani and Jones 1991b). Standards and models of lossless compression, such as JPEG-LS (Christopoulos et al. 2002; Weinberger et al. 1995), or CALIC (Wu and Memon 1997) and FELICS (Howard and Vitter 1993), have been created with specific assumptions for the data being handled. Generally, these methods operate in two independent phases, modeling and encoding (typically entangled). In the *modeling phase*, the image is studied under a predetermined sequential process in order to retrieve information for the creation of a probabilistic model that describes the data. In the *coding phase*, the creation of codes takes place, either in a sequential (adaptive) or a two-step process (training). Most of the compression methods, at their final stage, encode the error generated by the subtraction of the predicted data

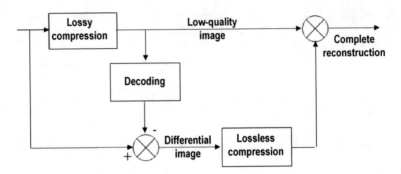

Fig. 5.2 Lossless compression in two stages

values from the actual values. An equivalent approach for lossless compression is a *lossy+lossless compression technique* in which (Fig. 5.2),

- the *initial image* (or, as usually referenced, the *original image*) is encoded with a lossy compression method, and
- the *differential image* produced by subtracting the encoded image from the initial, is losslessly encoded, so that complete reconstruction of the original data is possible

The *differential image*, as defined, has a significantly reduced variance compared with the initial image, with significantly less correlated pixel values. Figure 5.3 shows the normalized histogram of the luminance pixel values (a) of the known image 'Lena', (b) of the image after JPEG compression to 0.25 bpp and (c) of the difference between the first two images (the defined 'differential image').

Since the differential image, in this case, is the difference between two quantities with a resolution of 8 bpp, it should be represented by 9 bpp (i.e. its values are in the interval $[-256, 255]$). As shown in Fig. 5.3 the probability distribution for the differential image is substantially concentrated around zero (0) and can be approximated satisfactorily by a *double exponential distribution* with a probability density function $p = \frac{\alpha}{2} e^{-\alpha|x-\mu|}$, such that

$$\left. \begin{array}{l} p = \frac{\alpha}{2} e^{-\alpha|x-\mu|} \\ \mu = 0 \\ \alpha = \frac{1}{\sigma^2} \end{array} \right\} \Rightarrow p = \frac{1}{2\sigma^2} e^{-\frac{|x|}{\sigma^2}} \tag{5.1}$$

μ and α being the distribution parameters, which is also known as a *Laplace distribution* for the continuous case, or a *two-sided geometric distribution* for the discrete case, in which a zero mean ($\mu = 0$) and a σ^2 variance.

It is known that all the differential images of this type are approximately of the same form and differ only in the variance σ^2 (Habibi 1971; Jain 1981; Netravali and Limb 1980; O'Neal 1966). It is also known that the most suitable methods for the encoding of images with these statistical features include an entropy encoder. One approach is through the creation of *variable length codes*, such as *Huffman codes*,

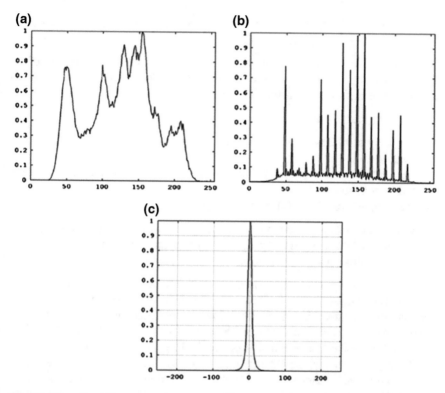

Fig. 5.3 Normalized histograms of the graylevel 'Lena' image: **a** original image, **b** JPEG compressed image at 0.25 bpp and **c** differential image

which can fit to the statistics of the image; usually, a dictionary of codes is constructed from the histogram and then the values are being encoded accordingly. Another approach would be to use *arithmetic coding*, in which context information from neighboring pixel values is being exploited. After the modeling of the probability distributions, an arithmetic coder can easily provide with efficient encoding. Given any discrete probability distribution, the arithmetic coder may be used to encode any random variable from this distribution with an average code-word length that approximates the entropy of the distribution. Unlike Huffman coding, arithmetic coding is efficient even when the probabilities are close to one (1).

There are several efficient compression algorithms available (based on DCT or DWT) that can generate the required differential images or the representation of the low quality + differential image (Chamzas and Duttweiler 1989), and the method presented here operates by

- using a standardized typical lossy compression scheme to generate a representation of an initial image into low quality + differential
- using a simple classification technique to partition the differential image into segments of arbitrary shapes

- encoding individually and sequentially the segments of the differential image by using an appropriate entropy encoder

This process is suitable for the progressive reconstruction of encoded portions of the differential image. Figure 5.4 shows a detailed block diagram of the key processes of the MSICT method. The method includes

- an encoder, responsible for
  - the lossy compression of the original image to generate information of low quality
  - the application of a segmentation using local statistical characteristics in the differential image
  - the compression of the differential image segments
- a decoder that reconstructs the original image through a multistep algorithm, in which the final image is formed progressively by sequential addition of differential data to the initial low quality image.

In experiments with this compression method, known algorithms were tested for the lossless coding stage (DCT (Pennebaker and Mitchell 1993) and DWT (Burrus and Guo 1998; Daubechies 1988, 1992; Mallat 1999; Rao and Bopardikar 1998), by applying the JPEG (Pennebaker and Mitchell 1993) and JPEG2000 (SantaCruz and Ebrahimi 2000a; Taubman and Marcellin 2002; Christopoulos et al. 2002; ISO-IEC 2000a) respectively), and an arithmetic encoder (Moffat et al. 1998) was used to produce the final compression codestring. The partition of the differential image was based on a Kohonen SOFM (Kohonen 1989, 1990, 1991, 2000; Haykin 1994), mainly due to its simplicity and effectiveness in solving similar classification problems.

The main contribution of this method was in the concept of the segmentation of the differential image by using criteria that aid the entropy encoder to increase its efficiency and thus the overall efficiency of compression. The method, by employing a separate coding of the various segments, it allows for the progressive reconstruction at the decoder. If, in addition, the encoded data are to be transmitted through a

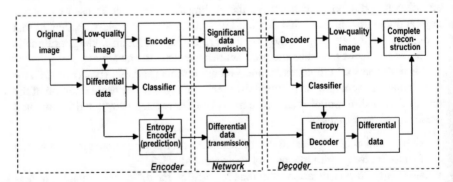

Fig. 5.4 Detailed block diagram of the MSICT method

channel with noise or congestion, the final formation of the code-stream may, indirectly, assist towards achieving error resilience. The method was tested on standard experimental graylevel images and resulted in compression rates of the order of those achieved with the state-of-the-art lossless compression methods (JPEG (Pennebaker and Mitchell 1993) and JPEG-LS (Christopoulos et al. 2002; Weinberger et al. 1995)). The advantage, as discussed, in relation to those standards is that the method supports progressive, lossless reconstruction through a low-complexity segmentation of the differential information and the restructuring of the final compression codestream that is transmitted in such a way that a reduction in quality due to data loss becomes less noticeable, as the errors are scattered throughout the surface of the image.

5.2.1 MSICT Encoding

The Multi Segment Encoder (MSEnc) of MSICT (Fig. 5.5) consists of two main stages: the stage of lossy coding and the stage of segmentation and classification of a differential image coupled with the final lossless entropy coding. The compressed image is decoded and the resulting image is subtracted from the original to create the differential image. The differential image is led to a neural classifier and a segment classification map is generated for the image (for 8×8 pixel blocks) based on local statistical characteristics. The choice of the size of the region was made for better recognition of correlations in small neighborhoods of pixels and a subsequent better clustering of the regions for the implementation of the adaptive context selection in the mechanism of the arithmetic encoder, at the final stage of the system. The lossy compressed image is individually being transmitted as the low-quality replica of the original, and along with the classification map, these two constitute the *essential information.*

Typically, differential images include large smooth areas with zero mean and low variance and areas with edges or 'activity', in which the variance grows and the dis-

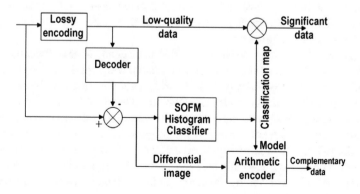

Fig. 5.5 The multi segment encoder

tribution exhibits skewness. Generally, the result of lossy coding, due to its inherent nature to compress the energy of the image and the fact that the images are treated as noise signals, is the distortion of the information for the edges. Information about the edges that is lost is expected to reappear in the differential image. Considering that the edges are usually thin pixel regions (regions representing abrupt changes) they are expected to be detected with high probability within a block of 8×8 pixels. By employing a classification technique on the differential image blocks, a classification map is generated that can be used for the segmentation of the differential image and to supply the arithmetic encoder with contextual information to accommodate the prediction model being used. The advantages of using such a classification scheme was already referenced in works such as (Pavlidis et al. 2001).

Two encoders were tested in this method for the lossy coding stage, JPEG, and JPEG2000. As a reminder,

- when encoding with JPEG, an image is divided into non-overlapping blocks of 8×8 pixels and each block is being independently encoded. The transform coefficients obtained by applying DCT are quantized by applying quantization table values corresponding to the desired final quality, and finally encoded by an entropy encoder to create the final codestream.
- when encoding with JPEG2000, the image may (or may not) be divided into blocks (called tiles in JPEG2000), and is transformed using the DWT. The coefficients are quantized by a scalar quantizer with a dead zone in accordance with the desired final data rate and grouped in the transform domain in regions that are coherent in the image domain. Entropy coding follows in two stages using significance coding and mechanisms for optimal data rate performance.

When considering the case of transmission through channels with noise, a medium-to-low quality would be preferable for the essential information being transmitted according to a scenario, in which the essential data are transmitted as soon as possible with low bandwidth requirements to ensure a proper transmission, and then the complementary data are transmitted progressively and gradually improve the reconstructed image at the decoder. Typical data rates for the essential data, ranging between 0.4 and 0.7 bpp, gave the best overall data rate result for the case of a JPEG encoding system, whereas in the case of JPEG2000, better results were achieved in terms of quality at even lower data rates.

In VQ (Gray 2003; Gray and Neuhoff 1998) an original image is divided into parts or k-dimensional vectors. Each vector x is compared with representative vectors (the codevectors), which are usually produced during a training process. The codevectors are also k-dimensional. The codevector which differs less from the vector being encoded is usually selected by the criterion of the least squared error. After selection of the most representative codevector, the original image vector is replaced by a pointer (index) p, which is of $log_2 N$ bits, where N is the number of codevectors being used. The decoder uses this index to restore the original image vector. In this way compression is achieved by replacing the image vectors with pointers (indices) in representative codevectors.

Fig. 5.6 Neural network
vector quantizer and its use
as a segmentation system

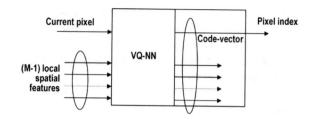

To avoid the complexity that accompanies a typical Vector Quantizer (VQ), an alternative simplified quantization method is to use neural networks; neural networks can be used for the creation of the codevectors (training) and for the final representation of the image with a reduced number of such vectors (as in (Papamarkos and Atsalakis 2000), in which neural networks are used for efficient color reduction). If the vectors to be used are the colors of the image the approach corresponds to a fast implementation of a color quantizer, which may exploit additional spatial or statistical characteristics. The basis of such a system may consist of a SOFM neural network (Fig. 5.6) and is named Neural Networks Vector Quantizer (NN-VQ) in MSICT. The NN-VQ operates either as an adaptive or a static training process: when the codevectors are generated from samples of the image being encoded it is adaptive, whereas when the training process and the creation of the dictionary of codevectors are made in advance using a set of representative samples, it is static. Static NN-VQ is obviously a faster implementation. However, acceleration can be achieved of the adaptive NN-VQ and a reduction in memory requirements. This can be done by employing techniques of *fractal sampling* (eg. following special curves that cover the surface of the image, such as the *Peano curves* or *fractal scanning*) to reduce the samples used in training, in a way, though, that the representation from all sections of the image is covered. After the process of training, the neurons of the output layer determine the codevectors of the NN-VQ. The result of the application of this quantizer is to ensure the existence of representative features (could be just colors) of the original image in the final quantized image along with texture information or other features that may have been used in the clustering. Spatial context is being captured by feature extraction in the pixel neighborhoods, so as to make effective utilization of local correlations. As shown in the following paragraphs, the result of applying this principle is to obtain data that can be encoded with high efficiency from a classic entropy coder (such as the arithmetic coder) and thus sufficient compression rates can be achieved, accompanied by some additional advantages (as progressive transmission and control of errors), which will become apparent during the following study.

Based on this analysis, the classifier chosen for grouping the parts of the differential image was an unsupervised neural network classifier. This classifier provides the system with progressive capabilities and with contextual information for exploitation by the entropy encoder. The specific SOFM classifier has been used repeatedly and successfully in respective applications of unsupervised classification of images (Papamarkos and Atsalakis 2000). In its operation,

- the differential image is divided into blocks of 8 × 8 pixels (or more)
- the histogram h of each block is calculated and a large table of histograms H is generated for all the image blocks
- the table of histograms H becomes the input to the neural classifier; the same table may also represent the training dataset, thus the network is adapted to the type of histogram at hand. Each block of the image is classified into a category according to the 'form' of its histogram
- the number of outputs was set to eight (8), although experiments were conducted with sixteen (16) outputs, which showed that this option may be preferable when more steps of progressive coding are required or needed; these outputs correspond to the number of parts the differential image is to be divided

As this system was tested with standard 8-bpp images (that produce 9-bpp differential images), the SOFM network used is one of a simple topology of 512 neurons in the input and 8 in the output, as shown in Fig. 5.7. The inputs take any real value, whereas the outputs are binary decisions (0 or 1). The input neurons i corresponding to the indices of the histogram data, with input neuron i representing the probability of the luminance value of i. At any given time t, the histogram h_t of an image block is presented at the input of the network.

This SOFM was trained by presenting all or part of the histograms of the image, until an error criterion (typically, mean square error) was satisfied. The training process was simple, performed within the network algorithm. After training, the network is ready to classify a histogram of an input image to the pre-defined classes. At the output, the network provides a single active output for a specific form of histogram, satisfying the model developed during the training. The end result of the classifier is a classification map that depicts a mapping of the differential image into an image of 3 bpp (corresponding to $2^3 = 8$ classes). This map can be used to segment the differential image and to supply the entropy encoder with contextual information. Figure 5.8 depicts 'average' histograms for two different classes after a classification and Fig. 5.9 shows a typical classification map, in which the luminance variations correspond to different classes. The particular classification map was produced by applying the method to the classic test image 'Lena', wherein the low-quality image

Fig. 5.7 The topology of the used neural network classifier

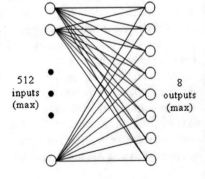

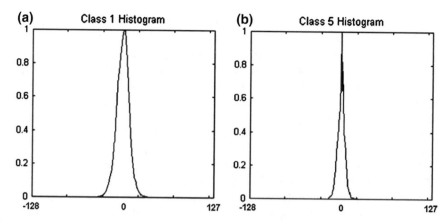

Fig. 5.8 Histograms of two different classes of the image blocks as obtained after segmentation and clustering

Fig. 5.9 Typical classification map produced by the neural classifier

was 0.25 bpp. It is important to observe here how the groupings in the differential image provide information relating to the original image. The portions of the image with similar perceptual characteristics can be found in the same class. The neural classifier operates essentially as a vector quantizer of high-dimensional data, which quantizes the image to a reduced resolution and color depth. The main advantage of using the SOFM network instead of a typical VQ lies in the significantly reduced complexity and computational cost.

Furthermore, restrictions in the VQ-NN input data may reduce the dimensions and therefore the computational complexity and processing time. As already mentioned, the differential image is described by a two-sided geometric probability mass function, which guarantees that most of the values are within an interval smaller (possibly significantly smaller) than what is theoretically expected. The same principle applies to each of the image blocks being processed. In a typical case, in which the original image is compressed to 0.25 bpp, the histogram of the differential image lies (on the average) in the range $[-50, 50]$. Since the histogram can be described, to a large extent, by the values within this interval, the other values can be discarded. This 'cropping' of the dynamic range leads to a lower number of input neurons in the classifier (101 in this case), leading to a dramatic reduction in training and classification times. The results presented in the following paragraphs correspond to a mild, 'adaptive' discarding only of those values that are zero at the limits of the theoretic interval.

Additionally, further restrictions could be applied in the training to improve the efficiency of the system, adopting the offline training scheme, in which training could be done once using a large volume of training samples. It could also be possible to fix the system without any training since all differential images are expected to exhibit similar characteristics of known probability distribution. Pre-fixing the network can be characterized as the static mode, unlike the case of the adaptive mode, where each input image provides the samples for the training. Nevertheless, in the general case, the static mode is expected to result in lower accuracy, whilst the adaptive mode demands for a higher computational cost. The selection of the appropriate mode of operation must be related to the nature of the application, as the training of a SOFM consumes valuable time at the beginning of the process.

Moving on to the entropy coding subsystem, a typical arithmetic coder is employed for the final encoding of the segmented differential data. A typical Arithmetic Coding (AC) system contains two basic stages of operation (Fig. 5.10), the 'engine' of the AC, where the input symbols and their probabilities are fed, and the 'model' that is responsible for assessing the probability of the current input symbol. The *model* can be described as the intelligent module of the AC, which tries to 'understand' the underlying structure of the input, whereas the *engine* is the engine-room of the encoder, responsible for converting this structure and each symbol to a code. A good model can generate symbol probabilities close to 0 or 1, which is very important to achieve an optimum response of minimum length codes. It is also noted that the engine does not depend on the model. The final stage of MSICT consists of such an AC. The AC uses contextual information from the classification map (it can also use contextual information from the low quality image) and encodes the

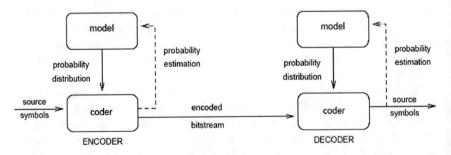

Fig. 5.10 Typical arithmetic codec diagram

differential image. The output can be either a single codestream or several partial codestreams, following the classification of the differential image into various classes by the neural classifier. Thus, the transmission may or may not be progressive. In progressive transmission, each transmitted segment represents the encoded portion of the differential image that belongs to a class. The main advantage of the partial coding method is that during transmission through noisy channels any possible information loss occurs within the class that is currently transmitted, which can be composed of image parts distributed in different regions of the original image. Therefore if each differential image class will consist of parts distributed over the entire surface of the

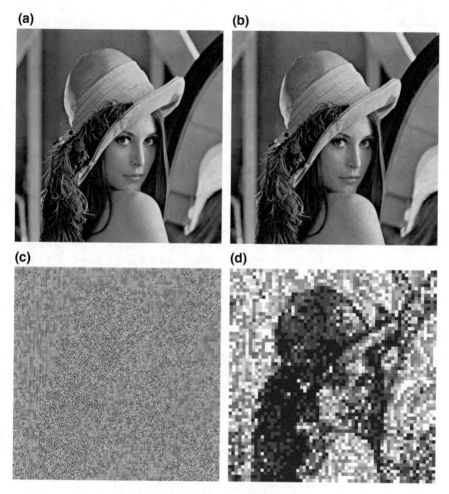

Fig. 5.11 Example of the application of MSICT **a** original image, **b** low-quality JPEG image at 0.33 bpp (32.95 dB PSNR), **c** differential image, and **d** classification map (blocks of 8 × 8 pixels, 8 classes)

original image, errors will not be as intense and the quality loss will not be strongly evident since the errors are distributed across the image surface.

Figure 5.11 presents images created by the application of MSICT, in which (a) shows the original image 'Lena' (8 bpp, 512 × 512 pixels), (b) shows the compressed image using JPEG at 0.33 bpp, (c) shows the differential image, and (d) shows the classification map after the application of the neural classifier on the differential image. JPEG compression for the production of the low-quality image was carried out via the conventional terminal-based encoder *cjpeg v.6b* provided by the Independent JPEG Group (2000), included in the distribution of Cygnus (Red Hat 2000) that was used for these experiments. The quality of this image was estimated at 32.95 dB PSNR. The differential image produced had a dynamic range of $[-55, 57]$ and was classified into 8 classes using blocks of 8×8 pixels. This way the classification map is an image with a color depth of 3 bpp and is 8 times smaller in dimensions than the original image in both directions. The classification map was losslessly encoded by a standard arithmetic coder, whereas the segments, resulted from the segmentation of the differential image, by an adaptive arithmetic coder, resulting in a total data rate of 4.35 bpp. It should be noted that the luminance values of the pixels in images (c) and (d) were scaled to allow a better visualization.

5.2.2   MSICT Decoding

The Multi-Segment Decoder (MSD) of MSICT is shown in Fig. 5.12; it consists of two main sequential decoding stages. Initially, the decoder receives the low-quality image and the classification map. This image is decoded by the appropriate decoder (e.g. JPEG) and the low-quality image is reconstructed and presented to the user. The subsequent data correspond to the differential image segments as described in the classification map. As the encoded image segments arrive, an arithmetic decoder decodes the data based on the map and contextual information from previously received data. Finally, each decoded segment is added to the reconstructed image progressively, and gradually improves the quality of the final image, as additional segments are sequentially being integrated.

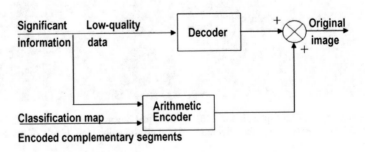

Fig. 5.12 The progressive multi-segment decoder

5.2.3 *Experimental Evaluation of MSICT*

In order to be evaluated, MSICT was implemented in MatLab and ANSI-C and was used in a series of different experiments, both to understand its characteristics through an experimental validation and to assess its theoretical advantages. Several experiments were carefully designed that targeted the

- confirmation of the theoretical utilization of the imposed segmentation by the arithmetic encoder
- confirmation of the superiority of this segmentation compared to simple tiling of the image
- exploration of various prediction models in the arithmetic encoder to achieve optimal compression
- exploration of various parameters and compression methods for the initial production of the low-quality image
- testing of a different number of segments of the differential image for optimal compression and transmission
- testing in transmission channels with different error rates and bandwidth
- confirmation of the possibility to use this method in real-world telecommunication systems and its inherent capacity to provide user-controlled cost policies

All experiments were performed with typical graylevel test images including either natural images or graphics, low quality photos and textual images (University of Waterloo 2002).

5.2.3.1 Verification of the Positive Effect of Segmentation

As has already been stated in previous paragraphs, the performance of the arithmetic encoder is based substantially on the form of the probability distribution of the data at its input. A probability distribution with a low variance is generally preferable. In terms of the higher order moments, such as the skewness and kurtosis of the probability distribution, a high value is preferable. Thus, in order to confirm the theoretical positive effect of the segmentation in MSICT by the arithmetic encoder, a simple experiment was carried out, in which the typical test image 'Lena' was encoded as follows: standard JPEG encoder was used to generate the low-quality image on the whole range of JPEG quality factors (from 5 to 95). Then the image was segmented and the classification map was created using the neural network classifier that classified image blocks of 8×8 pixels into 8 classes based on their histogram; finally arithmetic coding was applied to compress the data. The average values of the moments of the differential image along with the corresponding moments of the 8 image segments for low quality images with bitrates ranging within 0.17 and 2.8 bpp are listed in Table 5.1.[1] As is evident, the average variance in the segments of the

---

[1]It should be noted that the higher order moments skewness and kurtosis take real values, either positive or negative. The average values listed in the experiments were computed on the absolute

Table 5.1 Mean values of moments of the probability distributions in the overall differential image and in the segments produced by MSICT for the total range of low-quality images

	Variance	Skewness	Kurtosis
Overall differential image	25.051	0.054	2.817
Mean of all segments	5.528	0.674	48.960

Table 5.2 Comparison between the use of the MSICT and of a simple sequential segmentation

	MSICT segmentation	Sequential segmentation
Low-quality image (bytes)	10,860	10,860
Compressed classification map (bytes)	770	
Compressed differential image (bytes)	142,386	148,614
Total size [lossless] (bytes)	154,016	159,474
Total data rate (bpp)	4.70	4.87

differential image is much lower than the average variance of the total differential image. On the contrary, higher order moments (skewness and kurtosis) exhibit a significant increase. These facts indicate a better fit of the data of the differential image segments for arithmetic coding, confirming the theoretical superiority of using such a segmentation.

5.2.3.2 Verification of the Efficiency of the Segmentation

To confirm the effectiveness of the segmentation, the following simple test took place: after creating the low-quality and the differential image, segmentation was applied on one hand by the neural classifier and on the other by a sequential segmentation to the same number of non-overlapping rectangular regions of the same size. Table 5.2 presents the obtained compression values during the two tests. Results for the test image 'Lena' are reported here, with the image initially being compressed at 0.33 bpp. The results clearly show that the efficiency of compression is 4 % higher with the application of the MSICT classification. Figure 5.13 shows a comparative graph of results attained by applying the MSICT segmentation and classification and by a simple sequential segmentation. The curves present values of the overall data rate versus the data rate of the low-quality image. As shown in the graph, the application of classification always leads to a better overall average compression rate. Furthermore,

---

(Footnote 1 continued)
values. This was due to the fact that either negative or positive such moments, the result is always in favor of the coding efficiency as the absolute value increases.

Fig. 5.13 Average data rate
versus the data rate of the
low-quality image in MSICT
and sequential segmentation

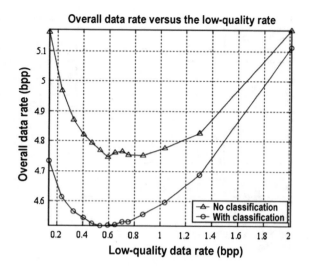

the graph shows that the best overall data rate is achieved for a range of low-quality image data rates in the interval 0.3 to 0.9 bpp, optimally around 0.6 bpp.

5.2.3.3   Various Prediction Models to Optimize Compression

As discussed in previous sections important role in the performance of an arithmetic coder plays the modeling of the probability distribution of the data at its input. To improve the performance during compression, various prediction models and context patterns have been tested in MSICT. Some of the most important are listed in Table 5.3, in which the letter 'x' corresponds to the current pixel and letters 'a' through 'h' represent pixels in the immediate 8-neighborhood as shown in Fig. 5.14. The models take into account the classification map and the low-quality image in various combinations. The models use at most 12 bits, giving $2^{12} = 4,096$ context combinations. This number is generally considered very large for an encoder that is expected to exhibit a quick adaptation.

This was also confirmed by the experimental results, which showed that better compression rate was achieved with less context combinations. Table 5.4 shows an example of a model used for experimentation. Further study of the nature of the images involved (both the differential and those images that contribute to contextual combinations) supported the idea of the application of a probability model with assumptions, such as model 7, which changes from model 4 to model 5 depending on the position and the expected changes in the probability distributions, something that happens in the marginal areas between different segments (where model 4 is more efficient) or within a single segment (where model 5 fits best). In summary, comparative results (average data compression rates) from the application of various prediction models across all the images of the experiments are shown in Table 5.5.

Table 5.3 Arithmetic coding prediction models and context patterns

Model	MSB	12-bit prediction pattern										LSB
	11	10	9	8	7	6	5	4	3	2	1	0
1					g	c	e	a	x	x	x	x
2	g	g	c	c	e	e	a	a	x	x	x	x
3	c	c	c	c	a	a	a	a	x	x	x	x
4					g	c	e	a	x	x	x	1
5	c	c	c	c	a	a	a	a	x	x	x	0
6								x	x	x	0	
7–1	g	c	e	a	x	x	x	1				
7–2	c	c	c	c	a	a	a	a	x	x	x	0

Fig. 5.14 Context 8-neighborhood

b	c	d
a	**x**	e
h	g	f

Table 5.4 Example of an arithmetic coding model: The case of model 3

MSB	12-bit prediction pattern										LSB
11	10	9	8	7	6	5	4	3	2	1	0
c	c	c	c	a	a	a	a	x	x	x	x

In total 12 bits are used

The template is filled with bits from the classification map starting with the least significant (LSB):

4 bits of the current pixel (the four most significant bits)

4 bits from the previous pixel (the four most significant bits)

4 bits from the top pixel (the four most significant bits)

Table 5.5 Average data rates bpp for the various models in all experiments

Prediction models						
1	2	3	4	5	6	7
4.63	4.52	4.63	4.54	4.52	4.63	4.50

5.2.3.4   Creation of the Low-Quality Image

Due to the design of the system, required to be based on typical coding methods, there is no direct dependence on the efficiency of the encoder to be used for the creation of the low-quality image. This independence does not refer to whether one or another coding method will eventually lead to a better overall data rate or image quality, as to the flexibility of the selection in itself provided by the system. The way

Table 5.6 Comparative results of two different encoders for the creation of the low-quality image

	Low-quality image data rate (bpp)	Classification map data rate (bpp)	Total rate of significant data (bpp)	Total data rate (bpp)
JPEG	0.40	0.01	0.41	4.02
JPEG2000	0.41	0.01	0.42	3.97

it works is essentially based on the good adaptation to the data of the differential image with the arithmetic encoder. As mentioned in the previous sections, almost all the differential images have similar statistical characteristics and therefore the low-resolution image, which contributes to their creation, does not play a significant role in the overall system. To confirm these observations experiments were performed with conventional JPEG and JPEG2000 encoders and a large range of target data rates for the low-quality image. The results showed that the use of JPEG2000 gives slightly better compression and image quality outcomes, but the significant computational cost due to the high complexity of the particular encoder does not justify the application of a system like this. Table 5.6 shows an average outcome of the system in using those two encoders. As shown, while the total amount of significant information (low-quality image+classification map) is almost the same, the final overall result is in favor of using JPEG2000, at a cost of a significant increase in runtime. Of course, as already pointed out in the analysis of the two algorithms in a previous chapter, the significantly faster execution of JPEG against JPEG2000 is considered as given.

5.2.3.5 The Number of Differential Segments

As part of the study for the optimum number of segments, in which the differential image may be segmented by the neural classifier before the arithmetic coding, a series of experiments was performed using two values, that is, eight (8) and sixteen (16) number of segments. The choice is related to the fact that these numbers are powers of two (2) (third and fourth power respectively) and therefore the image that corresponds to the classification map is a palette image with an integer number of colors (color depth of 3 or 4 bpp respectively). Of those experiments it turned out that there is no substantial difference in the compression efficiency. What differentiates the two cases relates to the progressive transmission and the fact that, by using more segments, transmission control may be more efficient, especially in conditions of occurrence of errors; in addition the setting of a possible cost policy becomes easier.

Furthermore, choices for segmentation into blocks of various sizes prior to classification were also investigated. Block sizes tested were of 8×8 and 16×16 pixels. The initial choice of 8×8 followed the partitioning used by the JPEG encoder. The idea was that the influence of the use of the particular encoder to create the differential image will be best captured within blocks of the same size (or multiples of that size). Experiments with 16×16 pixel blocks confirmed the original hypothesis giving slightly inferior compression results due to the classification imposed by the neural classifier in that case.

In both cases the results shown no significant deviations in the compression performance worth to be reported.

5.2.3.6 Noisy Channels and Cost Policy Issues

The last set of experiments of MSICT focused on a simulation of transmission over simple telecommunication systems, in which no error checking and concealment mechanisms exist. In this simulation approach it is easy to approximate ideal congestion situations. The most important contribution of MSICT in such a case stems from the basic idea behind its development, none other than the segmentation and characterization of the transmitted data into important and complementary. In these experiments the testing scenario was also complemented by aspects regarding possible cost policies; according to these aspects, the network service provider may

- guarantee the safe and lossless transmission of significant information packets with a specific cost, and
- provide options for the transmission costs of the additional information, depending on the rate of errors that is expected to occur in the non-guaranteed part of the overall network bandwidth

Two experiments were conducted for the simulation of these conditions: during the first experiment the neural network classifier was applied for the segmentation of the differential image, whereas during the second experiment a simple sequential tiling was applied in the same number of non-overlapping blocks. In these experiments, the classification was applied on blocks of 8×8 pixels into 8 classes. Then the significant data were transmitted through an ideal error-free communication channel and the complementary data through a channel with an error rate of 5×10^{-6}. The resulting images were reconstructed at the decoder and compared with the original. The whole simulation scenario was in agreement with a similar scenario used in the assessment of the error resilience in the standard JPEG2000 (Pavlidis et al. 2002, 2003). It should be noted that in these experiments the worst case scenario was considered, according to which when an error occurred in a packet of a transmitted encoded segment, the rest of the segment is considered 'lost' due to a simulated loss of synchronization at the decoder.

Figure 5.15 illustrates the results of the simulation for image 'Lena', for both cases of the experiment. Figures (a) and (c) show the differential image and the final reconstructed image at the decoder that corresponds to the case of segmentation with classification, whereas figures (b) and (d) show the corresponding results for the case of a simple sequential tiling. As apparent, the errors in the first case are dispersed across the surface of the differential and the final image, and are substantially less noticeable than in the latter case. Error measurements for both cases are summarized in Table 5.7. It is clear that, although the objective measurements show a slight

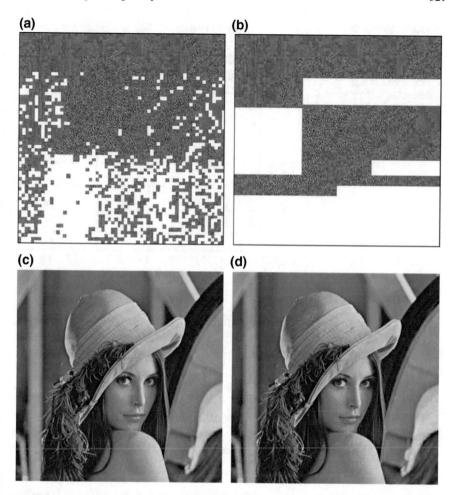

Fig. 5.15 Illustration of an image transmission over a noisy channel (random noise) for the case of segmentation with classification (**a, c**) and the case of simple sequential tiling (**b, d**)

Table 5.7 Comparative study of errors during transmission

	Segmentation with classification	Sequential segmentation
Maximum difference	112	133
RMSE	4.17	4.67
PSNR (dB)	35.72	34.74

improvement in favor of the segmentation with classification, the most significant advantage lies in the dispersion of the errors over the entire surface of the image and therefore in the significantly better subjective quality obtained in this way.

Fig. 5.16 Average final
reconstructed image quality
against the data rate of the
low-quality image attained
for differential image
segmentation with or without
classification

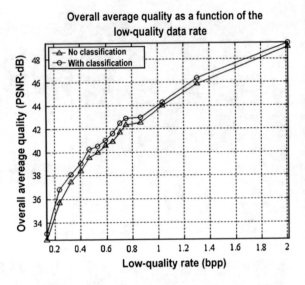

Fig. 5.16 Average final
reconstructed image quality
against the data rate of the
low-quality image attained
for differential image
segmentation with or without
classification

Figure 5.16 displays a graph relating to the quality of the final reconstructed image. It is a comparative graph of the average quality of the final reconstructed image as a function of the rate of the low-quality image for both cases of segmentation of the differential image. The results show that the use of classification is preferable (at least marginally), since it leads to better results across the total range of the data rates for the low-quality image.

The graph in Fig. 5.17 shows the average quality of the final decoded image as a function of the rate of errors in the transmission network. The low-quality image in this experiment was compressed to a pre-set 0.24 bpp, and the overall network error

Fig. 5.17 Average quality of
the final reconstructed image
against the rate of errors in
the transmission network
(low-quality image pre-set
data rate was 0.24 bpp)

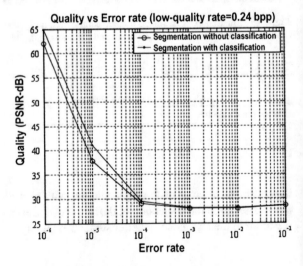

rate is within 10^{-6} to 10^{-1}. In high error rates (10^{-3} and above) the classification in the differential image segmentation does not contribute noticeably towards at least the objective quality. However, although the benefit seems marginal, it is the perceived image quality that still gives the advantage to the use of classification.

Furthermore, for the experimental confirmation of the cost policy flexibility offered by MSICT, the system was tested in the following virtual scenario, in which

- the network provider's cost policy imposes that significant data have a cost rate of one (1) unit per byte;
- the network provider's cost policy imposes a *cost ratio* (cr) for complementary/significant data, such that the pricing of complementary data is done with a smaller rate;
- the total transmission cost is, thus, defined as,

$$\text{Cost} = \text{significant bytes} \times 1 \, [\text{cost units/byte}] + $$
$$+ \text{complementary bytes} \times cr \, [\text{cost units/byte}] \quad (5.2)$$

- low-quality images are produced using a JPEG encoder with quality factors in the set {5, 10, 15, 20, 25, 30, 35, 40, 45, 50, 60, 70, 80, 90}; the respective average data rates of the significant data are accordingly {0.17, 0.29, 0.40, 0.49, 0.57, 0.64, 0.71, 0.77, 0.84, 0.89, 1.02, 1.21, 1.51, 2.13} bpp.

Figure 5.18 shows how the overall transmission cost is influenced based on the above simulation scenario. The three curves shown in the graph correspond to the cost of the significant, the complementary and the total data respectively. This particular graph depicts the situation when a cost ratio of 0.5 is used. By varying the cost ratio one may control the cost of the significant and the complementary data (and, of course, the total cost) depending, on the choice of quality, according to scenarios, in which,

Fig. 5.18 Normalized cost as a function of the significant data rate

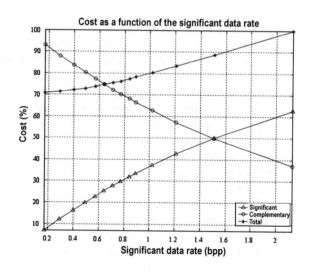

- using a cost ratio of 0.5, the total cost is split 50–50 % at significant and complementary data when the rate of significant data is 1.51 bpp (this appears in the graph of Fig. 5.18 where significant and complementary curves cross each other), and the total cost at this point is estimated to 89 % of the maximum cost in this case;
- using cost ratios lower than 0.5 (e.g. 0.1), the cost of significant and complementary data is split 50–50 % where the significant data rate is 0.4 bpp, and the total cost at this point is about 34 % of the maximum cost in this case;
- using cost ratios greater than 0.5 (e.g. 0.9), the cost of significant and complementary data is split 50–50 % where the significant data rate is 2.13 bpp, and the total cost at that point reaches the maximum cost;
- apparently, if a very low cost ratio is selected (i.e. cheap complementary data), then the average cost of the complementary information is less than the cost of the significant data, almost throughout the total data rate range (with the exception of very small data rates: 0.17 and 0.29 bpp), therefore this setting is an excellent choice for a consumer;
- if a cost ratio close to unity is selected (i.e. expensive complementary data), then the average cost of complementary data will always be higher than the cost of the significant data and the total cost will always be close to the maximum, therefore this scenario is a very bad choice for a consumer, as in any occurrence of network congestion the total cost will always be close to the maximum;
- if an 'average' cost ratio is selected (e.g. 0.5), the cost curves of significant and complementary data meet close to the average rate of significant data, providing a good choice for the consumer (concerning the costs) and the provider (concerning network bandwidth exploitation).

5.2.4   Using Other Statistical Features

The benefits of using the histogram-based classification over a simple sequential tiling became apparent by the experimental results of the previous paragraphs. Since this classification is based on vectorial data, the process can be simplified further by using scalar statistical characteristics such as the variance, variance coefficient, energy, entropy, skewness, kurtosis and activity (Saha and Vemuri 2000; Press et al. 1992). These features are mathematically expressed in Table 5.8. Combinations of these features can also be considered in order to increase the robustness of the neural network classifier, especially in situations where a single feature is not enough to adequately capture the structure of the distribution in the segments being classified. Figure 5.19 shows two possible topologies for the neural network classifier when the classification feature is the variance or when both the variance and the skewness are being used in combination. The case of a single feature as the input of the neural network may appear somewhat unusual and is not a common practice although there is no theoretical issue. It is known that neural networks are very capable in performing non-linear separation of multidimensional input signals, and therefore,

Table 5.8 Statistical features useful in classification ($\bar{x}$ = mean value)

Feature	Mathematical expression
Variance	$\sigma^2 = \frac{1}{N-1}\sum_i(x_i - \bar{x})^2$
Variance coefficient	$n = \frac{\sigma}{\bar{x}}$
Energy	$E = \frac{1}{N}\sum_i x_i^2$
Entropy	$E = -\sum p_x log_2(p_x)$
Skewness	$S = \frac{1}{N}\sum_i \left(\frac{x_i - \bar{x}}{\sigma}\right)^3$
Kurtosis	$K = \left\{\frac{1}{N}\sum_i\left(\frac{x_i-\bar{x}}{\sigma}\right)^4\right\} - 3$
Activity	$IAM = \frac{1}{M \times N}\left[\sum_{i=1}^{M-1}\sum_{j=1}^{N}|I(i,j) - I(i+1,j)| - \sum_{i=1}^{M}\sum_{j=1}^{N-1}|I(i,j) - I(i,j+1)|\right]$

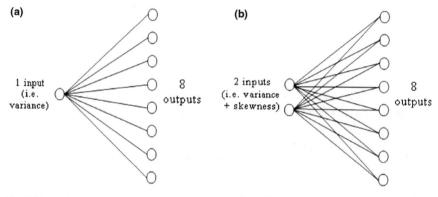

Fig. 5.19 Possible topologies of the simplified neural classifier

such classifiers are preferable in these cases over conventional vector quantization due to its reduced complexity. Nevertheless, the classification using a single or a simple combination of scalar features performs efficiently, even though the computational cost in this case may be considered as relatively high.

Figure 5.20 shows the average compression rate obtained by MSICT for various prediction models in the arithmetic encoder and various statistical features and their combinations for the classification in all experimental test images; in this figure

- the line that is apparently lower than the others, which corresponding to the best compression outcome, is the result of using classification based on histograms;
- the dashed-dotted line corresponds to the compression results obtained when the differential image is generated using a XOR operation and the classification is, again, based on the histograms;

$$\text{Differential image} = (\text{Original image}) \text{ XOR } (\text{Low} - \text{quality image}) \quad (5.3)$$

Fig. 5.20 Data rates against various prediction models for various classification features and strategies

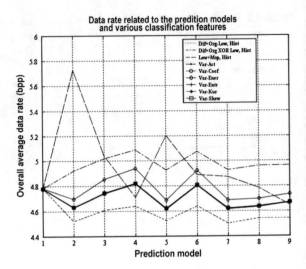

the dashed line represents the compression rate that corresponds to using the histogram-based classification and contextual information from the classification map and the low-quality image;

all other lines correspond to the compression results obtained by using classification based on pairs of scalar statistical features (such as the combination of variance and entropy).

Apparently, the results demonstrate that the use of only two scalar statistical characteristics (such as variance and skewness) can also yield satisfactory compression results while significantly reducing complexity.

5.2.5 MSICT Wrap-Up

MSICT was proposed as a compression and image transmission method, which is based on a lossy+lossless coding scheme, using typical coding standards. The heart of the system is a classifier of non-overlapping parts of a differential image that bases its classification on the histograms (probability distributions) of image tiles. The result of this classification is the segmentation of the differential image into partial images with inner statistical homogeneity, which may be efficiently encoded by a standard adaptive arithmetic encoder. Besides the efficient compression, such a segmentation can support effective progressive transmission and selective pricing. The advantages of MSICT may be summarized to

- the independence of the system from a specific encoder that produces the low-quality image
- the segmentation into segments on the basis of statistical similarity, results in more efficient compression of differential information
- the progressive transmission and reconstruction offered
- the improved subjective image quality in conditions of occurrence of transmission errors
- the possibilities for selective pricing and cost policies

It should be mentioned that since the system implements a vector quantizer of low complexity and the problem encountered is that of the *clustering of image parts* based on local characteristics, it is possible to apply other methods (at this stage) such as LBG, K-means, LVQ and fuzzy c-means (Jain and Flynn 1999; Milligan and Cooper 1985; Fasulo 1999; DeHoon et al. 2003; Tou and Gonzalez 1977; Trivedi and Bezdek 1986; Lim and Lee 1990; Kohonen et al. 1992). Finally, the system could be integrated into a single independent compression scheme around a wavelet-based encoder (DWT), with the application of segmentation taking into account the high-pass bands of the transformation (HL, LH, HH). The system is naturally progressive in resolution and quality and a decoder is not required at the side of the coding system. Of course, due to the larger dynamic range of the transform coefficients in the high-pass bands, histograms may not be the best choice as classification features and therefore alternative features should be studied, always balancing the complexity with accuracy and the required runtime. Additionally, the typical arithmetic encoder could be replaced by a binary arithmetic coder, following the example of the modern compression standards (like in JPEG2000). By incorporating schemes such as significance coding, the segments of the differential image can be further divided into quality layers within each segment. This allows the transmission of even smaller data packets with enhanced control and resilience against errors. Although MSICT was demonstrated with graylevel images, the system may be easily generalized for the case of color image encoding through the incorporation of a reversible color transform for the appropriate channel decorrelation.

5.3 Significance Segmentation in JPEG2000

This part of this chapter presents the application of the idea of segmentation into significant and complementary information in JPEG2000 encoding, to study the behavior of this standard in transmission through noisy and congested networks. The aim was both to study the noise resilience and correction mechanisms provided by the standard itself, and to explore the possibility of exploiting the separation and characterization of the transmitted information into significant and complementary for imposing and controlling costs and pricing policies. To that end, simulations were designed of two error patterns that occur either at the level of bits (channels with noise) or at level of packets (channel congestion); the first simulated error pattern involved two methods of error, corresponding either to loss or to reversal (or flipping) of bits, whereas the second involved various packet sizes.

It is reminded that the JPEG2000 image compression standard, came to fill modern efficient compression and progressive transmission needs, while providing error resilience and concealment mechanisms. Several studies present the behavior of the new standard under the occurrence of errors during transmission in order to compare it with older standards (Christopoulos et al. 2000a; Santa-Cruz and Ebrahimi 2000a, b; Santa-Cruz 2000). Here, the focus is to study the use of JPEG2000 in Layer-Resolution-Component-Position (LRCP) progressiveness, under various error transmission conditions. The overall outcome of experimentations is the general conclusion that the JPEG2000 can be efficiently used in a transmission scheme with information segmented into significant and complementary, enabling comprehensive cost policy agreements between a service provider and the users. An extensive description of the JPEG2000 standard, its functionalities and benefits, have already been given in a previous chapter, so the focus here is on the features relating to the ability to control and suppress errors during transmission.

In JPEG2000 variable length coding is being applied (through the MQ arithmetic coder) for the compression of the quantized wavelet transform coefficients. As known, the variable length codes are prone to transmission errors. An error at a bit level can lead to loss of synchronization at the decoder and the final image could end up significantly altered. To improve performance when transmitting encoded data through noisy channels, JPEG2000 includes and defines specific tools and syntax to the compression string to control errors. These tools encounter errors in various ways, that is by using data partitioning and resynchronization, employing error detection and concealment and by supporting QoS transmission based on priority (Christopoulos et al. 2000a; Moccagata et al. 2000; Liang and Talluri 1999). Error resilience is achieved at the entropy encoding level and at the transmission packets level. Table 5.9 summarizes the ways in which this is possible (Christopoulos et al. 2000a).

The entropy coding of the quantized transform coefficients takes place within code-blocks. Since the encoding (and decoding) is performed independently in each code-block, the occurrence of errors affects the information within an encoded block,

Table 5.9 Tools for error resilience in JPEG2000

Level of mechanism	Methodology
Entropy coding level	Partitioning into code-blocks
	Termination of the arithmetic encoder in each coding pass
	Reset of contexts after each coding pass
	Selective arithmetic coding
	Use of segmentation symbols
Packet formation level	Use of small-sized packets
	Use of resynchronization markers in packets

Table 5.10 Quality measurements (dBs PSNR) after the transmitting of the 'café' image through channels with different error rates

bpp	Error-free	Error rate 10^{-6}	Error rate 10^{-5}	Error rate 10^{-4}
0.25	22.64	22.45	20.37	16.02
0.50	26.21	26.01	23.35	16.20
1.00	31.39	30.25	25.91	16.52
2.00	38.27	36.06	25.80	17.16

providing a first potential isolation of the errors. Further increase of this isolation can be realized through the application of the arithmetic encoder termination after each coding pass, with accompanying reseting of the contexts. This allows the arithmetic decoder to continue decoding even after the occurrence of errors. The *lazy coding* mode of the arithmetic encoder can also aid in safeguarding against errors. This function relates to the optional function of the arithmetic encoder (the optional arithmetic coding bypass), in which some of the bits are not being encoded and are sent as raw data. This may be even more important in limiting the errors, to which the variable length codes are susceptible.

In terms of packets, each packet is accompanied by a resynchronization marker that allows the spatial partitioning and synchronization; practically this is implemented with the placement of an index (sequence) number in front of every packet, thus enabling an easy way to detect loss of a packet or packets during transmission. Table 5.10 presents average values of 200 coding tests on standard test image 'café', using reversible filters, and transmission through channels with variable error rates (10^{-6}, 10^{-5}, 10^{-4}) as described in (SantaspsCruz et al. 2000) and confirmed by the experiments that will be presented in the following paragraphs.

5.3.1   The Framework for the Experiments

First off, three widely used standard test images have been used to experiment with the JPEG2000 standard (no other than the images 'woman', 'café' and 'bike', illustrated in Fig. 5.21). The *kakadu* software implementation (version 3.0 (Taubman 2000b)) was used for the JPEG2000 encoding, which is described as the standard reference software, it is open source and has been proven by experiments that covers most of the functionalities described in the standard. To encode the images using the error resilience features the following kakadu syntax was used:

```
kdu_compress -i \$1.pgm -o \$1.jpc -rate
     -,.1,.25,.5,.75,1,.5,2 -full Cuse_sop=yes Cuse_eph=yes
     Creversible=yes Cmodes="RESET|RESTART|ERTERM|SEGMARK"
```

Fig. 5.21 The test images 'woman', 'café', 'bike'

where \$1 represents the filename; for the decoding with error control, the following kakadu syntax was used:

```
kdu_expand -i \$1.jpc -o \$1.pgm -resilient_sop
```

The produced encoded images were stored using progressive by quality (layer) priority with the ability of perfect reconstruction (with intermediate data rates of 0.1, 0.25, 0.5, 0.75, 1, 1.5 and 2 bpp), including noise resilience markers. Obviously, these additional markers increase the data rate to about 1 % approximately, which has already been reported in other works (ISO-IEC 2000a; Santa-Cruz and Ebrahim 2000a, b; Santa-Cruz et al. 2000). The encoded images pass through a simulated transmission channel with noise or congestion, and finally are decoded and checked to assess the robustness of the standard in suppressing errors, in order to produce comparative data and data for cost analysis.

The channel simulation was implemented via a software, which was responsible for injecting noise onto coded image data according to two basic error models,

model (1) *bit errors—simulation of a noisy channel*: the encoded data are affected at bit level, using two probability parameters,

    i. a probability of error occurrence (error rate) p_e;
   ii. a probability of occurrence of multiple consecutive errors (burst error rate) p_b.

The probability p_e is a typical error rate with values 10^{-6}, 10^{-5}, 10^{-4}, whereas the probability p_b introduces an additional error probability, simulating conditions of burst errors or momentary outage, which could very often occur in actual telecommunication systems. The values for probability p_b are given by $p_b = 1 - q$ where q is a burst factor with values in the set {1, 0.2, 0.1, 0.02, 0.01, 0.002, 0.001, 0.0002, 0.0001, 0.00001,

0.000001, 0.0000001}, in which smaller values represent a higher probability of error. Thus, a set of values $p_e = 10^{-6}$ and $q = 10^{-2}$ mean that
there is a probability $p_e = 10^{-6}$ to get an flipped or dropped sequence of
bits of an average length equal to $\frac{1-q}{q^2} = 9900$ bits. Typically, the error
probability introduces random errors, whereas the probability of burst
errors imposes a sequence of errors immediately after the occurrence of
the random error. In this scheme, two error modes, namely "0", and "1",
are being used with the following corresponding effect on the data:

0 the values of bits are reversed or flipped (random errors)
1 the values of bits are lost or dropped (outage)

model (2) *packet loss—simulation of a channel with congestion*: data are affected at
the packet level. The errors introduced in this scheme lead to packet dropping. Typical packet sizes used in experiments range from 100 to 2000
bytes, and the packet loss probability p_d takes values $10^{-3}, 10^{-2}, 10^{-1}$,
$2 \times 10^{-1}, 4 \times 10^{-1}, 6 \times 10^{-1}, 8 \times 10^{-1}$. During a first phase of experimentation, burst errors are not taken into account ($q = 1$). Thus, $p_d = 10^{-3}$
means that with a probability of 10^{-3} a packet is lost. Additionally, since
the images are encoded with eight (8) quality layers, the scheme takes into
account that it can distinguish significant and complementary information
among these quality layers. The simulation in this case considers that the
first packets of encoded data always arrive securely at the receiver (losslessly), simply by ensuring that the key information in the main header
of the file always arrive at the decoder.
To simulate more realistic communication channels (similar to the wireless packet-based transmission networks), a second packet loss scenario
was developed, in which a probability of burst errors $p_b = 1 - q$ was
introduced, with q representing (as in the first scheme) the burst error
factor. This possibility took values in the set {0.2, 0.1, 0.02, 0.01, 0.002,
0.001, 0.0002, 0.0001, 0.00001, 0.000001}. It should be noted that, as
in the first scheme, the final packet loss rate varies dramatically with the
introduction of this parameter.

The overall simulation scenario involved that the kakadu JPEG2000 codec,
encoded an image with quality priority, creating eight (8) quality layers using
reversible filters and all necessary markers to ensure maximum resilience to transmission noise. Then, during testing with scheme 1, the generated encoded image
was corrupted by noise during transmission according to two types of errors and
using various random error rates. Symbols affected were either flipped or dropped.
During the application of scheme 2, the encoded data were affected by a network
congestion. Afterwards, the kakadu JPEG2000 decoder decoded the corrupted data,
taking advantage of the built-in error resilience and reconstructed the image. At the
end, an estimator (the *imgcmp* software, part of the *jasper* (Adams 2002) system,
another free distribution included in the JPEG2000 standard), was used to assess
the quality of the reconstructed image. For the first error model, 100 experiments

were performed on each of the test images, using the two error modes, the three random error rates and the twelve burst error rates. Additionally, one to seven of the quality layers were successively characterized as significant. This simulation led to 151,200 results. For the second error model, 100 experiments were performed on each test image, using transmission of packets of sizes selected from twenty different size options ranging from 100 to 2000 bytes (in increments of 100 bytes). Seven packet error probabilities were tested. Further, one to seven of the eight quality layers were used as significant. In overall 336,000 results were produced. In all cases, the first transmission packets were kept intact during transmission to ensure proper transmission of the basic header data.

5.3.2 Experiments with Bit Errors

In the experiments with error model 1 (bit errors), the *error mode 1 (loss of bits or network outage)*, had the worst impact on the reconstructed images on the overall average, in the sense of measuring distortion regardless of the error rate. One explanation to this phenomenon could be that in conditions of bit dropping (loss of information), in addition to the loss of image data (which results in loss of synchronization until the occurrence of the next intact packet head), there is a potential loss of packets corresponding to tile-part headers, leading to possible loss of a large data segment, since many of the subsequent data may lose their meaning in the overall codestream.

The graph in Fig. 5.22 shows the average percentage of data corruption for the three error rates and the two error modes ("0" and "1"). These values are calculated over all burst error rates. The effect of the occurrence of burst errors is, of course,

Fig. 5.22 Average percentage of data distortion as a function of the error rate

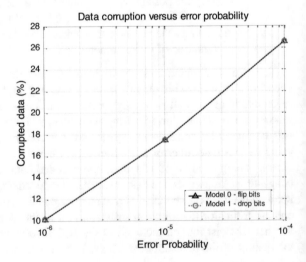

Fig. 5.23 Average image
quality as a function of the
error rate

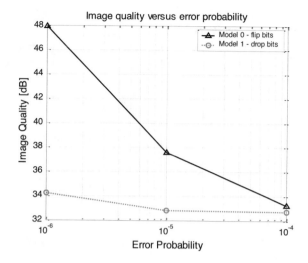

evident in these measurements since, for example, when the error rate is 10^{-6}, the
average distortion measured ultimately approaches the significantly higher value of
10^{-1}. As expected, the data distortion is independent of the error modes. Henceforth,
all references to an error rate will refer to the assessment over all burst error rates
(averaged), and therefore, this rate will substantially correspond to the probability
of initiating a sequence (block) of errors.

The graph in Fig. 5.23 shows the average reconstructed image quality (dBs PSNR)
as a function of the error rate for the two error modes, again, over all burst error rates.
As shown in the graph, the bit-dropping errors lead to higher distortion and the quality
is reduced in comparison to the occurrence of bit-flipping errors. This difference in
the effect of the two error modes is more apparent in lower error rates (say 10^{-6}) but
it losses its significance when the error rates go high (like in the case of 10^{-4}).

In Figs. 5.24 and 5.25 specific graphs show the effect of burst errors in image
quality and the distortion of data respectively. Figure 5.24 shows a graph of the
quality of the reconstructed image as a function of the burst error rate q for the
three rates and both the error modes. It is apparent that bit-dropping errors result in
significantly degraded quality, especially in the case of rates of the order of 10^{-6}, in
which the difference from the bit-flipping errors approaches the noticeable amount
of 20 dBs. At higher error rates, the differences are significantly reduced and error
rates of the order of 10^{-4} approach zero. Figure 5.25 depicts the percentage of data
distortion as a function of the burst error rate q. As is evident here, the choice of the
error mode does not affect the result. Both modes give exactly the same estimates for
the data distortion. The measurement of distortion here is defined as the percentage
of symbol changes under the influence of random and burst errors.

Figure 5.26 shows the reconstructed image quality as a function of the percentage
of data distortion. It is essentially a combination of the graphs of Figs. 5.24 and
5.25. In this graph it can be observed that the quality decreases abruptly for an

Fig. 5.24 Average image
quality as a function of the
burst error rate

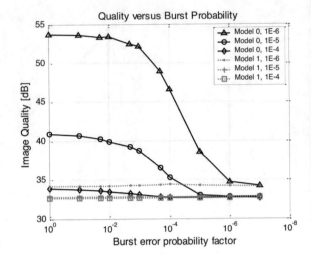

Fig. 5.25 Average
percentage of data distortion
as a function of the burst
error rate

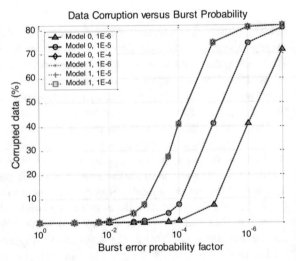

estimated percentage of data distortion of about 10 % and remains nearly constant
for a percentage of data distortion around 75 %. Indicatively, in the first 10 % the
quality degradation reaches 38 dB for error rate of 10^{-6}, and then up to about 75 %
the degradation is approximately 3 dB maximum. As expected from the previous
graphs, the worst case occurs when applying the bit dropping mode (error mode
1), where the curves are almost linear and nearly horizontal, imposing a low image
quality throughout the whole range of data corruption. On the other hand, application
of the bit flipping mode (error mode 0) results in a reconstructed image quality that
is proportional to the rate of random and burst errors in the network.

Another important aspect of the experiments is illustrated in Figs. 5.27 and 5.28,
which show the average data corruption and the reconstructed image quality in rela-

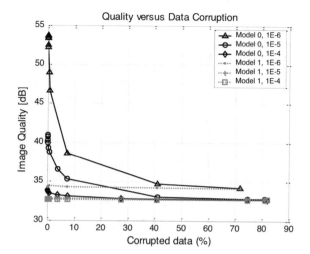

Fig. 5.26 Reconstructed image quality as a function of the percentage of data corruption

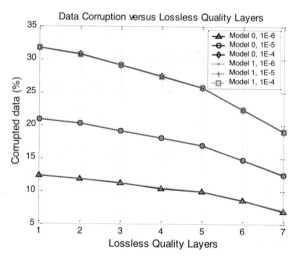

Fig. 5.27 Percentage of data distortion as a function of the significant quality layers

tion to the number of quality layers of encoded data that are marked as 'significant' (and transmitted losslessly). Specifically, if the system can guarantee the lossless transmission of some of the quality layers, then the result of the application of noise to the remaining layers is depicted as the percentage of data corruption in Fig. 5.27, and as the average reconstructed image quality in Fig. 5.28.

Apparently the curves in Fig. 5.27 can be linearly approximated. For example, in the case of error rate 10^{-5} (middle curves), the line approximating the curve can be expressed by

$$y = -1.42x + 22.86 \tag{5.4}$$

Fig. 5.28 Average
reconstructed image quality
as a function of the
significant quality layers

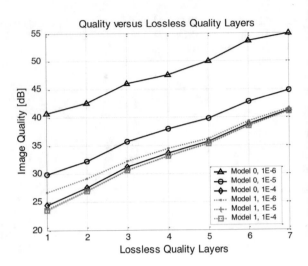

that corresponds to a slope of -1.42 (% corruption per quality layer). Correspondingly, Fig. 5.28 illustrates the fact that quality is directly related with the error rates and the error modes. In fact, these curves can be linearly approximated with a slope of around 2.5 (dB per quality layer) and expressed by

$$y = 2.5x + b \qquad (5.5)$$

where the parameter b takes values depending on the error rate and error mode (eg. for a rate of 10^{-5} in error mode 0, b equals to 27.34).

What emerges is a linear response for the quality as a function of 'significant' quality layers in the encoded image data. Knowing the image quality obtained for the first of the significant quality layers, as well as the error rate and error mode, it is possible to predict the final quality of the reconstructed image as the additional layers reach the receiver.

5.3.3 Experiments with Packet Errors

The following paragraphs present the results of simulations for the error model 2, which corresponds to the *packet loss scenario in a network with congestion*. What is called here an error rate reflects the probability of packet loss on the network.

First off, the graph in Fig. 5.29 shows the average percentage of data corruption as a function of error rate (seven error rates averaged over all packet sizes in the simulation). As shown, the actual percentage of data corruption is slightly less than that expected in relation to the error rate, and that is because the first codestream packets are always considered significant and their lossless transmission is guaranteed,

Fig. 5.29 Average percentage of data corruption as a function of the error rate

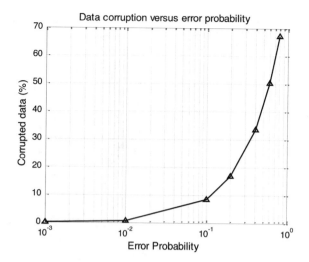

Fig. 5.30 Average reconstructed image quality as a function of the error rate

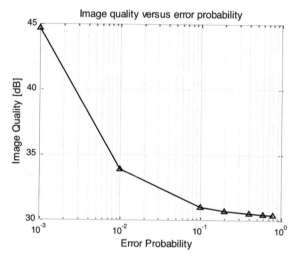

also because an averaging over all cases of significant-complementary information segmentation is considered.

Following a sharp increase in the corruption of data, the average reconstructed image quality also decreases rapidly with the increase of the error rate, as shown in Fig. 5.30, where the average reconstructed image quality is depicted as a function of the packet loss rate. Again, the effect is expected, as the increasing (in powers of 10) rate of packet loss, truncates increasingly larger parts of the original codestream and, therefore, the decoder becomes rapidly and increasingly unable to understand and restore the received data (even when full error resilience is imposed during the encoding).

Fig. 5.31 Reconstructed image quality as a function of the percentage of data corruption

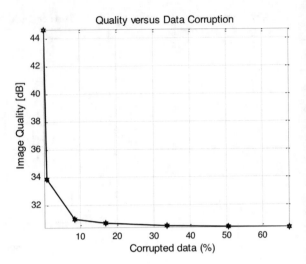

Following the paradigm of the presentation of results for the case of error model 1, a combined chart from graphs in Figs. 5.29 and 5.30 should be given, to present the rate-distortion picture obtained; this combination is shown in Fig. 5.31, which depicts the average reconstructed image quality as a function of the percentage of data corruption for the seven error rates of the experiments. It is evident that the curve is exponentially decreasing, corresponding to a behavior similar to that met in error model 1 (Fig. 5.26).

Another interesting aspect of the experimental results is shown in Figs. 5.32 and 5.33, which depict graphs of the average data corruption and the average reconstructed image quality as a function of the size of the packets used during the transmission and the number of significant (losslessly transmitted) quality layers. Figure 5.32

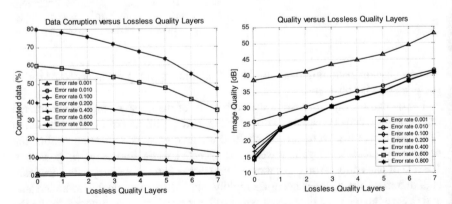

Fig. 5.32 Data distortion and corresponding reconstructed image quality as a function of the number of the significant quality layers

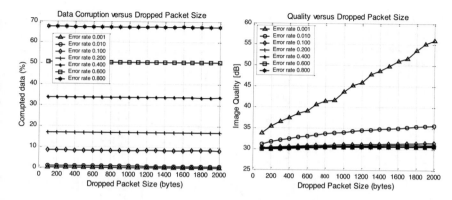

Fig. 5.33 Data distortion and corresponding reconstructed image quality as a function of the size of the transmission packet

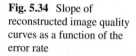

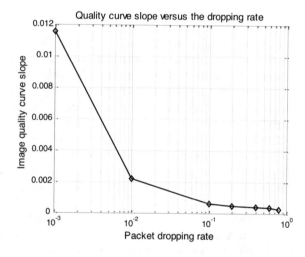

shows the graphs relating to the distortion and the image quality as functions of the number of significant quality layers, and Fig. 5.33 shows the corresponding quantities as functions the transmission packet size. An important consequence of the results is the observation that the reconstructed image quality follows the size of the transmission packet, whereby a larger packet size results in a linearly increasing quality. Another conclusion is that the slope of these curves decreases with the error rate and this is shown in Fig. 5.34.

The introduction of burst errors in the scenario of packet dropping significantly alters the overall error rate (as observed in the case of bit errors in error model 1). In this case, the error impact assessment tends to describe the assessment in the case of random errors but at a much higher rate. Note here that the multiplicity of experimentation in the packet dropping scheme is expressed through a burst error probability that controls essentially the number of consecutive information packets

Fig. 5.35 Average
percentage of data distortion
as a function of error rate in
the presence of burst errors

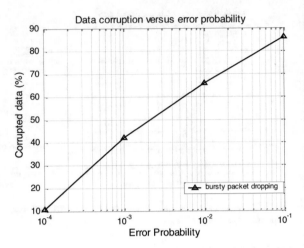

Fig. 5.35 Average percentage of data distortion as a function of error rate in the presence of burst errors

that are lost during transmission. This case can be considered as a worst case in the simulations and corresponds to the network conditions at which packet dropping (expressed as the network congestion) affects simultaneously multiple users.

Figure 5.35 depicts a graph of the average percentage of data corruption as a function of the four packet loss rates (calculated on average of all transmission packets, significant quality layers and burst error rates). The actual distortion is greater than that demonstrated by the error rate due to the presence of burst errors.

Following this sudden increase in corruption of data, the average reconstructed image quality also decreases rapidly with the increase of the error rate, as shown in Fig. 5.36 (calculated as the average of all the parameters of the experiments). Again, the result is largely expected, as the packet loss at a rate that increases in powers of 10 and the additional introduction of burst errors at the worst case simulation scenario, leads to significant truncations of the original codestream and makes the decoder unable to understand and restore the received data.

The influence of burst errors is illustrated in the graph of Fig. 5.37, where the average percentage of data corruption is plotted as a function of the burst error rate. Figures 5.38 and 5.39 present graphs expressing the effect of the characterization of some of the quality layers in the encoded data as significant. Specifically, Fig. 5.38 depicts the graph of the average percentage of data corruption and Fig. 5.39 the graph of the average reconstructed image quality as a function on the number of significant quality layers (which, obviously, are losslessly transmitted).

5.3.4 Cost Analysis

Another aspect that may be highlighted through the simulations on the previous sections concerns the capacity to use similar scenarios for designing cost policies;

Fig. 5.36 Average
reconstructed image quality
as a function of the error rate
in the presence of burst errors

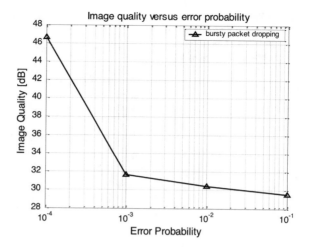

Fig. 5.37 Average
percentage of data corruption
as a function of burst errors
rate

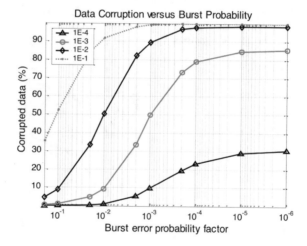

as a matter of fact, the results of this analysis were derived by the same experiments previously presented. To analyze the costs involved in transmitting data using such scenarios, one should, initially, establish a kind of policy in order to allow the pricing of various parts of the transmitted information from the overall codestream. To this end, the following scenario (assumptions) was adopted as the policy of the transmission network:

• An image codestream is divided into a significant and a complementary part, by imposing a progressive by quality JPEG2000 encoding. This way, the codestream contains a sequence of quality layers (1–8 in the experiments conducted), which can be distinguished and classified individually as significant and complementary. In this sense, the network is able to guarantee the lossless transmission only of the significant data (in this case, only of the first quality layers in the codestream).

Fig. 5.38 Average percentage of data corruption as a function of the number of significant quality layers

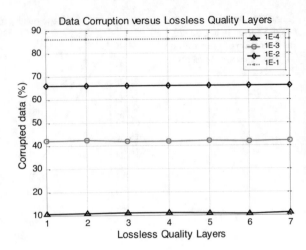

Fig. 5.39 Reconstructed image quality as a function of the number of significant quality layers

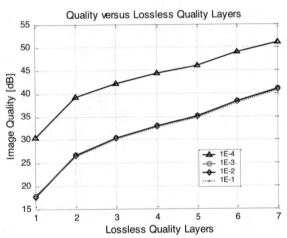

The complementary part of the information is considered non-essential (thus droppable) and is transmitted via a part of the bandwidth that can only be guaranteed to be within a specific error rate.

• Significant and complementary data are assigned a cost ratio C_r on a byte-wise basis. The cost ratios used in the presented experiments ranged from 1:1 to 10:1 using a 1:10 step (i.e. 90 cost ratios). The significant data are always assigned a cost of 1 cost unit per byte and thus the cost ratios express the significant: complementary cost units per byte.

Apparently, under these assumptions, the total cost can be estimated as

$$C = N_{nd} + C_r \times N_d \tag{5.6}$$

Fig. 5.40 Cost of unaffected
data as a function of the burst
error rate

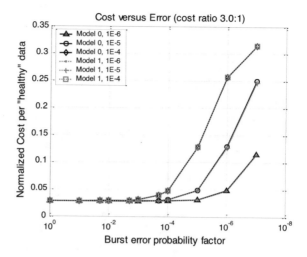

where C is the total cost in arbitrary cost units, N_{nd} is the amount of significant data
bytes, C_r is the cost ratio and N_d is the amount of complementary data bytes.

In the experiments carried out, due to the large size of the test images, cost
calculations gave values in the interval $[0.5 \times 10^7, 10^7]$ cost units, and, therefore,
were normalized in the interval $(0, 1]$ for a better demonstration. In most cases, the
cost is shown or expressed in terms of other quantities, such as the reconstructed
image quality or the amount of significant data, thus being a relative quantity.

The graph in Fig. 5.40 depicts the cost of unaffected data as a function of the burst
error rate, averaged over all of the random error rates, for a cost ratio of 3:1. The
term *cost of unaffected data* expresses the cost associated with data not corrupted
by any means by the channel errors, regardless of whether they are significant or
complementary; this term signifies, essentially, how much do the data arriving intact
at the receiver cost.

Figure 5.41, which is presented as the main evaluation chart of a channel, shows
the average cost per dB of the reconstructed image quality at the receiver as a function
of the reconstructed image quality at the receiver, for seven quality layers and three
of the tested cost ratios 2:1, 3:1 and 4:1. The points of the curves were produced
by increasing the number of significant quality layers from 1 to 7 (average data
rate from 0.1 to 2 bpp). By this graph, one may determine if the cost of each dB
of image quality received is 'worth' the reconstructed quality obtained, along with
an estimate of how many significant quality layers are needed (or are enough) so
that this can be achieved. For example, if the cost ratio is 3:1, error mode is 0 and
error rate is 10^{-5}, the cheapest dBs of quality are those at around 36 dB (medium
in terms of the HVS), when three quality layers are marked as significant.; the cost
at that point is 0.08 normalized cost units per dB of quality. For this case to hold,
the encoded image (codestream) has to be losslessly (error-free) transmitted up to a
part that corresponds to 0.5 bpp, leaving the rest of the codestream unprotected in
the occurrence of transmission errors.

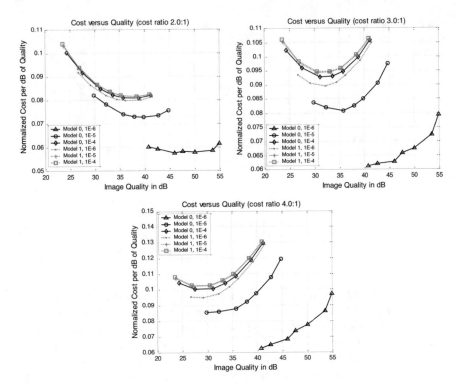

Fig. 5.41 Average cost per dB of reconstructed image quality as a function of the reconstructed image quality at the receiver for cost ratios 2:1, 3:1 and 4:1

Engineers and network designers can plan similar graphs for special cases and conditions, policies and cost ratios. In the experiments presented here, only cost ratios between 2:1 and 4:1 yielded *interesting results* concerning the cost policy options; it was only in this interval, in which the cheapest dBs of quality do not appear at the ends of the curves. The use of ratios less than 2:1 results in getting a minimum when all but one quality layers are characterized as significant, which is rather impractical. In contrast, ratios greater than 4:1, result in getting a minimum when only one of the quality layers is regarded as significant.

In addition to the graph in Figs. 5.41 and 5.42 represents the number of significant quality layers to enable a minimum cost as a function of the cost ratio. This graph makes clear how many of the quality layers should be designated as significant to achieve cheap dBs of reconstructed image quality across the overall range of cost ratios 1:1–10:1. Taking into account what was learned by the graph in Fig. 5.41, it can be deduced that for a cost ratio 3:1, error mode 0, error rate 10^{-5}, the number of significant quality layers is 3. Thus, the graphs in Figs. 5.41 and 5.42, taken together, reveal the best cost option for any given conditions regarding the errors and the significant information.

Fig. 5.42 Number of significant quality layers to enable a minimum cost per dB of reconstructed image quality as a function of the cost ratio

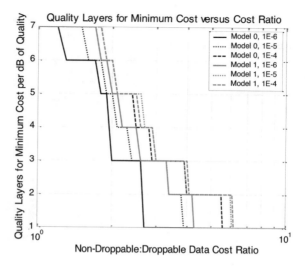

The conclusions that can be drawn from the experiments on error model 1 (errors in bits) are summarized as follows:

- When the error rate of the channel is high (such as 10^{-4}), then the reconstructed image quality at the receiver is expected to be low (33 dB on the average). At error rates of the order of 10^{-5} the quality ranges between 28 and 44 dB (36 dB on the average). At error rates of the order of 10^{-6} the quality is quite high between 40 and 55 dB (47 dB on the average).

- When errors of mode 0 occur (flipping of bits) then the costs per dB of quality are lower than in the case of mode 1 error (loss of bits). From the cost graphs it can be drawn that, typically, less significant quality layers are required to achieve cheap dBs of quality at the receiver. It is also evident that the reconstructed image quality at the receiver would be much better. Cheap dBs of quality in error mode 0 can be achieved using one less significant layer with respect to the application of error mode 1, which means that a smaller amount of lossless data would be needed to achieve a better reconstructed image quality at the receiver.

- In the case of error mode 1 (loss of bits−outage), the cost per dB of quality is high regardless of the error rate. It is noted that the reconstructed image quality at the receiver will be low even when all quality layers but one are being characterized as significant. For a cost ratio 2:1, cheap quality dBs can be obtained with the use of five to seven significant quality layers, meaning that a large amount of information needs to be losslessly transmitted, with the ultimate result, anyway, expected to be of low quality at the receiver.

- Cost ratios in the range of 2:1–4:1 qualify as 'critical' in the sense that the curves that are produced when these ratios apply are concave with at least one critical point (minimum). Minimums of cost are found typically in the middle of these curves. In all other cases, the curves exhibit a monotonic increase or decrease and thus correspond to unrealistic or 'unfair' cost policy options.

Fig. 5.43 Cost of unaffected data as a function of the number of significant quality layers

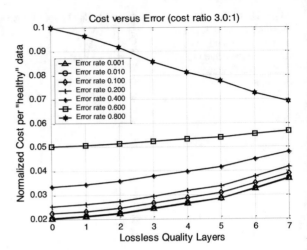

When assessing the cost in the case of the occurrence of packet loss errors, the segmentation of significant and complementary data may be imposed on packet level. The same conventions used in the previous presentation of the case of bit-level errors also apply in this case. In addition, in this case it was pre-supposed that an end-user is able to mark some of the packets as droppable, in an attempt to pre-set a lower price for them. Fixing the size of the droppable packets to a wide range (100 to 2000 bytes with 100 bytes increment), and varying the size of non-droppable data, interesting results and conclusions were drawn.

Figure 5.43 shows a graph of the cost of unaffected data to have reached the receiver as a function of the number of significant quality layers (that were, again, losslessly transmitted). The cost is expressed as the amount of data actually making it to the receiver, averaged over all packet sizes, error rates and for a cost ratio 3:1. As shown the cost increases for low error rates and reduces for high rates.

Figure 5.44 shows the cost per dB of quality as a function of the reconstructed image quality at the receiver, for cost ratios 2:1, 3:1 and 4:1. Each curve represents a different error rate in the network. The results are similar to those obtained in experiments on error model 1, with higher error rates requiring more significant quality layers and resulting lower reconstructed image quality. The fact is also illustrated in Fig. 5.45, where the number of significant quality layers required to achieve cheap dBs of image quality at the receiver is expressed as a function of their cost. Overall in these experiments similar results were obtained for the case in which burst errors were introduced.

A network administrator can take advantage of these results when designing a communication system and its cost policy according to the needs and requirements of the users and the network capabilities. On the other side, users are able to review the services provided and choose from possible cost scenarios depending on their needs. If a 'fair' cost policy is defined as the policy in which the dBs of image quality provided to the user are equally priced, then curves (in Figs. 5.41 and 5.44) which

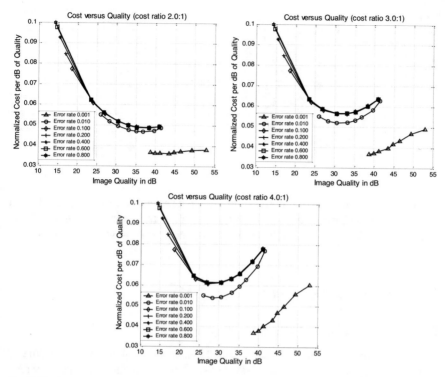

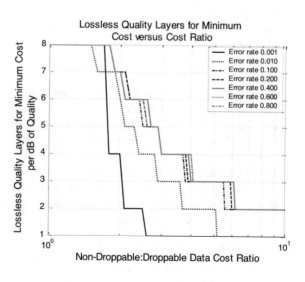

Fig. 5.44 Average cost per dB of reconstructed image quality as a function of the image quality at the receiver for cost ratios 2:1, 3:1 and 4:1

Fig. 5.45 Number of significant quality layers required to achieve a minimum cost per dB of reconstructed image quality as a function of the cost ratio

Table 5.11 'Fair' cost ratios
for a congested network with
various error rates

Error rate	Cost ratio
0.001	1.9:1
0.01	2.4:1
0.1	4.0:1
0.2	4.5:1
0.4	5.0:1
0.6	5.4:1
0.8	5.5:1

are 'balanced' around the minimum value should be selected. Using this definition it is possible to summarize the overall cost policy for a 'fair' communication channel as shown in Table 5.11, which shows the desired cost ratio for various error rates.

5.3.5 Simulation Conclusions

Summarizing, in a channel with noise (flipping of bits occurs),

- the distortion of data and the reconstructed image quality at the receiver estimated by averaging over all burst error rates is logarithmically related with the error rate;
- the reconstructed image quality at the receiver as a function of the data distortion during transmission exhibits a parabolic behavior, similar to the known typical rate-distortion curves;
- the distortion of data and the reconstructed image quality as a function of the number of significant quality layers exhibit a linear behavior with constant slope for the case of the quality, and slopes depending on the error rate in the case of distortion;
- the cost of unaffected data is proportional to the error rate and increases with increasing rate of burst errors;
- the cost per dB of quality exhibits maxima and minima in closely related with the error rate; appropriate cost ratio selection can lead to optimized 'fair' operation of a network;
- when the cost-quality curves are complemented by significant information-cost ratios curves, it is possible to draw very important conclusions for selecting the mode of segmentation of the information into significant and complementary so that desired costs may be achieved;

In a channel with outage (loss of bits occurs),

- the reconstructed image quality at the receiver is lower than that estimated in conditions of bit flipping and exhibits less correlation with burst error rate;

- the reconstructed image quality as a function of the data distortion does not exhibit a parabolic behavior; rather, it can be expressed as a linear relation, with the quality being almost constant for all values of data corruption percentage;
- the distortion of data and the reconstructed image quality as a function of the number of significant quality layers are also linearly approximated with constant slopes in the case of quality, or slopes that depend on the error rate for the data distortion, similar to the case of bit flipping errors; again, quality is significantly lower than in the case of bit flipping errors;
- the cost of unaffected data, the cost per dB of quality and the curves of significant information as functions of the cost ratio, exhibit the same characteristics as those in the case of bit flipping errors; although the qualitative results are similar, the image quality is lower and the costs per dB of quality is higher;

In a congested channel (loss of packets occurs),

- the quality of the reconstructed image as a function of the unaffected data exhibits the same parabolic behavior encountered in the case of a noisy channel (bit flipping);
- the quality of the reconstructed image as a function of the size of transmission packets confirms that in conditions of a high rate of packet loss a small packet size is preferable;
- the cost per dB of quality as a function of the amount of significant data exhibits similar characteristics to those found in the case of a noisy network (flipping of bits);

References

Adams, M. A. (2002). *The JasPer project home page.*
Burrus, C. S., Gopinath, R. A., & Guo, H. (1998). *Introduction to Wavelets and wavelet transforms: A primer.* Prentice Hall, ASIN: B00EB0QNWY.
Carpentieri, B., Weinberger, M., & Seroussi, G. (2000). *Lossless compression of continuous-tone images.* Hewlett-Packard: Technical report.
Chamzas, C., & Duttweiler, D. L. (1989). Encoding facsimile images for packet-switched networks. *IEEE Journal on Selected Area in Communications, 7*(June), 5.
Christopoulos, A., Skodras, A., & Ebrahimi, T. (2000a). The JPEG2000 still image coding system: An overview. *IEEE Transactions on Consumer Electronics, 46*(4), 1103–1127.
Christopoulos, C., Ebrahimi, T., & Lee, S. U. (2002). JPEG2000 special issue. In *Elsevier signal processing: Image communication* (Vol. 17). Elsevier.
Daubechies, I. (1988). *Orthonormal bases of compactly supported wavelets.* XLI: Communication on Pure and Applied Mathematics.
Daubechies, I. (1992). Ten lectures on wavelets. *SIAM.*
DeHoon, M., Imoto, S., & Miyano, S. (2003). *The C clustering library.* Technical report: Institute of Medical Science, Human Genome Center, University of Tokyo.
Fasulo, D. (1999). *An analysis of recent work on clustering algorithms.* Technical report and Departmaent of Computer Science & Engineering: University of Washington.
Gray, R., & Neuhoff, D. (1998). Quantization. *IEEE Transactions on Information Theory, 44*(6),
Gray, R. M. (2003). *Fundamentals of vector quantization.*

Habibi, A. (1971). Comparison of nth-order DPCM encoder with linear transformations and block quantization techniques. *IEEE Transactions on Communication Technology, 19*, 948–956.

Haykin, S. (1994). *Neural networks: A comprehensive foundation.* New York: Macmillan Co., ASIN: B00EKYJLHA.

Howard, P. G., & Vitter, J. S. (1993) (April). Fast and efficient lossless image compression. In *Proceedings of data compression conference* (Vol. 30, pp. 351–360).

Independent JPEG Group, IJG. (2000). *JPEG reference software.*

ISO-IEC. (2000a) (December). Information technology—JPEG. In: *Image coding system—Part 1: Core coding system, ISO/IEC international standard 15444-1.* ISO/IEC: Technical report.

ISO-IEC., (2000b). *JPEG2000 Part I final draft international standard (ISO/IEC FDIS15444-1) JTC1/SC29/WG1 N1855.* ISO/IEC: Technical report.

Jain, K. (1981). Image data compression: A review. *Proceedings of the IEEE, 69*, 349–389.

Jain, M. N. M. A. K., & Flynn, P. J. (1999). Data clustering: A review. *ACM Computing Surveys, 31*(3), 264–323.

Kohonen, T. (1989). *Self organization and associative memory* (3rd ed.). Berlin-Heidelberg-New York-Tokyo: Springer. ISBN 978-3540513872.

Kohonen, T. (1990). The self organizing map. *Proceedings of the IEEE, 78*(9), 1464–1480.

Kohonen, T. (1991). Self organizing maps, optimization approaches. In *Proceedings of the international conference on artificial neural networks* (pp. 981–990)

Kohonen, T. (2000). *Self-organizing maps* (3rd ed.). New York: Springer. ISBN 978-3540679219.

Kohonen, T., Kangas, J., Laaksonen, J., & Torkkola, K. (1992). LVQ_PAK: A program package for the correct application of learning vector quantization algorithms. In *Proceedings of the IEEE international joint conference on neural networks* (Vol. 1, pp. 725–730).

Langdon, G. (2001). *Lossless image compression—general overview.* Technical report: University of California.

Lim, Y. W., & Lee, S. U. (1990). On the color image segmentation algorithm based on the thresholding and the fuzzy c-means techniques. *Pattern Recognition, 23*, 935–952.

Liang, J., & Talluri, R. (1999) (January). Tools for robust image and video coding in JPEG2000 and MPEG-4 standards. In *Proceedings of the SPIE visual communications and image processing conference* (Vol. 3653, pp. 40–51)

Mallat, S. (1999). *A wavelet tour of signal processing.* 2 edn. Academic Press, ASIN: B01A0CGJF2.

Milligan, G. W., & Cooper, M. C. (1985). An examination of procedures for determining the number of clusters in a data set. *Psychometrika, 50*, 159–179.

Moffat, A., Neal, R. M., & Witten, I. H. (1998). Arithmetic coding revisited. *ACM Transactions on Information Systems, 16*(3),

Moccagata, I., Sodagar, S., Liang, J., & Chen, H. (2000). Error resilient coding in JPEG2000 and MPEG-4. *IEEE Journal of Selected Areas in Communications (JSAC), 18*(6), 899–914.

Netravali, N., & Limb, J. O. (1980). Picture coding: A review. *Proceedings of the IEEE, 69*, 366–406.

O'Neal, J. B. (1966). Predictive quantization differential pulse code modulation for the transmission of television signals. In *Bell System Technical Journal, 45*, 689–721.

Papamarkos, N., & Atsalakis, A. (2000). Gray-level reduction using local spatial features. *Computer Vision and Image Understanding, 78*(3), 336–350.

Pavlidis, G., Tsompanopoulos, A., Atsalakis, A., Papamarkos, N., & Chamzas, C. (2001) (May 14–18). In *IX Spanish symposium on pattern recognition and image analysis SNRFAI: A vector quantization-entropy coder image compression system*

Pavlidis, G., Tsompanopoulos, A., Papamarkos, N., & Chamzas, C. (2002) (Sep. 22–25). JPEG2000 over noisy communication channels: The cost analysis aspect. In *IEEE International conference on image processing ICIP 2002.*

Pavlidis, G., Tsompanopoulos, A., Papamarkos, N., & Chamzas, C. (2003). JPEG2000 over noisy communication channels: Thorough evaluation and cost analysis. *Elsevier Signal Processing: Image Communication, 18*, 497–514.

Pavlidis, G., Tsompanopoulos, A., Papamarkos, N., & Chamzas, C. (2005). A multi-segment image coding and transmission scheme. *Signal Processing, 85*(9), 1827–1844.

Pennebaker, W. B., & Mitchell, J. L. (1993). *JPEG still image compression standard.* New York: Springer.

Press, W. H., Teukolsky, S. A., Vetterling, W. T., & Flannery, B. P. (1992). *Numerical recipes in C: The art of scientific computing, 2nd edition* (pp. 609–650). Cambridge University Press.

Rabbani, M., & Jones, P. W. (1991b). Digital image compression techniques, Volume II. *SPIE optical engineering press.*

Rao, R. M., & Bopardikar, A. S. (1998). *Wavelet transforms: Introduction to theory and applications.* Prentice Hall, ASIN: B01A65JU7W.

Red Hat, Inc. (2000). *Home of the Cygwin project.*

Saha, S., & Vemuri, R. (2000). How do image statistics impact lossy coding performance? In *Proceedings of international conference on information technology: Coding and computing (ITCC'00).* March 27–29: Las Vegas.

Santa-Cruz, D., & Ebrahimi, T. (2000a) (September). An analytical study of JPEG 2000 functionalities. In: *Proceedings of IEEE International Conference on Image Processing—ICIP 2000.*

Santa-Cruz, D., & Ebrahimi, T. (2000b) (September). A study of JPEG 2000 still image coding versus other standards. In *Proceedings of X European signal processing conference* (Vol. 2, pp. 673–676)

Santa-Cruz, D., Ebrahimi, T., Askelof, J., Karsson, M., & Christopoulos, C. A. (2000) (August). JPEG2000 still image coding versus other standards. In *Proceedings of SPIE, 45th annual meeting, applications of digital image processing XXIII* (Vol. 4115, pp. 446–454)

Taubman, D. S. (2000b). *Kakadu software—A comprehensive framework for JPEG2000.*

Taubman, D. S., & Marcellin, M. W. (2002). *JPEG2000 image compression fundamentals, standards and practice.* Kluwer Academic Publishers, ASIN: B011DB6NGY.

Tou, J. T., & Gonzalez, R. C. (1977). *Pattern Recognition Principles.* 2 edn. Addison-Wesley, ASIN: B011DART5O.

Trivedi, M. M., & Bezdek, J. C. (1986). Low-level segmentation of aerial images with fuzzy clustering. *IEEE Transactions on System Man and Cybernetics, 16,* 589–598.

University of Waterloo, Canada. (2002). *Image repository.*

Weinberger, M. J., Seroussi, G., & Sapiro, G. (1995) (July). *LOCO-I: A low complexity lossless image compression algorithm.* Technical report ISO/IEC JTC/SC29/WG1 document N203.

Wu, X. (1996). An algorithmic study on lossless image compression. In *Proceedings of the IEEE data compression conference* (pp. 150–159)

Wu, X., & Memon, N. (1997). Context-based, adaptive, lossless image coding. *IEEE Transactions on Communications, 45*(4), 437–444.

Wu, X., & Memon, N. (2000). Lossless interframe image compression via context modelling. *IEEE Transactions on Image Processing, 9*(5).

Index

A
Arithmetic coding, 140, 141, 143, 145, 155, 184, 305, 315, 316, 318, 319, 328

B
Basis functions, 71, 77, 85, 92–95, 110
Basis images, 71, 75, 76, 85–89, 93, 94, 96–98
Binary encoding, 136

C
Codec, 151
Color quantization, 118, 253, 264
Color space, 22, 30–32, 34–39, 41–44, 68–70, 118, 152, 163, 167, 172, 174, 253–255, 282, 296, 302

D
Data
  coding, viii
  compression, vi, 49, 50, 128, 146, 147, 206, 317
  encoding, 147
  transmission, 183, 196, 197

E
Expectation maximization, 248

G
Geometric distribution, 135
Golomb coding, 135

H
Histogram clustering, 254
Histogram equalization, 40, 236, 241
Huffman coding, 132, 135, 155, 163, 277, 282, 304, 305
Human
  eye, 2, 6, 10, 12, 14, 23, 26, 30
  observer, 1, 69
  perception, 1
  retina, 10
  vision, 1, 2, 121
  visual perception, viii, 1, 6, 23, 24, 216
  visual system, 2, 5, 10, 16, 17, 19, 44, 65, 66, 69, 117, 121, 150, 152, 173, 180, 181, 186, 214–216, 220, 264, 343

I
Image
  analysis, 216, 223, 225, 233, 244, 261, 262
  coding, 86
  compression, vii, 49, 65, 66, 72–74, 77, 97, 104, 116, 117, 140, 150, 170, 213, 261, 262, 270, 301, 302, 328
  encoding, 197, 327
  processing, 2, 6, 58, 71, 81, 274
  segmentation, 217, 222, 223, 225, 233, 236, 237, 239, 240, 242, 245, 252, 302, 323
  transmission, viii, 188, 197, 302, 326
Image transform, 70
  discrete cosine transform, 81
  discrete Fourier transform, 78
  discrete Hartley transform, 93
  discrete sine transform, 89
  Haar transform, 100

© Springer Nature Singapore Pte Ltd. 2017
G. Pavlidis, *Mixed Raster Content*, Signals and Communication Technology,
DOI 10.1007/978-981-10-2830-4

Printed in the United States
By Bookmasters